BOSTON STUDIES IN THE PHILOSOPHY OF SCIENCE

VOLUME 244

TURKISH STUDIES IN THE HISTO
AND PHILOSOPHY OF SCIENCE

TURKISH STUDIES IN THE HISTORY AND PHILOSOPHY OF SCIENCE

Edited by

GÜROL IRZIK
Boğaziçi University, Istanbul, Turkey

and

GÜVEN GÜZELDERE
Duke University, Durham, U.S.A.

Springer

A C.I.P. Catalogue record for this book is available from the Library of Congress.

ISBN 10-1-4020-3332-X (HB) Springer Dordrecht, Berlin, Heidelberg, New York
ISBN 10-1-4020-3333-8 (e-book) Springer Dordrecht, Berlin, Heidelberg, New York
ISBN 13-1-4020-3332-X (HB) Springer Dordrecht, Berlin, Heidelberg, New York
ISBN 13-1-4020-3333-8 (e-book) Springer Dordrecht, Berlin, Heidelberg, New York

---

Published by Springer,
P.O. Box 17, 3300 AA Dordrecht, The Netherlands.

Printed on acid-free paper

Printed in the Netherlands.

We dedicate this book to our colleagues and friends
Robert Cohen, Arda Denkel, Berent Enç and İlham Dilman

TABLE OF CONTENTS

PART III. PHILOSOPHY OF LANGUAGE AND MIND

PART IV. CAUSES AND ACTION

PART V. OTTOMAN SCIENCE STUDIES

ABOUT THE CONTRIBUTORS

Murat Aydede received his Ph.D. from the Philosophy Department of Maryland University, College Park. After teaching at the University of Chicago for a while, he became an associate professor of philosophy at the University of Florida where he is currently teaching now. His main scholarly interests are philosophy of mind with emphasis on pain and cognitive science. His articles appeared in such journals as *Consciousness and Emotions*, *Philosophy and Phenomenological Research*, and *Philosophy of Science*.

Samet Bağçe received his Ph.D. in Logic and Scientific Method from the Philosophy Department, London School of Economics. Currently, he is an associate professor of philosophy at Middle East Technical University. His areas of specialization include history and philosophy of science, especially philosophy of geometry. He has published articles in the philosophy and methodology of science and geometry in various journal and collections.

Osman Bahadır completed a DEA (Diplôme d'Etude Approfondi) Degree at Denis Diderot (Paris 7) University. He was the editor-in-chief of *Bilim Tarihi* (History of Science) Journal for its 30 issues from 1991 to 1994. His published books include *Osmanlılarda Bilim* (Ottoman Science), *Bilim Cumhuriyeti'nden Manzaralar* (Scenes from the Republic of Science), *Cumhuriyet'in ilk Bilim Dergileri ve Modernleşme* (Republic's First Science Magazines and Modernization) and *Elektriğin Kısa Tarihi* (A Short History of Electricity).

H. H. Günhan Danışman is an associate professor of History of Technology at Boğaziçi University. He received his Ph.D. from University of London in the same field. He is the author of numerous articles in various journals and the co-author or editor of several books in Turkish and English. In June 2002, he was elected as a council member of the International Oral History Association.

Arda Denkel received his Ph.D. in philosophy at Oxford University. He taught at the Philosophy Department of Boğaziçi University, Istanbul until his untimely death at the age of 51 and published widely in analytical ontology, epistemology and philosophy of language. He wrote more than thirty articles and a dozen books, three in English, including *Object and Property* and *The natural Background of Meaning*.

Sun Demirli is an assistant professor in the Department of Philosophy at Bogazici University. He received his Ph.D in philosophy from Syracuse University. His areas of interest include the history of modern philosophy, metaphysics and ontology.

Berent Enç received his Ph.D. at Oxford University and was an emeritus professor of Philosophy at the University of Wisconsin-Madison. His publications ranged from philosophy of science and philosophy of mind to metaphysics and action theory. His articles appeared in *Journal of Philosophy*, *Mind*, *Philosophy of Science*, *Philosophical Studies*, *Noûs*, *Canadian Journal of Philosophy*, and *History of Philosophy Quarterly*. He passed away in 2003 at the age of 65.

David Grünberg is an associate professor of philosophy at Middle East Technical University, Ankara. He received his Ph.D. from the Philosophy Department of Middle East Technical University. He specializes in logic and philosophy of science. His publications include "A Wittgensteinian Approach to Truth" (*20th International Wittgenstein Symposium*), and "Bootstrapping and the Problem of Testing Quantitative Theoretical Hypotheses" (*Proceedings of the Twentieth World Congress of Philosophy, Volume 10*).

Teo Grünberg is an emeritus professor of philosophy at Middle East Technical University, Ankara. He received his Ph.D. from the Philosophy Department of Istanbul University. His areas of specialization are logic and philosophy of science. He is the author of numerous books in Turkish, including a three-volume book on logic. His publications appeared in such journals as *Journal of Symbolic Logic*, *Logique et Analyse*, *British Journal for the Philosophy of Science*, and *Studies in History and Philosophy of Science*.

Güven Güzeldere is an assistant professor of psychological and brain sciences at Duke University. He received his Ph.D. degree from Philosophy Department at Stanford University. He specializes in philosophy of mind, cognitive science, history and philosophy of psychology and neuroscience. He has published more than forty articles and coedited *Nature of Consciousness: Philosophical and Scientific Debates* (with N. Block and O. Flanagan). He is also the founding editor of *Psyche: An Interdisciplinary Journal of Research on Consciousness*.

Ilhan Inan is an assistant professor of philosophy at Boğaziçi University, Istanbul. He received his Ph.D. from the Philosophy Department of UC Santa Barbara. His areas of specialization include philosophy of language, epistemology and free will and determinism. Currently, he is working on a book on paradoxes.

Gürol Irzık is a professor of philosophy at Boğaziçi University, Istanbul. He received his Ph.D. degree from History and Philosophy of Science Department at Indiana University, Bloomington. He specializes in causation and the relationship between positivist and postpositivist approaches to science. He has contributed to many volumes and published in such journals as *British Journal for the Philosophy of*

Science, Philosophy of Science, Studies in History and Philosophy of Science, and *Science and Education*.

Ekmeleddin Ihsanoğlu is currently the president of the Division of History of Science of the International Union of History and Philosophy of Science. He received his Ph.D. in Ankara University and taught at many institutions, mostly at Istanbul University as a professor of History of Science. He is the author and editor of numerous books and articles in many languages in science, history of science, Turkish culture and the relationship between Moslem and the Western world. His most recent publication is *Science, Technology and Learning in the Ottoman Empire*.

Berna Kılınç is an assistant professor of philosophy at Boğaziçi University, Istanbul. She received her Ph.D. in Conceptual Foundations of Science at the University of Chicago and specializes in history and philosophy of science, statistical reasoning and cognitive psychology. Her articles appeared in *Studies in History and Philosophy of Science* and *British Journal for the History of Philosophy*.

Yalçın Koç is an emeritus professor of philosophy at Boğaziçi University, Istanbul. He received his Ph.D. from the Philosophy Department of Istanbul University and specializes in philosophy of quantum mechanics, Plato and Kant. He is the author of several books in English in Turkey and published in such journals as *Physics Letters, Nuovo Cimento* and *Isis*.

Erdinç Sayan received his Ph.D. in philosophy from the Ohio State University. He is currently an associate professor in the Department of Philosophy at the Middle East Technical University, Ankara. His areas of interest include philosophy of science, philosophy of mind, and analytic metaphysics. He has published in *Philosophical Studies, Philosophy Research Archives* and *Studies in East European Thought*.

Karanfil Soyhun is an assistant professor of philosophy at Boğaziçi University, Istanbul. She received his Ph.D. from the Philosophy Department of University of Rochester. Her areas of specialization include philosophy of language, applied ethics and theories of rationality. Currently, she is working on a book on applied ethics.

A. M. Celal Şengör is a professor of geology at Istanbul Technical University, Istanbul. He is considered to be the world's foremost authority on the plate-techtonic evolution of Eurasia. He has published many books and hundreds of articles on geology and its history and been honored as a member of National Academy of Sciences (USA) and European Academy of Sciences.

Ümit Yalçın is an associate Professor at East Carolina University. He received his Ph.D. degree in Philosophy from UC Berkeley and specializes in epistemology and metaphysics. He is the coeditor of *Supervenience: New Essays* and published in *Philosophical Papers* and *Philosophical Studies*.

Hüseyin Yılmaz specializes in general relativity theory and is the author of numerous articles in books and in such journals as *Physical Review Letters*, *Annals of physics*, *Hadronic Journal* and *Nuovo Cimento*.

PREFACE

The academic world is a strictly hierarchical world. Administrators and policy makers continuously devise all kinds of criteria in order to construct, and subsequently preserve, the hierarchies of the institutions. But in attempting to assess its own state, academia has dispensed with one of its basic constitutional values, and succumbed to the fatal attraction of quantification, since such hierarchies appear to undermine what academics so passionately preach: that what should be a qualitative assessment cannot be expressed quantitatively.

There is, nevertheless, another kind of hierarchy whose criteria are at the antipodes of the quantifiable standards. This other hierarchy is dependent on strictly subjective criteria: it concerns the way each member of the scientific community views his or her colleagues, with standards which sometimes may have the consensus of the community, but very often do not. Having the opportunity to get acquainted with the works of scholars who are members of emerging communities of historians and philosophers of science, has been one of the aims of the sub-series about the state of history and philosophy of science in national contexts, started so successfully by the untiring efforts of Bob Cohen.

The volume *Turkish Studies in the History and Philosophy of Science* shows in no uncertain terms that there is an active community of philosophers and historians of science in Turkey with an impressive scholarly agenda. The volume, also, includes a rather unique piece: the interview with Maria Reichenbach and David Kaplan, which – together with the particularly informative comments of the editors – brings to surface many aspects of academic life in Turkey in the past, unknown to many scholars. All the contributors to this volume are tackling problems lying squarely within the mainstream *problematique* of philosophy and history of science, thus being engaged in a critical dialogue with those who have the relevant expertise in these fields. The editors have done an excellent job in presenting the articles and the overall rationale behind the structure of the volume.

Let me bring up a number of issues, spurred by this collection of well-argued articles.

The hierarchies mentioned above have progressively brought about a stricter division between center and periphery, strengthening what for many people constituted the very basis of this dichotomy: almost everything that is considered "best" is to be found in the center, and, alternatively, the very notion of center is often understood to be the space, which almost exclusively accommodates what is considered to be "best." The center is, thus, affirming its role as the producer of what is new and novel and the periphery strengthens its image of having acquired its identity exactly because of its role in applying or consuming these new intellectual or material products. The link between center and periphery has been codified as a relationship based on the processes of a unidirectional transmission – of ideas and techniques from the center to the periphery. But this particular form of received wisdom does not appear to bear the brunt of recent (re)considerations of the notion of "transmission." The relationship between center and periphery is recently being examined within the framework of the dynamic co-existence, on the one hand, of cultural affinities and dispositions for adoption and, on the other, of the potent proclivities to resistances in the receiving culture. Hence this relationship is viewed in terms of processes of *creative appropriation* of ideas and practices which have been initiated in the center, rather than in terms of the notion of transfer or transmission

In the discussions about the transmission of ideas and techniques from the center to the periphery, what had been glossed over was that the periphery plays an intriguingly creative role, since ideas and practices are never received in a passive manner: the receiving culture almost always *appropriates* what has been coming from the centers. And appropriation is a creative process. One must always recognize that ideas are not simply transferred like, as it were, material commodities. They are always transformed in unexpected and sometimes startling ways as they are appropriated within the multiple cultural traditions of a specific society during a particular period of its history. The scholars of the periphery are not passive agents whose only function is to distribute locally the well-packaged goods delivered to them from the centers, but they act as subjects who receive many goods with no particularly clear directions on how to dispose of them locally and that it is their role to chart such local strategies.

Thus the concept of the transfer of ideas, used extensively by those who have discussed these issues in the past, is found to be ultimately inadequate in contextualizing the dissemination of ideas and practices in the societies of the periphery. The notion of appropriation appears to be a more coherent and fruitful analytic instrument. Appropriation directs attention to the measures devised *within the appropriating culture* to shape the new ideas within the local traditions which form the framework of local constraints – political, ideological as well as intellectual constraints. To examine such issues requires discussing the ways in which ideas that originate in a specific cultural and historical setting are introduced into a different milieu with its own intellectual traditions as well as political and educational institutions. Indeed, a major challenge for whoever examines the processes of appropriation across boundaries is precisely to transcend the merely geographical, and to concentrate on the character of the specific receiving culture.

That such an approach can be particularly useful in the history of science does not need to be further qualified. But what about philosophy? For philosophy, generally, and philosophy of science, in particular, there is a series of questions, which I feel, preoccupy many of those who work in academic institutions of the periphery and often contemplate about redetermining their role by trying to combine scholarly excellence with local particularities. Let me mention some of these questions: Should we continue to be denying so emphatically that locality has no role in the further discussion and analysis of the mainstream problems in philosophy? Is it possible to have further insights in the discussion of many of these problems, if attempts are made to start approaching the standard problems by drawing homologies with analogous themes related to the local cultural framework? Is the cultural diversification that is so pronounced in so many societies and its effects so intense in so many aspects of everyday life, to be ignored when it comes to the examination of many of these philosophical problems? Can these aspects of localities, be made to have a theoretical implication in the ways scholars discuss the philosophical problems? Is the aim of the educational programs, research agendas and organizational structures of the relevant academic Departments in the periphery to be good replicas of their equivalents in the institutions of the center or is there an unexplored range of possibilities which, if realised, will complement the already successful programs?

There are well argued objections to what constitute the presuppositions of such queries. In philosophy, especially, it is claimed that the problems are independent of localities. Everyone agrees that these problems appear repeatedly in the history of philosophy, but many regard this history as the sum total of the different attempts in dealing with these problems, again, independent of localities. But do such convictions continue to have their validity, if examined within the new realities of center and periphery? Might it be the case that the new emerging communities of philosophers may become a contributory factor in rejuvenating our overall philosophical *problematique*? What I am trying to argue is that it might be worthwhile, to re-examine these issues, and try to sense whether it would be possible to formulate some non-trivial answers – there may not be any, but it may be worth the discussion. The present appears to be an opportune moment for such discussions, since in many of the countries of the periphery there is an emerging community of particularly well-versed philosophers and historians of science and who often indirectly touch upon these issues in their attempts to (re)define their role as scholars, teachers and intellectuals.

Europe is presently in the throes of its most dramatic transformations since the end of the Second World War. New nation-states have come into being, new borders emerged, new institutions appeared, and old institutions restructured themselves. These changes will force many historians and other scholars to look again at the past. The work that has already been done, as well as newly available sources, combined with (comparatively) open intellectual environments and increases in funding for trans-cultural contacts offer an unprecedented opportunity for a critical re-examination of the historical character of science in Europe. It is obvious that such a re-examination, will not be tried on solely its scholarly merits, and that there will be many attempts to assimilate such reconsiderations of the past within the aspiring ideologies in Europe. Let me give an example of the dangers involved.

In a 1995 White Paper on the question of unemployment and on the ways young people can gain as many skills before finishing high school, the European Union proposed that history of science and technology be included in the school curricula. It was no doubt a good recommendation but for the wrong reasons.[1] The White

[1] White Paper published by the European Commission titled *Teaching and Learning: Towards the Learning Society* (Luxembourg, 1995). See sections II.B and C.

paper suggested that by learning the history of science, and especially the history of technology, young people will acquire knowledge of a variety of skills and techniques and will become aware of the boundlessness, as it were, of human inventiveness. The recommendation of the report, however, is embedded in one of those interesting mental summersaults that the bureaucrats in Brussels are so fond of performing. It was noted that science had been a European phenomenon, that modern science has been born in Europe and that it should be taken as our common European heritage. From these, it is but a short step, to be confronted with the elusive notion of European Science.

Here is one of those instances where there is such a dichotomy between the bureaucracy's goals and the aims of an academic pursuit. Never mind that historians of science have been trying to articulate local differentiations and trying to bring to surface the deviations from the view that holds the scientific enterprise to be an all inclusive homogeneous practice. European integration as planned in Brussels needs "European" notions like European Science and the specter of Europeanizing everything will be continuously finding justification. Nevertheless, the dynamics of these processes in Europe will offer new opportunities for academic pursuits, despite the fact that they take place within a framework full of contradictions and struggles for hegemonies – ideological, political, of research agendas, of educational priorities etc.

So here is another dimension about the sciences in the European periphery: Talking about the periphery will result in articulating differences and not in seeking identities. The view which considers the sciences at the European periphery as the out-of-focus reflections of what has been happening at the center is mostly for ideological use. The history of the sciences at the periphery is not an attempt to chart the map of the watered down version of what happened at the center. The study of the sciences in the periphery will bring forth interesting philosophical and historical issues only if such divergences in the European context are understood. Otherwise it would be trivial to study it: after all we do know that in the countries of the periphery there were no Newtons, no Laplaces, no Leibnizes, and no Eulers.

Thus, perhaps, one of the most intriguing challenges for philosophers and historians of science is to chart their own thematic atlas within this geographically expanded and culturally diverse Europe, whose present configuration provides a unique opportunity for symbiosis between established and emerging communities of historians. Members of newer communities will soon have to decide how to recast what have often, and for many years, been local topics in ways that can be linked to contemporary historiography of science, devise convincing methodologies of analysis and legitimate the attempts for the new syntheses.

Let me be clear in disavowing two possible misunderstandings: Firstly, I do not think that whatever new and refreshing will be coming from the work of the scholars from the periphery. Quite the opposite is what I want to stress: that there is an untapped potential in the cultural diversity of the international community of scholars which is being strongly bolstered and further consolidated by the ever more assertive presence of the emerging scholarly communities in the periphery. And such a diversity may be a contributory factor in the various ongoing attempts for new syntheses either in philosophy or history of science. Secondly, all of the above is not another password for agreeing with the excesses of social constructivism. Unfortunately, the strongly partisan discussions of the past decade between those who uncritically acclaimed that social constructivism will be the new catharsis from the sins of the past and those who believed that such discussions were undermining the traditional holy values, did not create a milieu where researchers would talk about culture and localities without having to apologize and take distances from methodological prescriptions. But social and cultural history and its theoretical and philosophical considerations, has been an undertaking with a long history of its own and which has given many excellent samples of scholarly work.

Unknown to many, Turkey commands a uniqueness in the history of history of science: The first graduate student of the Harvard professor of History of Science George Sarton, was a Turk, Aydın Sayılı, whose subsequent work has ranked him among the experts of Islamic science and, specifically, astronomical observatories in the Ottoman Empire. He returned to Turkey after his studies and was initially appointed to a junior post at the University of Ankara, eventually becoming a professor of history of science there. In an exasperated letter to his teacher, soon after his return, he describes to him his academic loneliness in Ankara where none of his colleagues, there or anywhere he went in Turkey, could understand, let alone appreciate, what he was doing. He asked for his teacher's help, in case his teacher had some kind of ready and convincing answer to help him change the mood of indifference. A little over half a century later, our Turkish colleagues can rightly boast that they managed to come a long way from the conditions Sayili was describing, and the present volume is a rather convincing evidence of their achievements. And, personally, I feel deeply honored to have been asked by the editors to write the preface.

Kostas Gavroglu,
Department of History and Philosophy of Science,
University of Athens

GÜROL IRZIK & GÜVEN GÜZELDERE

INTRODUCTORY REMARKS

The Republic of Turkey was founded in 1923 upon the ruins of the Ottoman Empire, which lasted more than six hundred years. The founders of the Republic explicitly denied the heritage of the Ottomans in all spheres and aimed to construct a modern nation-state based on Western values and principles. This was obviously no easy task and turned out to be more difficult than imagined. Turkey since 1923 is therefore best described as a country in continuous transition, that has given rise to striking similarities and differences, obvious continuities and ruptures between the old and the new, between a traditional, Islamic culture and a modern, secular one.

In this context the writing of the history of philosophy of science in Turkey can be either a too easy or a too difficult task. From one perspective, it is all too easy because there was simply no philosophy of science in the standard sense until the formation of the Turkish republic; all that exists is contemporary philosophy of science. From another perspective, this is denialism pure and simple because since at least the 19th century during which modernization (i.e. Westernization) attempts gained momentum in all spheres including education, there emerged a conspicuous philosophical interest and activity in logic, mathematics, physics and social thought. During this period a number of Ottoman young men were sent to Europe, especially France, to study the state of the art in the sciences.[1] But the writing of the history of philosophy of science from this perspective is not at all a simple task, the major reason being the language barrier. Although the lay people spoke Turkish, the official language of the Ottoman Empire was Ottoman, which was an Esperanto consisting of Persian, Arabic and Turkish, written in Arabic script. In 1928, the Latin alphabet was accepted and the language was drastically purified into Turkish, a process which continued well into the seventies, and the teaching of Ottoman language was excluded from the educational system. The dramatic result was that few could read anything written before 1928. Consequently, even if there is a distinctively philosophy of science heritage, it is mostly buried in old texts which are not so easy to find either.

Just like the history of philosophy, a detailed history of philosophy of science (including the Ottoman and the Republican period) is yet to be written as well. In this introduction we cannot do justice to this complex issue nor are we equipped to. What we can at best do is to mention some of the key figures since the thirties.

G. Irzık & G. Güzeldere (eds.), Turkish Studies in the History and Philosophy of Science, 1-6.

In the year 1933 Hans Reichenbach came to Turkey and began teaching in the Faculty of Letters at Istanbul University. He was invited by the Turkish government to establish a modern philosophy department and was appointed as its first chairperson. This was part of a much bigger movement of reforming Istanbul University as a whole, a reform that began in 1933 and was carried out with the help of about eighty-five non-Turkish academicians, scientists and technical personnel. A vast majority of them were German and Austrian refugees who escaped from the Nazi regime. Among them were philologists and professors of literature Leo Spitzer, Eric Auerbach and Helmut Ritter; professor of economics Fritz Neumark; professor of mathematics Richard von Mises; and professor of philosophy Ernst von Aster. The Turkish government employed them with the explicit purpose of turning Istanbul University into a modern institution of higher education (see Widmann 2000). It also adopted the reverse policy of sending, with government scholarship, promising young students and scholars to study abroad. Thus, the first generation of philosophers, philosophers of science, and historians of science in the early years of the Turkish Republic are to a large extent the outcome of these two policies.

Hans Reichenbach taught mostly logic and scientific philosophy, and even some history of philosophy since students were not knowledgeable about Western philosophy. Later he recruited von Aster to teach history of philosophy not only because the latter was a very good historian of philosophy, but also because he was sympathetic to scientific philosophy. Reichenbach's logic notes were published in Turkish under the title *Logistic*, and several of his lectures appeared in Istanbul University publications, which Reichenbach used in his *The Rise of Scientific Philosophy*.[2] This is not much, given that Reichenbach taught at Istanbul University until 1938. It appears that during his five-year stay Reichenbach focused on his book *Experience and Prediction* and did not seriously think of spending the rest of his life in Turkey. There are several reasons for this. He felt badly isolated from the community of philosophers of science. He had helped recruiting Richard von Mises to teach in the Mathematics Department, but obviously one sparrow did not make a summer. The libraries were extremely poor for doing research. Few of his students knew any foreign language, so his lectures had to be consecutively translated into Turkish by his assistants, Macit Gökberk and Nusret Hızır. Even Macit Gökberk, who later became a well-known historian of Western philosophy in Turkey, had trouble following Reichenbach's lectures because he had no mathematical background, and, as he himself confessed, auditing von Mises' mathematics classes did not help either (see Kaynardağ 1986). Finally, the law permitted a contract for only five years with no retirement benefits (Çağlar 1999).

These are not the only reasons why Reichenbach's influence was extremely limited. The 1933 Istanbul University reform meant that many Turkish faculty members either lost their jobs or were relegated to a second-class status. Most of them were trained in Islamic philosophy and therefore knew little about the recent developments in Western philosophy, especially, scientifically oriented philosophy. When Reichenbach became the chair of the Philosophy Department, he wrote negative reports about the competence of some of his Turkish colleagues. All of this caused much envy and hostility among not only philosophers but also other academicians, which must have made Reichenbach uneasy and become an additional obstacle against philosophy of science's taking root in Turkey (see Kaynardağ 1986).

Reichenbach's scientific philosophy made its limited impact through his Istanbul University colleagues such as Vehbi Eralp, Hilmi Ziya Ülken, and Nermi Uygur. But the person who was most instrumental in this respect was Nusret Hızır. Hızır studied physics, mathematics and philosophy in Germany and served as Reichenbach's assistant from 1934 to 1937. At the time he was probably the only one who really understood the new logic and philosophy, which he adopted enthusiastically. But, unfortunately, in 1937 he was appointed as a researcher at the Institute of Turkish History in Ankara, and this meant, when coupled with Reichenbach's departure for USA in 1938, that no courses in modern logic and philosophy of science were to be offered for years to come. Philosophy at Istanbul University turned heavily historical, almost exclusively German, under various influences. Nevertheless, Hızır was able to return to teaching philosophy at Ankara University from 1942 until 1968. He also taught briefly at Ecole Normal Superieur in Paris and Middle East Technical University in Ankara after his retirement and disseminated the ideas of scientific philosophy.

From late sixties onward philosophy of science and logic began flourishing in Ankara, especially after Hüseyin Batuhan and Teo Grünberg joined Middle East Technical University and established a philosophy of science and logic program there. It is interesting to note that although both of them began their philosophical career in Istanbul University, they "discovered" the existence of modern logic and the analytic philosophy of Russell, Wittgenstein, Austin, Ayer and Quine on their own. It is also worth noting that five of the contributors to this collection have taken courses at some level with either Grünberg or Batuhan or both. We should add, however, that the impact of the refugee scholars on the Turkish philosophers and their role in the academic and intellectual life in general has not been fully explored.

Let us now say a few words about the history of science in Turkey, which originally had a better footing than philosophy of science. Two names stand out in recent history: Adnan Adıvar, an intellectual medical doctor who wrote the first treatise on the history of Ottoman science, *La Science chez les Turcs Ottomans*, published in France in 1939; and Aydın Sayılı, the first person to receive a PhD under George Sarton at Harvard University in 1942. While Adıvar's book represents popular, narrative, and ahistoric historiography, Sayılı's works represent the analytical, academic historiography, which emphasizes the indispensability of original sources. Sayılı was handpicked by Mustafa Kemal Atatürk, the founder of the Turkish Republic, to be given government scholarship to study history of science abroad. The choice turned out to be more than appropriate; Sayılı became an internationally distinguished historian of science. His major work is *The Observatory of Islam and its General Place in the History of the Observatory*, published in 1960. He formed the first history of science program in Turkey at Ankara University in 1952 and began producing PhDs. The most well known of them is Sevim Tekeli who published on the nature of instruments used in the observatories in the East and in the West in the sixteenth century. There is now a strong tradition of Ottoman science studies at Ankara University. Despite Sayılı's heritage, however, historical studies of

Ottoman science and technology have not completely purged themselves of an ahistoric perspective especially when treating scientific concepts.

This volume begins with an interview with Maria Reichenbach, Hans Reichenbach's wife. One of the interviewers is David Kaplan, who took courses with Reichenbach at UCLA, so the interview gives us an insight into Reichenbach's life both in Istanbul and LA.

Most of the articles in this volume are written by scholars who have done their graduate work abroad. Some of them continue to teach abroad. In this sense the articles reflect a universalist and cosmopolitan character. Part I deals with the philosophy of logic and physics. Part II is concerned with epistemological and methodological issues in science such as confirmation, objective evidence and relativism. Part III contains articles in philosophy of language and mind. Part IV deals with the topic of causality in relation to analytical ontology and action. Finally, Part V is devoted to Ottoman science studies. The volume does not aim to represent an exhaustive survey of philosophy of science, much less of history of science, by Turkish philosophers and historians of science. It merely aims to give the reader an overall idea about their work, which we hope to be of interest to the international community of scholars with similar concerns.

The idea for this book was suggested to us by Robert Cohen. We are grateful to him for the initial impetus. However, the publication of this volume was delayed for a number of reasons. During its preparation, first Arda Denkel in 2000 and then Berent Enç in February of 2003 passed away unexpectedly. It is of some consolation to us that they were able to complete the writing of their articles before their untimely death. Shortly after Enç'es death, another well-known Turkish philosopher, Ilham Dilman, too died. We regret not being able to include a contribution by him. It is to their memory we dedicate this volume.

We received generous help from a number of colleagues and students. We thank Stephen Voss for his suggestions and proofreading, and Rob Bowers and Melis Erdur for the transcription of the interview with Maria Reichenbach. We thank them all. Our greatest debt is to Bürkem Cevher. Without her technical assistance this volume would not have been possible. Finally, we would lilke to thank Charles Erkelens and especially Ingrid Krabbenbos from Kluwer for their generous support, guidance and infinite patience. It was a pleasure to work with them.

Department of Philosophy, Boğaziçi University
Department of Philosophy, Duke University

NOTES

[1] Just to give two examples, Salih Zeki (1864-1921), a distinguished mathematician, translated several books by Poincare including *Science and Hypothesis*, and lectured extensively on the philosophical meaning of the discovery of non-Euclidean geometries. Kerim Erim (1894-1958), another well-known mathematician, wrote on the philosophical aspects of relativity theory, the issue of determinism and probability, and the foundations of mathematics; he was also an active participant in Reichenbach's seminars in Istanbul University.

[2] These include the following: 1) "Felsefe ve Tabiat Ilimleri" (Philosophy and the Natural Sciences), *Üniversite Konferansları* 1933-1934, Istanbul Universitesi Yayınları, 1934. In this opening lecture of the

"General Philosophy" course for the 1933-34 academic year, Reichenbach discusses the relationship between the natural sciences and the "system philosophies" of Descartes, Hume and Kant. 2) "Ilmi Felsefenin Bugünkü Meseleleri" (Today's Issues of Scientific Philosophy), *Üniversite Konferanslari* 1936-1937, Istanbul Universitesi Yayınları, 1937. In this paper Reichenbach describes scientific philosophy as the analysis of knowledge, of the language of science. 3) "Tabiat Kanunu Meselesi" (The Problem of Law of Nature), *Üniversite Konferansları*, Istanbul Universitesi Yayınları, 1937-1938. This lecture seems to have formed the essence of Chapter 10-Laws of Nature of *The Rise of Scientific Philosophy*. 4) "Illiyet ve Istikra" (Causality and Induction), *Felsefe Semineri Dergisi*, Istanbul University Yayınları, 1939. Here, Reichenbach discusses Hume's and Kant's views on causality and induction and argues that a probabilistic approach promises to solve the problem of induction.

REFERENCES

Çağlar, A. *Türkiye Cumhuriyeti'nin 75. Yilinda 1933 Üniversite Reformu* (1933 University Reform in the 75th Year of Turkish Republic). Ankara: TÜBA Yayinlari, 1999.

Kaynardağ, A. "Üniversitelerimizde Ders Veren Alman Felsefe Profesörleri" (German Philosophy Professors who Taught at our Universities). In *Türk Felsefe Araştırmalarında ve Üniversite Öğretiminde Alman Filozofları* (no editor), Ankara: Türkiye Felsefe Kurumu Yayinlari, 1986, 1-31.

Widmann, H. *Atatürk ve Üniversite Reformu* (translation of *Exil und Bildungshilfe*). Istanbul: Kabalci, 2000.

AN INTERVIEW WITH MARIA REICHENBACH AND DAVID KAPLAN

(Conducted by Güven Güzeldere)

M: Maria Reichenbach
G: Güven Güzeldere
K: David Kaplan

M: The first year in Istanbul, everybody invited everybody else among those people who could or would no longer teach in Germany under Hitler. There was not only a Jewish community, but there were also lots of other people who for political reasons did not want to stay or could not stay in Germany. There were about 40 people with their families, and they got a very good contract of five years at academic institutions and had assistants and interpreters. In Hans's case, depending on what the interpreter spoke–if the interpreter talks in German, then he talks in German and if in French, then he talks in French and if in English then in English. After every sentence the interpreter would translate it into Turkish, the students could write down the whole lecture verbatim, sentence after sentence. He was also taking the students–also something new at the time in Turkey–for skiing at Mount Uludağ on the Anatolian side. The faculty lived either in Pera or Bebek or Kadıköy. I was teaching for the children of the professors, and I traveled around and toured into Turkish nursery schools and WYCE, where I taught English and French. I must have taught from 8 in the morning to 8 in the evening. Hans got an offer from UCLA during his five-year contract, but they did not let him go, so he had to finish the five-year contract. During his stay, he instituted some kind of interdisciplinary discussions, as he had done in Berlin also. Of course, that died immediately after he left. After five years almost everybody left. Politically, things became more chauvinistic somehow. Gazi [Mustafa Kemal Atatürk] was dead. So maybe one or two families stayed, but otherwise people came to USA, and later after the war, some returned to Germany, also to Switzerland, I do not know. The generation that knew these people probably is dead now.

G: In Turkey, I found people who are students of students of this generation. Concerning the relation of these academicians with German officials in Turkey... Probably, the German government must have been pressuring Turkey to...

G. Irzık & G. Güzeldere (eds.), Turkish Studies in the History and Philosophy of Science, 7-24.

M: If you had anything to do with them, you had to make use of the people in schools. There was a German school, there was a French lycee, there was American high school for boys and girls—so there were lots of them, and we knew these people.

G: I am wondering if the German government was making life difficult for the immigrants.

M: Yes, but they [the German government] did not have much to do with them. People became stateless after 1938, I think. They did not renew your passport, you know. But if you had an offer from here [USA], you could come in.

G: How big was the whole community?

M: There were 40 families, I think. But there were others who went to Ankara. Not everybody went to Ankara. Gazi was building so many new ministries there, and there were many architects who went to Ankara. All sorts of people went to Ankara. But we stayed in Istanbul.

G: What about Fritz Neumark?

M: He was one of the professors from Germany.

G: Is he still alive?

M: I doubt it. He said in one of his letters that he has passed the Biblical age. The Biblical age is 70, which is today not so terribly old. But I don't know, I have no contact anymore with anybody.

G: How about your family when you first came to Turkey?

M: I came alone.

G: Oh, you came by yourself.

M: Yeah.

G: I wonder, if you came because any other friends came.

M: No, I came with any other family,

G: And you already had a PhD?

M: Yeah, I had my PhD. February 1933. The Nazi party was already there, but I had won it [the PhD degree] just . . .

G: Just before you came to Turkey. You must have had doubts when you were thinking about going to Turkey. I mean, it was probably a place that you had never been before.

M: No, I had never been there before. But it was beautiful. And the people all were very nice, you know.

G: But before you went, was it an easy or a difficult decision?

M: I had all kinds of conflicts and complications. I came with a family who wanted some kind of companion. So, they took me alone. I had been in Italy for nine months. Somehow, that did not work out so well. So, when I came back to Berlin, I heard about this and then I went [to Turkey]. I made friends very quickly in the academic community.

G: And then you taught children of some of these professors.

M: Yes.

G: So, I guess there was a bit of an adventure to go to a place that you had never been before.

M: But I came in 1934 and I married in '35 but with somebody else, not Hans. Hans was already there [in Istanbul], but I think he was married to somebody else.

G: Was that person also an academician?

M: No, he was in business there. But I knew him from Berlin. His mother and my grandmother were friends.

G: Did you speak Turkish in Istanbul?

M: Yes, yes, we knew enough Turkish to get around to do the bargaining, to do shopping, to talk a little bit. And I was back in Istanbul four or five years ago. I do not remember exactly. We were in Cyprus. I was with a friend, and we went also to Istanbul and I retaught myself some Turkish. But, you know, Istanbul had changed so much. There was still the Galata Bridge over the Golden Horn, there was now another one. But it was still very beautiful.

G: Did you recognize some of the places you've been through?

M: Yes, yes, of course. All the mosques, Pera, the Bosphorous, the Golden Horn. O yes, Süleymaniye mosque, the Grand Bazaar...

G: Where did you live in Istanbul?

M: Pera. Everybody lived in Pera, Bebek, or Kadıköy on the Anatolian side.

G: How were relationships among the professors themselves? I think that some liked each other and some probably didn't like each other.

M: Of course, there were mysteries, secrets.

G: I read that Reichenbach was politically very involved before he came, as a student.

M: Well, I tell you. He was involved as a student, but not as a professor. He tried to get a position in Berlin. They held that against him–that he was in socialist student groups, but he was not active later.

G: Is it because he lost interest?

M: No, no. Because he wanted to use his time for writing. I mean. He gave up other things. He gave up chess, because it occupied his mind so much, and he did not want it. He wanted to concentrate and write. And then he was teaching. He did completely different things sometimes. I wrote about it. He was a very good photographer, he did skiing and ice-skating. He was many-sided. He could talk to many levels. He could talk to colleagues, and he could talk to ordinary people, to children... He never had any feeling of condescension. Everybody loved him. The students, very much. I know people who are 70 years old and say "I was in his class". Sometimes I meet people, I give my name and they say they remember him from his famous logic classes. He was a great amuser, great relationer.

G: There was one professor in Duke who is fairly old. He said he was in his class when he was a graduate student.

M: What is his name?

G: Martin Golding. He does philosophy of law.

M: Did he like him?

G: Yeah, everybody I talked to had a great time.

M: Yeah, he was a very good teacher, and he never read anything, you know. He spoke almost automatically. Even at meetings. He spoke freely. Even when people speak to him very rough, you know. Some people often came very rude and some people got kind of confused. Then Hans would stand up and summarize everything and put a little order, you know.

G: Professor Kaplan, were you a graduate student then?

K: No, I was an undergraduate. Then I was a graduate student.

G: Are you from Los Angeles?

K: I was born in LA. So, when I went to UCLA, I took a logic class from Don Kalish. As a freshman I took most of the logic that was being offered.

K: Reichenbach taught two courses. He taught a course that was called 'inductive logic' or something like that at the lower division level. It was a popular course, with a large number of people. And he was a good lecturer to a large number of people. He held your interest, he had lots of facts, he had stories to tell. If you read his books, you get the sense of a person who has a kind of easy, conversational way–even in some of his writings, I think. So he taught this course on inductive logic, and it was kind of a beginning course, not really quite on scientific method but sort of on rationality, looking at evidence, how to weigh evidence and so on. But the main part of the course was devoted to the Bruno Hauptmann and Lindbergh baby kidnapping case. Lingbergh was such a figure, hero in the United States, and the baby was kidnapped and killed actually. And this guy, Bruno, who was an unemployed carpenter or something, was ultimately arrested, convicted and executed, I guess, based all on circumstantial evidence. He always denied that he had done it, but it was a very complex case.

M: But Hans used it as a schema.

K: He used it as a schema, right, but there were a lot of details. It was kind of fascinating.

M: You have the evidence, and then you deduce something.

K: Yeah, and what was good evidence and what was bad evidence and so on.

M: Both deductive and inductive inference.

K: Well, deductive and inductive inference, but also something about how scientists operate, how you try to think rationally. And at the end of the course, he actually did a little bit of technical stuff. Kolmogorov axioms, and a little bit of probability. Then he taught an upper division course on his probability book. That was a very serious course. So I took the first course and then I took the second course. He actually died at the beginning of the term.

M: Then I think Salmon took over.

K: No, Richard Montegue. They actually brought Richard Montegue, who was still being tortured by Tarski, and not being allowed to get a Ph.D. So he was brought down, and he came down and picked up that course.

M: Why wasn't he allowed to get his Phd?

K: That is another story. So there was this course, Philosophy 30, we still teach 31, it's deductive, 30 is inductive. So we were in this course. I had a girlfriend, an opera singer, and she introduced me to this friend of hers who was in psychology, who happened to be taking this course. She was very smart and a very good psychologist. She is my present wife. So we studied together, my wife came to every session of class, and she would kind of listen listen listen, and then she'd start taking notes like mad, and then she'd kind of listen listen listen . . . And I kind of came to class and didn't come, but when Hans got to the real stuff at the end, you know, the Kolmogorov axioms, then I really started to take notes. My wife, my present wife, professed she didn't really understand the technical part at all, so I spent most of my time explaining her. But we didn't study for the final exam together. We took the exam, she got an A, and I got a C. The teaching assistants went to Hans and said "you can't give this boy a C in your inductive logic course, he's the teachers' pet of the department and logic". I was a sophomore, but I was so into doing logic. I knew all the TA's, they all hung out in one place, and all the students taking logic course would go to that coffee shop. And I was always there doing problems. Reichenbach raised my grade to a B, so in the end I actually got a B just on the basis of this appeal. So, in the final exam there was only question on the Kolmogorov axioms, one small question, and all the rest was on other stuff. As I've learned later on, when I talked to my wife, as I said, who is a clinical psychologist, she did it in all her courses. Reichenbach would lecture, and every once in a while, she would think to herself "ok, this is what he's going to ask on the exam", and she would write it down. I never could tell. But she could just tell from the body language and the like, when the person was talking about something that was really important to him. So then she would start to take notes, notes on exactly what was asked on the exam. But that's really a story about me. Hans was an entertaining lecturer. He held your attention, you got something out of his course, it was intellectual, it was serious. It was very accessible. Then when we moved into the upper division, then he started talking about theorems. I would say that stuff was relatively inaccessible, but he did a good job of explaining it. It was just a technical course. And I think you see that in a lot of his other writings, you know, there is *The Rise of Scientific Philosophy*, he kind of reaches out and does something that's almost ideological, he wants to bring people to adopt a certain world view. And then there's all that technical stuff on space and time that is also philosophical.

G: Well, when I read some of his writings, it's easy to see that there's a lot of excitement and a sense of things changing. I guess that's the sense you call they have something ideological: the whole scientific worldview.

K: Yes, yes.

G: How do you think that vision translated into the next decade and into today's philosophy? I don't see, for example, that kind of an excitement these days.

K: There was a big change. There's an interesting question there. I don't know very much about Hans' politics. I know that one of his first couple of publications were for probably what was a student kind of a political movement.

M: As a student, he was a member of the socialist student movement.

K: Yeah, he was quite a political.

M: But as a student.

K: As a student, yeah. I think, if you see, that's true of a lot of them, I mean Carnap, for example. Carnap was also very political all the way to the end of his life.

M: He called himself a socialist.

K: He called himself a socialist all the way through.

M: Hans gave up, concentrating on philosophy. But, you know, the most terrible time here was the early fifties, there was this terrible loyalty oath.

K: It was the McCarthy era, and in the university California, because we were a state university, there was this oath you had to sign, that you did not belong to any organization that aimed to overthrow the government by force.

M: It was not because people were socialists, it was because of principle they didn't want to sign.

K: What they asked you to sign was not something that most people would object to. What they objected to was the speech act. It was being forced to sign a loyalty oath, you know.

G: Was that only for California?

K: This was in California, but there were similar things that went on in other places. At that time University of California was one of the great universities of the world. If you didn't sign, you didn't get a paycheck. So there was a lot of fuss about it. In the end there were only one or two people who did not sign. One of them was a young physicist at UCLA. He didn't sign, he lost his job, he got a job working for the bureau of numerical analysis. It is interestingly enough a federal agency working on campus. So he kind of moved his office out of one building into another. Subsequently he became the president of the University of California, which is quite interesting, I mean, it says something about the change. When he retired as a president of University of California he became the chairman of the board at MIT. And then he retired from there, he's back at UCLA. After a period, they abandoned the oath, and I guess it was ruled unconstitutional. But it took a long time. It was a terrible period. I was getting picked up on campus. But it was kind of benign. They had a rule at that time on campus called "Rule 23"–I'll never forget it–that said that you could not distribute political literature on campus. It was supposed to be a politics-free zone. Of course, basically the constitution now says that the governmental areas are where we can participate in such activities. But they had this campus rule.

M: They wanted to put something on Hans too, and he was very much afraid. The man who saved him was a lieutenant commander who was a part of the department.

K: I used to get picked up distributing socialist literature. But they just took me to the dean of students, and the dean of students would just say "David, you're not supposed to be doing this". But it was a very benign thing, and then off I went, and a couple of months later they took me up again. It was a very graceful arrangement. But the loyalty oath was a different matter because it went through bureaucracy.

G: Did it apply to all state workers in the state of California?

K: Yes, everyone who got a pay check. I'm trying to remember why the fuss was at the University of California. But it must have applied to all state employees. But maybe the university adopted some special strong form, or stronger form of it. What happened was that over time they modified it so that it was less offensive, and then they abandoned it. It was a funny period. Eisenhover was president, so this kind of bland politics, with this ugly underside of McCarthyism. Anyway, the thing that's interesting is that all these guys, Carnap and Reichenbach, had a very radical philosophical vision. You know, Carnap used to talk about overcoming metaphysics. He took it very seriously. So they had a radical intellectual vision, and they had a relatively radical social vision also. I mean it all ran together. I think there was this new way of doing philosophy and thinking of philosophy. I think it's plain that Reichenbach and Carnap and others wanted to persuade people that that was the way that philosophy should be done. There were a lot of excesses in the past, a lot of pseudo-problems, you know, that people had struggled with and worried about.

M: But it was not only ideology, though. They thought that they could give some better answers on the basis of physics and mathematics. I was a believer too.

K: So was I. When I came to UCLA as an undergraduate first and then as a graduate student, Russell was at UCLA, Reichenbach was in UCLA, and then came Carnap and Alonzo Church, who did not belong to that movement, but was another great kind of logician philosopher. They would axiomatize the two systems and then compare the Platonism and nominalism, and we could compare the systems from the meta-linguistic point of view. And there was Richard Montegue, of course. He was like a burning flame, you know. I think many of us in the graduate school really saw us as part of the movement. Then there was also the Wittgensteinian tradition that was dominant in Oxford and the so-called ordinary language philosophers.

M: Analytical philosophers...

K: Well, they are analytical philosophers, but they are not in the logic tradition. There's this famous paper by Strawson "On Referring", which is a response to Russell's "On Denoting". It says at the end that there's no precise logic of ordinary language, that there can't be. Ordinary languages are too subtle. The positivist movement wanted to push the rest of the philosophers aside. These ordinary language philosophers were saying "you can occupy a space if you want to, but it's mathematics, it's not philosophy". So in the 50's, when I was a graduate student, there was the ordinary language vs. logic struggles, kind of going back and forth.

G: Was that part of the UCLA tradition as well?

K: Well, there wasn't much ordinary language tradition at UCLA. The tradition was Reichenbach, Carnap...

M: But don't forget that there was enormous amount of philosophy of science that Hans taught, it is not only logic.

K: Yes. Originally the positivist movement and then the logical empiricist movement, and all of the stuff that flowed from that, with all the technicality attached to it. Actually when I was a graduate student, the UCLA philosophy department required that in order to get a Phd, you had to get an MA in another field.

M: But that was after Hans's doing.

K: Was that Hans's doing? I didn't know where it came from because it was in place when I came. They abandoned it later on. To me it was a reflection of a certain ideological position; that is, learn some science, and then you'll be able to do some philosophy about it, because philosophy is the foundations of science. It isn't a domain in itself. So there was a lot of fervor about all that stuff at that time, which I think has really died. There isn't somebody doing philosophy of science in UCLA.

M: I think that's very sad.

K: But philosophy of science is still an important field in philosophy, and it's being pursued well.

M: But not at UCLA.

K: Not at UCLA, but it's being pursued well at a lot of places.

M: Boston...

K: Irvine, San Diego, Stanford and so on. However, I don't know what the views are in these departments concerning methodology and history of science. But going [back] to philosophy, my sense is that generally speaking the idea was that by bringing science and technicality together all the problems of philosophy will be solved. I think that idea has burned out some time around 1970 or so.

G: Yes, that's what seems to have happened.

M: And Carnap died, too.

K: I don't attribute it to any particular death. It's a cultural change.

M: I don't know whether that was something global. There are also foundational issues.

K: People are still interested in foundations of physics.

M: Biology...

K: There's a lot of interest in foundations of biology. As I said, philosophy of science is an extremely interesting topic, and it raises issues in metaphysics and epistemology that we're all interested in. But the overcoming of the old, putting in of a new perspective... We had a lecture recently. You know who Kit Fine is?

G: Yes.

K: He's a logician and a metaphysician. He's an extraordinarily brilliant person and a very nice guy, too. He does metaphysics, but uses a lot of mathematical machinery. He does Aristotelian kind of metaphysics that I think Reichenbach, Carnap, and probably Montegue and so on would all think: "What is this? This is speculative metaphysics". You know Reichenbach's book *The Rise of Scientific Philosophy* opens up with this wonderful quote from what he calls "speculative metaphysics". I read it out in some meeting, I can't remember what it was. I just loved it. So what Kit is doing obviously, it isn't like that. But it's more of a variety of that than it's a variety of foundations of science or something like that. [Anyway], in terms of the movement, I don't know if there's a strong, powerful kind of ideology that's going on in philosophy [now]. I think there's been a globalization of analytic philosophy. I don't see great ideological battles [anymore]. When I was a graduate student, I thought it was fun. There was an enemy out there. You read their articles and made fun of them, and it was all very adrenalin rich. Now there's just these hard problems, and I don't know what do you think about them.

M: What should be done about them?

K: There is a way in which what these guys were doing has won. Analytic philosophy, which was a part of their tradition, with the use of techniques from logic, from the formal fields, is the way philosophy is done everywhere. There's a few places in Europe, as you're surely aware, in France, and there are some other places where they have a slightly different tradition, although they want to get themselves lined up with the analytic tradition. They don't want to stand up and say, the way that Richard Rorty does, "we're against analytic philosophy". They want to say "we have our own way of doing it that we want to call attention to these problems, we want to be in communication with you".

M: But what should be the aim?

K: I don't know what Hans and Carnap and people like them would be saying and doing now. Look, the point of the rise of scientific philosophy is "get out of the way, here we come", right? So Hans opens up [*The Rise of Scientific Philosophy*] with this passage and say "Let me tell you what the new way of things is".

M: But this is a popular book, you know.

K: I know, I know. But Carnap believed in it, Reichenbach believed in it, and Montague believed that all philosophy was a definitional extension of set theory. And these are all great, world-class minds that had huge accomplishments. But along with the accomplishment was this kind of challenge to the way, a kind of disdain for what either a majority or a minority of the people in the discipline were doing.

M: What do we do now? After you said all this, what should we aim for?

K: To do analytic philosophy, and to do it well. I think that's what the legacy is, but there isn't an ideological war that's going on now, and I think it's because in some sense they won, you know, although nobody now says that they are logical positivists or logical empiricists.

G: I think it's the continuation of that legacy that shaped today's philosophy, right?

K: I think they won really.

M: So where do we go now?

K: We'll see.

M: That's so vague.

G: I want to come back to that question, but I also think that the zeal is also no longer around. That's not because old metaphysics came back. It's not that kind of...

K: I think it's gone. But the interesting thing is that you do see people like Kit, with his powerful mathematical mind, coming back and trying to do things that are deeply metaphysical. You see Saul Kripke, coming in and talking about de re modal properties, right? Doing deeply metaphysical kinds of things.

M: But that's nothing new.

K: Metaphysics has always been going on, but what you see is that there's more dipping into substantive metaphysics, kinds of problems Aristotle worried about, and maybe even the kinds of problems that Hegel and other people worried about, not in that way. And I think Carnap and Reichenbach would not have done things the way people are doing now. Carnap tried to stay at the inside questions, outside questions, questions that are internal to a theory, external to a theory. How do you evaluate the theory? Well, the way you evaluate any scientific theory, productivity, efficiency, and so on. Stay at the meta-language level, at the principle of tolerance; you can use whatever object-language you want, but use first order logic in the meta-language to describe the semantics and the syntax. There was a lot of, kind of, keeping your hands out of these dirty substantive metaphysical problems and trying to recast them in a way that they can actually be resolved by science and mathematical methods.

M: They did that.

K: That's the way they did it. But I think many of their heirs, and I think of people like Kripke, Kit, myself, are dipping more deeply into the substantive metaphysics and arguing about issues that are not going to be resolved by mathematics. Maybe it took a lot of cleaning in order to be able to do that.

G: What do you think will dominate the philosophical agenda in the near future?

K: Ask me what the stock market is going to do in two years!

G: Any speculations?

K: I don't really have one. I've seen how things change. There are still a lot of people that are interested in philosophy of language. But a lot of philosophy of language has merged with linguistics. A lot that went on in logic, foundational issues in logic has merged so much with mathematics. Even my colleague, D. A. Martin, who is a set theoretician says that he really cannot read everything in the literature. Philosophy over the ages has spun off other subjects. At the present time there seems to be less spinning off and more sort of interdisciplinary stuff going on. Now, you're a philosopher, you think of yourself as a professional philosopher.

G: I do.

K: But you're going to these conferences working in areas such as brain research...

G: Yeah, I do some brain research myself.

K: So there you are.

G: I am very much trained in the Reichenbach tradition, my teachers were physicists and mathematicians turned philosophers, had worked with people like Tarski. And I learned some philosophy of language as an undergraduate. Now I do philosophy of psychology, philosophy of neuroscience, and there I very much think one has to know what is really going on in the particular sciences.

K: But you also think that as a philosopher you bring something to the table.

G: Oh yes.

K: I think that in a lot of these areas, there isn't going to be a spinning off, you know, in the way that psychology sort of sprung out of philosophy, although there's still some philosophical psychology. Some other fields, I think will be merging together. When that happens, the philosophers are going to have a tough time because they're not very good at doing interdisciplinary stuff. One of the things about the positivists was that they were all trained in science, and so they all came to philosophy comfortable with the science, with the mathematics, physics, whatever it was, right? There are now a lot of philosophers around who are not comfortable with these.

M: Now they should be in biology and genetics and all that stuff, it's a fantastic field.

K: Yes, there's a lot to learn, and there's a lot of turning in. In my own department there are some reasonably intense disagreements that are sort of tapered over by good human relations about looking at work and saying "But this isn't philosophy, this person did some philosophy, this is philosophy and that was philosophy, but this isn't really philosophy, this is really mathematics, it's linguistics or it's something else".

M: That's what happened to Hans. It was so difficult for him to get a position in Berlin because [theorizing about] probability is not philosophy!

K: Oh yeah, it's not philosophy, but it's not mathematics either.

M: But the philosophical questions are there, right there.

K: It is actually one of the few things that really drives me up to the wall. Here people say about the work of other people, people whose philosophical work they admire–so we're not talking about people who are third rate, we're talking about people who are first rate–"well, this work, that's philosophy, and that's very important work, and they're extremely good at it. But their interests have drifted over to this kind of interdisciplinary kind of subject, and what they are doing now is not philosophy anymore". This just drives me wild, it's such a wrong attitude.

M: Maybe they are not trained to do that kind of work.

K: I'm not sure exactly what the source of it is. I think there are people who work in the interdisciplinary areas who say that they are doing philosophy, and I would say that what they are doing is not philosophy. Not because it's something else, but because it's nothing. They're not really bringing something to the table at all. They're carrying some jargon back and forth between two disciplines . . . I believe that there are issues that are almost pure philosophical matters that don't have to do with the foundations of some other science. But I also think that [philosophy] is continuous [with other disciplines], and I don't think we can draw a line. I think of myself as a philosopher, I don't identify myself as a mathematician. Sometimes I identify myself as a logician, that's dear to me also . . . I came to hear Maria talk about the time in Turkey.

M: We did that a bit before you came.

G: What's your nicest memory that was impressed on your memory from the days between 1934 and 1938?

M: It was a strange atmosphere. Due to their contract people had a very comfortable life, but everybody knew we were not going to be able to put our roots down there, you know. After the five years were up, the Turks wanted Hans to become a Turkish citizen, and that he didn't want to do. Quite a few learned enough Turkish to get around, but only two spoke fluently. Everybody spoke six or seven languages. I think that on the whole they got along themselves very well. They were excellent people. We went hiking, sailing and swimming, and they stared at us because swimming on the Bosphorus wasn't common. We kept the German conventions and customs. I don't think there was very much social interaction with the Turks.

G: Even though Turkey at the end declared war against Germany, there was I think some people who were sympathetic towards the German government before and during the war in Turkey. Was that a problem for you?

M: I don't know enough about that, I really don't.

K: Were most of you [immigrants] Jewish?

M: No, some were, of course, but others were there [in Turkey] because politically they didn't want to stay, they couldn't or they wouldn't stay [in Germany]. And they were very decent people.

K: What percentage was Jewish? Could you make any estimate?

M: Oh, maybe one third was not Jewish. Quite a few were, and some of them went back to Germany later.

K: And probably many of the ones who were Jewish weren't religious.

M: No, they were not.

K: Like yourself?

M: No, no. I was brought up as a Protestant, you know.

K: You were brought up as a Protestant?

M: My parents were Protestant, and they were baptized and I was baptized.

K: Did you consider yourself Jewish?

M: Of course, because of descendancy, racially also, but I have never been in a Synagogue. The Jews, especially in Berlin or in Germany, were very assimilated. But then there was the idea that maybe we could escape anti-Semitism if we became Protestant. Also you could get positions as judges and so on, it was easier also for professors.

K: So when you say of the people who were in Turkey on these contracts in the academics, about a third weren't Jewish,

M: Nobody was accusing or anything...

K: I understand, but where did you put yourself, in the one third or in the two thirds?

M: Today?

K: No, no, then.

M: I knew I was Jewish.

K: So you would put yourself in the two thirds.

M: I would say so, absolutely.

K: I see. Though your parents were baptized, you were baptized...

M: Yeah, but actually they came from Jewish stock, both sides. But I know in my family there were intermarriages, so the children were half Jewish. They stayed in Germany and had to work in factories, etc., but they survived.

K: I asked the question because I was interested in how the whole group interacted with the Turkish culture, which was a totally different culture. So a lot of them were Jewish but they weren't religious at all, right?

M: Right.

K: And they hadn't been practicing Jews maybe for a generation, but they were all dropped down into a religious culture that was foreign to them.

M: As for Hitler if your grandparents were Jewish, you were Jewish.

G: Was there a group among the immigrants who attended synagogue or church in Istanbul ?

M: No, no.

G: Turkey has long had a Jewish population...

M: There were Spanish Jews...

G: And they are an influential community, financially, and so on.

K: So you don't know how they interacted with this group of people.

M: No. But in Istanbul there were Greeks, Arabs, Turks, Persians, there were Germans, French... It was such a mixture. Everybody spoke different languages, went to different schools...

K: One of the things that's interesting to me is the sense of a kind of island culture—these people on university contracts were dropped down in the middle of Istanbul. I don't know very much about history, and I don't know much about Istanbul, but just from popular culture I know enough to know that it was an intense focus of international concern. The place was filled with spies from every nation, and there was all kinds of political stuff going on. Maria's sense of it is an enclave of ex-patriots; they got together with one another, they talked old politics, etc.

M: Hans did the same thing as he did in Berlin. He tried to do some interdisciplinary thing going, lectures and so on, [also] asking students to study some kind of science outside of philosophy.

G: Was there any local anti-Semitism that you faced?

M: No, we didn't have enough contact.

K: It must have been an atmosphere in Turkey at this time that was extremely welcoming to these [activities]. When things like that happen, there are always strains in the population, there are always people that are fighting against them, and say "you're wiping out our culture". And these people were brought in and dropped down from above. Isn't that right, because a decision was made by one person basically that "we are going to move to modernity".

M: It was really revolutionary in that sense.

G: How were interactions with the students? Were there any Turkish students who later came to UCLA to study with Hans Reichenbach, or was there anybody who kept in touch with them?

M: He had one student. She taught somewhere in Northern California...

G: But was there a sense that philosophy students or the general outlook on philosophy was also being transformed in Istanbul?

M: Oh, yes, enormous influence.

G: So he also had the sense that he was really transforming...

M: Oh yeah. And he always had close relations with the students, he talked to them outside of lectures, and he took them skiing.

K: You know, Maria and Hans were not married.

M: No, we were not married. I have something here. This is the same picture but that is different.

G: Is there any picture of you two together?

M: I have so many. This one Hans took on the Bosphorus, wonderful picture. He was a photographer, yeah. These are pictures from his childhood...I was very ambivalent towards Germany, I still am very ambivalent. And I try to avoid who are still alive and were active under Hitler. I have very good friends, German friends, but they are too young to have been Nazis. I select them carefully.

G: The '52 visit was your very first time?

M: Oh, that was because Hans was invited to give a lecture and then we traveled around, visited his sister and her family, and then we went to Switzerland, my family from Brazil came there. And then there was a logic congress in August, and we went back to Paris.

G: Was it emotionally difficult to go back and visit?

M: Well, I had sort of kept it to myself, I still do that. I don't want to spend my money in Germany, no. I have a good friend there, but he is also too young to be a Nazi. He is my editor for the German edition of the collected works . . . I went back to Germany later, I also was in Berlin. And it's different because the streets are very different from what I remember. And then I was back in 91, I was back in East Berlin. This was the time that the people in East Germany lost all their academic positions because they were communists . . .

G: Was any part of your or Hans Reichenbach's family left behind in Germany?

M: Yes, his sister.

G: Did they have a hard time during the war?

M: Later, one of his nephews became a mayor in a little town. He had another nephew, who was Italian. One brother was here a mathematician and a physicist in the East, and one was, Bernard, he was in London. His wife was very good friend of Maria, Otto Neurath's wife. Hans by the way shared an office with Bertrand Russell when he was UCLA. He made pictures of him. In 62 I was in London, and I was invited and he [Russell] picked me up. He had just sat on the street against the bomb. He was 89 years old. He talked about that.

K: Did you say that Hans made a photograph of Russell at UCLA?

M: Yeah, I think I have a photo.

K: I'd love to be able to borrow that and make a copy of it. I've been searching around to get a picture of Russell when he was at UCLA. I haven't been able to locate one . . . Oh my goodness, that's Hans's writing, isn't it?

M: Yeah.

K: Will you let me get a copy?

M: Yeah, but swear that you will bring it back . . .

PART I

PHILOSOPHY OF FORMAL AND PHYSICAL SCIENCES

TEO GRÜNBERG

DEMARCATION OF THE LOGICAL CONSTANTS AND LOGICAL TRUTH IN TERMS OF ANALYTICITY[1]

1. INTRODUCTION

The definition of logical truth (or validity) in particular languages presupposes an enumeration of the logical constants. But a fully satisfactory solution of the problem of *demarcation* of the logical constants in general semantics, i.e., "the question whether and how 'logical' and 'descriptive' (in the sense of non-logical) can be defined in terms of other semantical terms"[2] has not been published until now. The purpose of this paper is to contribute to the solution of that problem by devising—at least for usual languages—a criterion of demarcation of the logical constants, and thus of logical truth, in terms of analyticity. The essential points of our criterion of demarcation can be roughly stated as follows.

1. Let $\mathscr{L}$ be a formalization of some portion of ordinary language endowed with a relation $\vdash_{an}$ of *analytic implication* and thus with a set of analytic sentences.[3] Γ being a vocabulary in the sense of a set of constants (i.e., primitive constant symbols) of $\mathscr{L}$, a Γ-*preserving interpretation* (or model) is an interpretation of that language which keeps invariant the meaning—determined by the intended interpretation—of (at least) the constants belonging to Γ. If L is the set of logical constants of $\mathscr{L}$, then the set of admissible interpretations (used to define the usual semantical concepts such as validity and satisfiability) consists precisely of the L-preserving interpretations.[4]

2. A Γ-*truth* is a sentence of $\mathscr{L}$ which is true under all Γ-preserving interpretations. The *logical truths* are the L-truths. We say that a set X of sentences Γ-*implies a* set Y of sentences, or that the sequent $<X, Y>$ is Γ-*valid*, iff for every Γ-preserving interpretation *a*, whenever every member of X is true under *a* then some member of Y is true under *a*. We write then $X \models_\Gamma Y$. A sentence A is Γ-true, iff the empty set Γ-implies the unit set $\{A\}$.[5]

3. An *analyticity-preserving* vocabulary is a set Γ of constants such that $\models_\Gamma \subseteq \vdash_{an}$, i.e., the relation of Γ-implication is included in the relation of analytic implication. It is a fact that the set L of logical constants is analyticity-preserving. In particular, all logical truths are analytic.[6]

4. We make the conjecture that (conversely) the largest analyticity-preserving set of constants, satisfying possibly some additional conditions, constitutes the set of logical

G. Irzık & G. Güzeldere (eds.), Turkish Studies in the History and Philosophy of Science, 27-38.

constants, at least in case of usual languages. Such a conjecture provides a *criterion of demarcation* for the logical constants.

2. FORMAL DEVELOPMENT

Let the meta-variables A, B range over sentences, X, Y over sets of sentences, c over constants, Γ over sets of constants, and *a*, *b*, *i* over interpretations. An *interpreted language* $\mathscr{L}$ is an ordered septuple of the form

$$<\mathscr{C}, \mathscr{S}, \mathscr{I}, \mathrm{i}^*, \mathrm{Pr}, \vdash_{\mathrm{an}}, \mathrm{L}>$$

where $\mathscr{C}$ is the set of constants, $\mathscr{S}$ is the set of sentences (or formulas), $\mathscr{I}$ is the set of possible interpretations, i* is the intended interpretation, Pr is a relation called the meaning-preservation relation such that $\mathrm{Pr}(a, \Gamma)$ expresses that the interpretation *a* preserves the meaning of the constants belonging to Γ, $\vdash_{\mathrm{an}}$ is the relation of analytic implication such that $\vdash_{\mathrm{an}} \subseteq \mathscr{PS} \times \mathscr{PS}$ (where $\mathscr{P}$ stands for "the power set of"), and L is the set of logical constants of the language $\mathscr{L}$. We assume that $\mathscr{L}$ satisfies certain axioms which will be introduced in due time.

Axiom 1.
$[\mathrm{Pr}(a, \Gamma_1) \,\&\, \Gamma_2 \subseteq \Gamma_1] \rightarrow \mathrm{Pr}(a, \Gamma_2)$ (for every $\Gamma_1, \Gamma_2 \in \mathscr{PC}$)

If *a* preserves the meaning of all members of Γ_1, it obviously preserves also the meaning of the members of the subset Γ_2 of Γ_1.

Definition 1.
$\mathscr{I}_\Gamma = \{a : \mathrm{Pr}(a, \Gamma)\}$
$\mathscr{I}_\Gamma$ is called the set of Γ-*preserving interpretations*.

Definition 2.
Γ-implies Y or $X \models_\Gamma Y$ for short, iff $\forall a[[a \in \mathscr{I}_\Gamma \,\&\, \forall \mathrm{A}(\mathrm{A} \in X \rightarrow \mathrm{A}$ is true under $a)] \rightarrow (\exists \mathrm{A}(\mathrm{A} \in X \,\&\, \mathrm{A}$ is true under $a)]$

Definition 3.
Sequent $<X, Y>$ is Γ-*valid* iff $X \models_\Gamma Y$

Definition 4.
A is Γ-*true* iff $\varnothing \models_\Gamma \{\mathrm{A}\}$
A is a *determinate truth* iff for some Γ, A is Γ-true.

Definition 5.
Set X is Γ-*satisfiable* iff not $X \models_\Gamma \varnothing$

Definition 6.
A is *analytic* iff $\varnothing \vdash_{\mathrm{an}} \{\mathrm{A}\}$

Definition 7.
A is *contradictory* iff $\{A\} \vdash_{an} \varnothing$

Axiom 2.
$\vdash_{an}$ has the properties of a consequence relation: it is reflexive, monotonic, cut-free, etc.

Axiom 3.
$\vdash_{an} \subseteq \models_{\mathscr{C}}$
In particular, if $\{A\} \models_{an} \{B\}$ and A is Γ-true, then B is Γ-true.

Proposition 1.
$\Gamma_1 \subseteq \Gamma_2 \rightarrow \mathscr{I}_{\Gamma_2} \subseteq \mathscr{I}_{\Gamma_1}$

Axiom 4.
Pr $(i^*, \mathscr{C})$

Proposition 2.
$i^* \in \mathscr{I}_{\Gamma}$

Proposition 3.
$\Gamma_1 \subseteq \Gamma_2 \rightarrow \models_{\Gamma_1} \subseteq \models_{\Gamma_2}$

Proposition 4.
$\Gamma_1 \subseteq \Gamma_2 \rightarrow$ (A is Γ_1-true $\rightarrow$ A is Γ_2-true)

Proposition 5.
A is Γ-true $\rightarrow$ A is $\mathscr{C}$-true.

Proposition 6.
A is a determinate truth iff A is $\mathscr{C}$-true.

Proposition 7.
There are non-determinate truths.

For example, the sentence

$$\exists x \exists y \sim x = y \tag{1}$$

of a first-order language with identity is a non-determinate truth.[7] It is indeed false in an interpretation with a domain having less than two elements.

Proposition 8.
A is analytic $\rightarrow$ A is $\mathscr{C}$-true.

Definition 8.
Γ is *analyticity-generating*, or Ag(Γ) for short, iff $\models_{\Gamma} \subseteq \vdash_{an}$.

Axiom 5.
$Ag(L)$

Proposition 9.
$[Ag(\Gamma_1) \& \Gamma_2 \subseteq \Gamma_1)] \rightarrow Ag(\Gamma_2)$

3. FIRST TENTATIVE CRITERION OF DEMARCATION: CD 1

By Axiom 5 every set of logical constants is analyticity-generating. We conjecture that, conversely, every analyticity-generating set of constants is logical. We obtain then the following criterion of demarcation of the logical constants.

CD 1. $\Gamma \subseteq L \leftrightarrow Ag(\Gamma)$ (for every $\Gamma \in \mathscr{P}\mathscr{C}$)

We derive from CD 1 the following propositions.

Conjectural Proposition 10.
$\forall i\,[i \in I \rightarrow Ag(\Gamma_i)] \rightarrow Ag \cup \{\Gamma_i\}_{i \in I}$

Conjectural Proposition 11.
$Ag(\cup \{\Gamma : Ag(\Gamma)\})$

Conjectural Proposition 12.
$L = \cup \{\Gamma : Ag(\Gamma)\}$

Unfortunately CD 1 is subject to a counter-example[8] and, therefore, must be abandoned. The counter-example consists in a first-order language $\mathscr{L}$ with a unique non-logical constant, viz. a monadic predicate P such that in the intended interpretation i*, P is true of infinitely many objects and false of infinitely many ones. More precisely the domain $|\,i^*|$ of i* includes a subset D such that the sets $D \cap i^*(P)$ and $D - i^*(P)$ are both denumerably infinite. ($i^*(P)$ is the extension of predicate P under i*). Then $\models_L = \models_C$ where L is the set of logical constants of $\mathscr{L}$ (say $L = \{\sim, \&, \forall\}$) and $\mathscr{C} = L \cup \{P\}$. Since $Ag(L)$, it follows that $Ag(\mathscr{C})$. Hence $\mathscr{C}$ is "logical" under CD 1 despite that it contains the non-logical constant P. Now in order to show that $\models_L = \models_C$ or for that matter $\models_C \subseteq \models_L$, it is sufficient to prove that every L-satisfiable set of sentences of language $\mathscr{L}$ is also $\mathscr{C}$-satisfiable. The proof is as follows.

Let X be any L-satisfiable set of sentences of $\mathscr{L}$. There is then an L-preserving interpretation *a* which satisfies X (i.e., such that every member of X is true under *a*). By the Löwenheim-Skolem theorem there is an interpretation *b* with a countable domain $|\,b\,|$ and which satisfies X. Then there is a 1-1 mapping f: $|\,b\,| \rightarrow D$ such that for every $b \in |\,b\,|$:

$$f(b) \in \begin{cases} D \cap i^*(P), & \text{if } b \in b\,(P) \\ D - i^*(P), & \text{if } b \notin b\,(P) \end{cases}$$

We can now uniquely define an interpretation i by means of the following two conditions.

(i) $|i| = f[|b|]$ (where $f[|b|]$ is the f-image of set $|b|$).

(ii) $i(P) = f[b(P)]$.

The defined interpretation i is a $\mathscr{C}$-preserving interpretation which satisfies X. Hence every set X of sentences which is L-satisfiable is also $\mathscr{C}$-satisfiable.

4. THE SECOND TENTATIVE CRITERION OF DEMARCATION: CD 2

In the counter-example mentioned in sec. 3,

$$\models_{L\cup\{P\}} = \models_L \tag{1}$$

or equivalently

$$\models_{L\cup\{P\}} \subseteq \models_L . \tag{2}$$

This means that the constant P does not contribute to the generation of additional valid sequents. We say, therefore, that c is not "non-idle" within the set L ∪ {P}, and we make the following definitions.

Definition 9.
c is *non-idle within* Γ, or Nd (c, Γ) for short, iff $\models_\Gamma \neq \models_{\Gamma - \{c\}}$

Definition 10.
Γ is *non-idle*, or Nd (Γ) for short, iff $\forall c(c \in \Gamma \rightarrow \models_\Gamma \neq \models_{\Gamma - \{c\}})$

Proposition 13.
$\mathrm{Nd}(\Gamma) \leftrightarrow \forall c(c \in \Gamma \rightarrow \mathrm{Nd}(c, \Gamma)]$

Proposition 14.
Nd (c, Γ) iff $\exists X \exists Y(X \models_\Gamma Y \ \& \ {\sim}X \models_{\Gamma - \{c\}} Y)$

The set of logical constants of usual languages is non-idle so that we can adopt the following axiom.

Axiom 6.
Nd (L)

For example consider the set $L = \{\sim, \&, \forall\}$ of logical constants of a first-order parametric language. We show below on the basis of Propositions 13, 14 that Nd (L) holds. ('p' is a sentential parameter, or place-holder.)

1. Nd (∼, L), since
   $\{p\} \models_L \{\sim\sim p\}$, but not $\{p\} \models_{L-\{\sim\}} \{\sim\sim p\}$

2. Nd (&, L), since
$\{p \& q\} \models_L \{p\}$, but not $\{p \& q\} \models_{L - \{\&\}} \{p\}$

3. Nd (∀, L), since
$\{\forall x \forall x p\} \models_L \{\forall x p\}$, but not $\{\forall x \forall x\ p\} \models_{L - \{\forall\}} \{\forall x p\}$,

therefore Nd (L).

We may now attempt to avoid counter-examples by revising our criterion of demarcation in the following way.

CD 2. $\Gamma \subseteq L \leftrightarrow [Ag(\Gamma) \& Nd(\Gamma)]$ (for every $\Gamma \in \mathscr{P}\mathscr{C}$)

However CD 2 is also confronted with a counter-example. Indeed consider a first-order language $\mathscr{L}$ containing just two non-logical constants, viz., the monadic predicates P and Q. Then $L = \{\sim, \&, \forall\}$ and $\mathscr{C} = L \cup \{P, Q\}$. Suppose now that in the intended interpretation, both P and Q are true of infinitely many objects and false of infinitely many ones. Then it can be shown, in a way similar to the proof in sec. 3, that

$$\models_{L \cup \{P\}} = \models_L \tag{1}$$

$$\models_{L \cup \{Q\}} = \models_L \tag{2}$$

so that the sets $L \cup \{P\}$ and $L \cup \{Q\}$ are not non-idle. Let us assume that the sentence

$$\forall x(Px \leftrightarrow Qx) \tag{3}$$

is analytic. Since (3) is analytic the set $\mathscr{C}$ is analyticity-generating.[9] The set $\mathscr{C}$ is also non-idle since

$$\varnothing \models_{\mathscr{C}} \{\forall x(Px \leftrightarrow Qx)\} \ \& \ \sim [\varnothing \models_{\mathscr{C} - \{Q\}} \{\forall x(Px \leftrightarrow Qx)\}] \tag{4}$$

holds. But by hypothesis P and Q are non-logical so that $\sim \mathscr{C} \subseteq L$. Hence we obtain

$$Ag(\mathscr{C}) \& Nd(\mathscr{C}) \ \& \sim \mathscr{C} \subseteq L \tag{5}$$

Obviously (5) contradicts CD 2 which is thus refuted.

In case sentence (3) is a determinate truth but not an analytic one, the set $\mathscr{C}$ is no more analyticity-generating, but it includes the analyticity-generating subsets $L \cup \{P\}$ and $L \cup \{Q\}$. This shows that the union of analyticity-generating sets may not be itself analyticity-generating, since the union of the analyticity-generating sets $L \cup \{P\}$ and $L \cup \{Q\}$ is not analyticity-generating.[10]

5. THIRD TENTATIVE CRITERION OF DEMARCATION CD 3

In the counter-example mentioned in sec. 4 we are confronted with the case of a non-idle set, (viz., the set $\mathscr{C}$) possessing subsets (viz., the sets $L \cup \{P\}$ and $L \cup \{Q\}$) which are not non-idle. In order to avoid such cases we shall attempt to formulate a new criterion of demarcation in terms of non-idle sets whose subsets are also non-idle. For this purpose we make the following definition.

Definition 11.
Γ is called *strongly non-idle*, or Snd (Γ) for short, iff $\forall \Gamma'[\Gamma' \subseteq \Gamma \rightarrow \mathrm{Nd}(\Gamma)]$

We obtain then the following propositions.

Proposition 15.
$\mathrm{Snd}(\Gamma) \rightarrow \mathrm{Nd}(\Gamma)$

Proposition 16.
$\mathrm{Snd}(\varnothing)$

Proposition 17.
$[\mathrm{Snd}(\Gamma_1) \,\&\, \Gamma_2 \subseteq \Gamma_1] \rightarrow \mathrm{Snd}(\Gamma_2)$

Proposition 18.
$\mathrm{Snd}(\Gamma) \rightarrow \forall c(c \in \Gamma \rightarrow \mathrm{Nd}(\{c\})$

Proposition 19.
$\mathrm{Snd}(\{c\}) \leftrightarrow \mathrm{Nd}(\{c\})$

The set of logical constants of usual languages is strongly non-idle so that we can adopt the following axiom.

Axiom 7.
Snd (L).

It follows from Proposition 18 and Axiom 7 that each member of L must be non-idle. Indeed the logical constants of usual languages are non-idle. In particular each member of the set $\{\sim, \&, \forall\}$ of logical constants of the parametric first-order language mentioned in sec. 4. is non-idle. This can be shown in the following way.

1. Nd ($\sim$), since
   $\{p\} \models_{\{\sim\}} \{\sim\sim p\}$, but not $\{p\} \models_{\varnothing} \{\sim\sim p\}$

2. Nd (&), since
   $\{p \,\&\, q\} \models_{\{\&\}} \{p\}$, but not $\{p \,\&\, q\} \models_{\varnothing} \{p\}$

3. Nd(∀), since
$\{\forall x \forall x\, p\} \models_{\{\forall\}} \{\forall x\, p\}$, but not $\{\forall x \forall x\} \models_{\varnothing} \{\forall x\, p\}$

We could propose the following tentative criterion of demarcation.

CD 3. $\Gamma \subseteq L \leftrightarrow [\mathrm{Ag}(\Gamma) \,\&\, \mathrm{Snd}(\Gamma)]$ (for every $\Gamma \in \mathscr{PC}$)

However CD 3 is subject to a counter-example suggested by Quine's criticism of Carnap's tentative criterion of demarcation of the logical constants.[11] The latter criterion can be reconstructed in our system in the following way.

Given that voc (A) is the vocabulary of A in the sense of the set of constants occurring essentially in the sentence A of a given language, and L is the set of logical constants of that language:

(i) There is a sentence A such that voc (A) ⊆ L.

(ii) For every sentence A, if voc (A) ⊆ L then A is analytic or contradictory.

Quine criticizes this criterion by remarking, in essence, that if the sentences of a language whose vocabulary is purely logical can be divided into analytic or contradictory ("in purely syntactical terms"), then the adjunction of a non-logical constant such as the binary predicate "heavier than" will not change such a division. Indeed the sentences formed from first-order logical constants and the predicate "heavier than" such as the sentences "∀x ∼ x is heavier than x", "∀x∀y∀z [x is heavier than y & y is heavier than z) → z is heavier than z]" are analytic and their negations are contradictory. Hence the non-logical predicate "heavier than" would be "logical" under Carnap's criterion of demarcation.[12] Note that one can also object to Carnap's criterion of demarcation on the basis that although the vocabulary of the sentence "∃x∃y ∼ x = y" is included in the set of logical constants, this sentence is neither analytic nor contradictory.

Now Quine's criticism of Carnap's criterion of demarcation can be directed also towards our CD 3. Indeed let us consider the set of L ∪ {"heavier than"}-valid sequents, i.e., sequents <X, Y> such that

$$X \models_{L \cup \{\text{``heavier than''}\}} Y \tag{1}$$

holds. Quine's criticism suggests that (1) implies

$$X \vdash_{an} Y \tag{2}$$

so that the set L ∪ {"heavier than"} is analyticity-generating (since $\models_{L \cup \{\text{``heavier than''}\}} \subseteq \vdash_{an}$). Furthermore we can show that this set of constants is also strongly non-idle. Hence L ∪ {"heavier than"} is "logical" under CD 3 although this set contains a non-logical constant. It follows that CD 3 must also be abandoned.

Note that the true sentence

$$\exists x \exists y \; x \text{ is heavier than } y \tag{3}$$

is not analytic. It is a non-determinate truth which is false in a domain with less than two elements. Therefore (3) cannot be used for showing that L ∪ {"heavier than"} is, after all, not analyticity-generating.

6. THE CRITERIA OF DEMARCATION CD 4 AND CD 4′

Quine has remarked that Carnap's criterion of demarcation ultimately avoids the difficulty we have mentioned in sec. 5 by using Cartesian co-ordinates. Indeed Carnap assigns to each object (or event) E a set K_E of quadruples of real numbers which are the spatio-temporal co-ordinates of the point-events constituting that object. Let $K_E[t]$ be defined as follows.

$$K_E[t] = \{<x, y, z>: <x, y, z, t> \in K_E\} \quad (1)$$

$K_E[t]$ characterizes the momentary state at time t of object E.[13] Now the constants of the form K_E and $K_E[t]$ are logical proper names. We shall add these new constants to the vocabulary of any language as *auxiliary logical constants*. We denote the set of these constants by $\mathscr{K}$. Then the set of logical constants of the *extended* language is $L \cup \mathscr{K}$.[14] The members of L are the *proper* logical constants. We can show that although the set L ∪ {"heavier than"} is analyticity-generating, the set $L \cup \mathscr{K} \cup$ {"heavier than"} is not analyticity-generating. Indeed

$$\varnothing \models_{L \cup \mathscr{K} \cup \{\text{"heavier than"}\}} \{K_1[t] \text{ is heavier than } K_2[t]\} \quad (2)$$

holds but

$$\varnothing \vdash_{an} K_1[t] \text{ is heavier than } K_2[t] \quad (3)$$

is false. Hence we are no more forced to conclude that "heavier than" is logical. This suggests the following criterion of demarcation of the logical constants formulated in the extended language.

CD 4. $\Gamma \subset L \leftrightarrow [\mathrm{Ag}(\Gamma \cup \mathscr{K}) \,\&\, \mathrm{Snd}(\Gamma)]$ (for every $\Gamma \in \mathscr{PC}$)

CD 4 is supported by the fact that in usual languages both Axiom 6 and the following one hold.

Axiom 8.
$\mathrm{Ag}(L \cup \mathscr{K})$

Furthermore non-logical monadic predicates such as "is blue", or for that matter the monadic predicate involved in the counter-example against CD 1 and the binary non-logical predicate "heavier than" involved in the counter-example against CD 3 are all non-logical under CD 4. For example let us show that the negation sign "~" is indeed logical under CD 4. We must show

$$\{\sim\} \subseteq L \tag{4}$$

and, therefore,

$$Ag(\{\sim\} \cup \mathscr{K}) \,\&\, Snd(\{\sim\}) \tag{5}$$

Now by Proposition 19, Snd ({~}) iff Nd ({~}). But we have shown in sec. 5 that Nd ({~}). Therefore we have Snd ({~}). We must therefore show only Ag ({~} ∪ $\mathscr{K}$). For this purpose we must show

$$X \models_{\{\sim\}\cup\mathscr{K}} Y \rightarrow X \vdash_{an} Y \qquad \text{(for every } X, Y \in \mathscr{P}\mathscr{S}) \tag{6}$$

We have

$$X \models_{\{\sim\}\cup\mathscr{K}} Y \rightarrow X \models_{\{\sim\}} Y \tag{7}$$

since the meaning of the members of $\mathscr{K}$ does not effect the validity of sequent <X, Y>. For example,

$$\text{since } \{Fk_1k_2\} \models_{\{\sim\}\cup K} \{\sim\sim Fk_1k_2\}, \text{it is the case that } \{Fk_1k_2\} \models_{\{\sim\}} \{\sim\sim Fk_1k_2\} \tag{8}$$

On the other hand we have

$$X \models_{\{\sim\}} Y \rightarrow X \vdash_{an} Y \tag{9}$$

Indeed in case sequent <X, Y> is {~}-valid it is valid in all {~}-preserving interpretation so that it depends only on the meaning of the constant "~". Hence X $\vdash_{an}$ Y. We derive then (6) from (7) and (9).

As a second example let us show that the empirical predicate "is blue" is indeed non-logical under CD 4. We must show

$$\sim(\{\text{"is blue"}\} \subseteq L) \tag{10}$$

and, therefore,

$$\sim Ag(\{\text{"is blue"}\} \cup \mathscr{K}) \vee \sim Snd(\{\text{"is blue"}\}) \tag{11}$$

We shall show that the first disjunct of (11) is true. Indeed we have

$$\varnothing \models_{\{\text{"is blue"}\}\cup\mathscr{K}} \{\text{"k is blue"}\} \;\&\; \sim(\varnothing \vdash_{an} \{\text{"k is blue"}\}) \tag{12}$$

on the assumption that k ∈ $\mathscr{K}$ and the location k is occupied, in the actual world, by a blue object. "k is blue" is a determinate truth, since it is true in all ({"is blue"} ∪ K)-

preserving interpretation. But it is not analytic since one can't know a priori that a location is occupied by a blue object.

CD 4 implies that the set

$$\cup\{\Gamma : Ag(\Gamma \cup \mathscr{K}) \,\&\, Snd(\Gamma)\}$$

is both analyticity-generating and strongly non-idle. If this fails we can still adopt in place of CD 4 the following restricted and weakened criterion of demarcation.

CD 4′ L is some maximal element of $\{\Gamma : Ag(\Gamma \cup \mathscr{K}) \,\&\, (\Gamma)\}$.

7. NOTES

[1] I acknowledge my great gratitude to Quine who has inspired my work on logical constants and logical truth. The present paper is a drastically modified version of my paper "Logical Constants", *Felsefe Araştırmaları Enstitüsü Dergisi* (Journal of the Institute of Philosophical Researches, University of Ankara), Vol. X, 1976. Quine, who has read that paper, encouraged me to prepare a new version.

[2] See R. Carnap, *Introduction to Semantics*, Cambridge, Mass., 1948, p. 59.

[3] We use here "analytic" not as opposite to "synthetic" but rather in the sense of necessary and a priori (or, in the context of pragmatics, relatively immuned from revisions). We agree with Quine that analyticity is a vague and elusive concept which cannot be defined in terms of logical truth. The notion of analyticity has a very wide range including, in particular, classical logical truths, modal-alethic truths, epistemic truths, deontic truths and set-theoretical truths. Thus the choice of a set of analytic sentences of a particular kind will determine a set of "logical" constants of a particular kind (say truth-functional, quantificational, modal, epistemic, deontic, set-theoretical constants).

[4] The L-preserving (i.e. the admissible) interpretations keep invariant not the reference, or extension, of the logical constants, but only their meaning. Indeed the extensions of such logical constants as the identity and the quantification signs vary with the domain of the interpretation. Cf. S. T. Kuhn "Logical Expressions, Constants, and Operator Logic", *Journal of Philosophy*, Vol. LXXVIII (1981), p. 491.

In general Γ-preserving interpretations keep invariant not the extension but only the meaning of the members of Γ. E.g., if the constant "is blue" belongs to Γ, then the extension of "is blue" in a Γ-preserving interpretation with domain D consists in the set $\{d : d \in D \,\&\, d \text{ is blue}\}$.

[5] The concept of Γ-truth is a semantical reformulation of Quine's idea of a truth with respect to a particular vocabulary Γ, i.e., of a true sentence in which only the constants belonging to Γ occur essentially. See W. V. Quine, *Mathematical Logic* (Cambridge, Mass., 1958), p. 2.

The concept of Γ-truth establishes a parallelism with Kripke's theory of "Naming and Necessity" in *Semantics of Natural Language* (Boston, 1972). Indeed we assume that the Γ-preserving interpretations (for any arbitrary Γ) correspond to Kripke's possible worlds. Then the elements of Γ are the rigid constants (inc. the rigid designators), the Γ-truths are the necessary truths, the analytic Γ-truths are the necessary a priori truths, and the non-analytic Γ-truths are the necessary a posteriori truths.

[6] This is a disputed philosophical point, but we take it here for granted.

[7] Quine has pointed out to me that the formula "$\exists x \exists y \sim x = y$" "is indeed anomalous, expressing seemingly extra-logical content through purely logical notation."

[8] I am indebted for this counter-example, as well as for the ingenious proof involved in that counter-example, to an anonymous referee of *The Journal of Symbolic Logic*.

[9] We can show that $\models \subseteq \vdash_{an}$ in the following way. Let $X \models_{\mathscr{C}} Y$, i.e., $X \models_{L \cup \{P,\ Q\}}$. Let then X' and Y' result from substituting respectively in X and Y the formula "Px" for the formula "Qx". Then $X' \models_{L \cup \{P\}} Y'$. But $\models_{L \cup \{P\}} = \models_L$ Hence $X' \models_L Y'$. Since Ag(L), i.e., $\models_L \subseteq \vdash_{an}$, we obtain $X' \vdash_{an} Y'$. $\forall x\,(Px \leftrightarrow Qx)$ being analytic we have $\varnothing \vdash_{an} \{\forall x\,(Px \leftrightarrow Qx\}$. We derive from the latter and $X' \vdash_{an} Y'$ the result that $X \vdash_{an} Y$.

[10] I am indebted for the latter point to the anonymous referee mentioned in n. 8.

[11] See R. Carnap, *The Logical Syntax of Language*, (London, 1937), § 50.

[12] See W. V. Quine, "Carnap and Logical Truth" in *The Ways of Paradox and Other Essays* (New York, 1966) pp. 116-117.
[13] See Quine, *ibid.*, p. 117.
[14] Note that if $k_1 \in K$, and $k_2 \in K$, where k_1 and k_2 are distinct, then

$$\sim k_1 = k_2 \models_{L \cup \mathscr{K}} \exists x \exists y \sim x = y$$

holds in classical quantificational logic (with existential presupposition). But the left-hand side of the implication is analytic whereas the right-hand side is not so. In order to avoid this difficulty we should have recourse to free logic. Indeed in free logic the above-mentioned implication does not hold. Such a recourse is natural since the auxiliary logical constants denote locations which may be unoccupied. Hence we may introduce in the extended language the predicate "E" of existence. "Ek" is true iff location k is occupied by an object.

HÜSEYİN YILMAZ

GENERAL THEORY OF RELATIVITY AND THE 5TH TEST

Abstract

Einstein's special theory of relativity is established as securely as any theory can be but his general theory of relativity dealing with gravitation is quite far from reaching a comparable status. About his field equations of general relativity Einstein himself have said: "My equation is like a house with two wings; the left-hand side is made of fine marble, but the right-hand side is perishable wood." The purpose of this article is to indicate that there exists a basic modification of general relativity which renders the right-hand side of the equations fine marble as well.

1. INTRODUCTION

It is known that the only possible modification of general relativity that passes the four classical "test-body" tests, namely 1) the gravitational red-shift, 2) the light-bending, 3) the relativistic perihelion advance and 4) the time-delay of radar signals, is [1, 2, 3]

$$\frac{1}{2}G^{\nu}_{\mu} = \tau^{\nu}_{\mu} + \lambda t^{\nu}_{\mu} \tag{1.1}$$

where G^{ν}_{μ} is the Einstein-Hilbert tensor, τ^{ν}_{μ} is the Einstein "matter tensor," and the t^{ν}_{μ} is the Yilmaz "gravitational field stress-energy tensor" (here written in Cartesian coordinates and in its Newtonian slow-motion limit)

$$t^{\nu}_{\mu} = -\partial_{\mu}\phi\partial^{\nu}\phi + \frac{1}{2}\delta^{\nu}_{\mu}\partial^{\rho}\phi\partial_{\rho}\phi \tag{1.2}$$

The λ is a numerical parameter passing through the values of 0 and 1. It is also known that $\lambda = 0$ corresponds to the usual "Einstein General Relativity" (EGR) and $\lambda = 1$ to what we here call (for convenience) the "Yılmaz General Relativity" (YGR)

G. Irzık & G. Güzeldere (eds.), Turkish Studies in the History and Philosophy of Science, 39-58.

$$\lambda = \begin{cases} 0 & \text{Einstein General Relativity} \\ 1 & \text{Yılmaz General Relativity} \end{cases}$$

The difference between the two theories is sometimes stated as that of a paradigm shift $\tau_\mu^\nu \Rightarrow \tau_\mu^\nu + t_\mu^\nu$, that is, in EGR the right-hand sides of the field equations, and of the equations of motion, contain "matter alone," whereas in YGR the right-hand sides of both the field equations and of the equations of motion contain "matter plus the gravitational field stress-energy tensor." In other words, gravity t_μ^ν participates in the generation of space-time curvatures by being in the field equations and responds to the curvatures by being in the equations of motion, both on equal footing with the matter stress-energy. That is, "gravitational energy gravitates" like any other form of energy and the basic equations of YGR (written in background Lorentzian coordinates) are of the form [3]

$$\frac{1}{2} G_\mu^\nu = \tau_\mu^\nu + t_\mu^\nu \tag{1.3}$$

$$\sigma \frac{du_\mu}{ds} = \frac{1}{2} \partial_\mu g_{\alpha\beta} (\tau^{\alpha\beta} + t^{\alpha\beta}) \tag{1.4}$$

$$\partial_\nu (\sqrt{-g} g^{\mu\nu}) = 0 \tag{1.5}$$

where (1.3) and (1.4) are the two postulates of the theory following from the paradigm and (1.5) is a gauge condition in which the physical quantities have their proper (locally Lorentzian) interpretations. Moreover, it is only under conditions (1.5) that quantum mechanics and gravity appear to be compatible [3]. These equations are derivable from a variational principle within the Riemannian geometry where the field equations and the equations of motion arise from different variations, hence, in general, they are not obtainable from each other. The t_μ^ν in Equation (1.2) is the field stress-energy in the Newtonian slow motion limit with $\lambda = 1$. General forms of τ_μ^ν and t_μ^ν can be found in [3]. Note that the symbol τ_μ^ν has a double meaning. Mathematically it is a differential operation on ϕ_μ^ν (the gauge potential) but physically it is the "matter tensor." In analogy to the Newtonian theory, matter is where (and when) the gauge D'Alembertian of ϕ_μ^ν does not vanish [4].

This theory exhibits a full gauge theory analogy to the electromagnetic gauge-field theory in curvilinear coordinates as [3]

A^ν	ϕ_μ^ν
$j^\nu = {}^2A^\nu - \dfrac{\partial_\alpha(\sqrt{-g} g^{\nu\beta} \partial_\beta A^\alpha)}{\sqrt{-g}}$	$\tau_\mu^\nu = {}^2\phi_\mu^\nu - \dfrac{\partial_\alpha(\sqrt{-g} g^{\nu\beta} \partial_\beta \phi_\mu^\alpha)}{\sqrt{-g}}$
$\partial_\nu(\sqrt{-g} j^\nu) \equiv 0$	$\partial_\nu(\sqrt{-g} \tau_\mu^\nu) \equiv 0$
$\partial_\nu A^\nu = 0$	$\partial_\nu \phi_\mu^\nu = 0$
$j^\nu \Rightarrow \rho u^\nu$	$\tau_\mu^\nu \Rightarrow \sigma u_\mu u^\nu$

The arrow $\Rightarrow$ means j^{ν} and τ_{μ}^{ν} are being identified as hydrodynamic type charge and mass distributions as sources of electromagnetic and gravity fields. In this article we shall take them mostly as distributions of point-like charges and masses. More generally, of course, the physics problem is to find models of charge and matter-energy compatible with the equations, hence other forms of mass-energy (for example, the energy in a Maxwell field) are, in principle, also possible.

The connection between ϕ_{μ}^{ν} and $g_{\mu\nu}$ is a functional relation already contained in my 1969 seminar at the Boston Studies in the Philosophy of Science, namely [5]

$$\begin{gathered} dg_{\mu\nu} = 2\left(g_{\mu\nu}d\phi - g_{\mu\alpha}d\phi_{\nu}^{\alpha} - g_{\nu\alpha}d\phi_{\mu}^{\alpha}\right) \\ \partial_{\nu}\phi_{\mu}^{\nu} = 0, \quad \partial^{\mu}\phi_{\mu}^{\nu} = 0 \end{gathered} \tag{1.6}$$

where $\partial^{\mu} = g^{\mu\nu}\partial_{\nu}$ and ϕ is the trace of ϕ_{μ}^{ν}. The iteration solution of (1.6) is the functional exponential

$$g_{\mu\nu} = \left\{\eta e^{2(\mathbf{1}\phi - 2\mathbf{\Phi})}\right\}_{\mu\nu} \tag{1.7}$$

where η is the Lorentz metric, $\mathbf{1}$ is the unit matrix and $\mathbf{\Phi} = \phi_{\mu}^{\nu}$. In the slow motion limit where the dominant field is ϕ_0^0 (which in this limit equals ϕ) we immediately get

$$g_{00} = e^{-2\phi}, \quad -g_{ik} = \delta_{ik}e^{2\phi} \tag{1.8}$$

$$ds^2 = e^{-2\phi}dt^2 - e^{2\phi}(dx^2 + dy^2 + dz^2) \tag{1.9}$$

$$\phi = \sum_A \frac{m_A}{r_A} + C \tag{1.10}$$

$$\partial_{\nu}(\sqrt{-g}g^{\mu\nu}) = 0 \tag{1.11}$$

where the N-body solutions exist because in the limit

$$^2\phi = \sigma, \quad ^2 = -(\sqrt{-g})^{-1}\nabla^2 \tag{1.12}$$

with ∇^2 the ordinary Laplacian, so that one can write

$$\nabla^2\phi = -\sqrt{-g}\sigma = -\sum_A m_A\delta^3(\mathbf{x} - \mathbf{x}_A) \tag{1.13}$$

whose solution is the N-body solution of Eq. (1.10) with $r_A = |\mathbf{x} - \mathbf{x}_A|$. Note that by virtue of the differential character of (1.6) the potentials will have integration constants as in $\phi_{\mu}^{\nu} \Rightarrow \phi_{\mu}^{\nu} + C_{\mu}^{\nu}$ which can be used to interpret the potentials as potential differences starting with the observation point. This allows the kinematics to be interpreted as locally Lorentzian.

Finally, we may consider the Hamilton-Jacobi equation, namely $g^{\mu\nu}p_\mu p_\nu = m^2$ (where m is the rest-mass), and find the N-body Hamiltonian as

$$H = P_0 = \sum_A{}' \left\{ e^{-\phi}\sqrt{m^2 + e^{-2\phi}p^2} \right\}_A \tag{1.14}$$

with which we shall make all of our calculations in this article. (′) means potential at a particle, excluding its own field, as a particle does not accelerate by its own field.

2. SOLUTIONS OF THE RESPECTIVE FIELD EQUATIONS

It is known that the time-independent solution of EGR in isotropic coordinates is the usual (1-body) Schwarzschild metric ($\lambda = 0$)

$$ds^2 = \left[\frac{1-\frac{\phi}{2}}{1+\frac{\phi}{2}}\right]^2 dt^2 - \left(1+\frac{\phi}{2}\right)^4 (dx^2 + dy^2 + dz^2) \tag{2.1}$$

$$\phi = \frac{m}{r} \tag{2.2}$$

$$\partial_\nu(\sqrt{-g}g^{\mu\nu}) \neq 0 \tag{2.3}$$

where m is the Schwarzschild mass. In YGR the corresponding solution is, as we have seen, the so-called Yılmaz N-body exponential metric $\lambda = 1$

$$ds^2 = e^{-2\phi}dt^2 - e^{2\phi}(dx^2 + dy^2 + dz^2) \tag{2.4}$$

$$\phi = \sum_A \frac{m_A}{r_A} + C \tag{2.5}$$

$$\partial_\nu(\sqrt{-g}g^{\mu\nu}) = 0 \tag{2.6}$$

where $r_A = |\mathbf{x} - \mathbf{x}_A|$. This is a most remarkable situation: We have complicated the Einstein equations by adding the Yılmaz tensor t_μ^ν to the right-hand side. One would expect that the solutions would get immensely more complicated. Instead, a) they simplify into just two exponentials, b) they become exact N-body metrics with inter-particle symmetry, hence go beyond the EGR "test-body" theory, c) the potential admits an arbitrary integration constant as in the Newtonian theory, d) it fully predicts the four classical tests, and e) it automatically satisfies the harmonic (gauge) conditions. Thus the theory already becomes a prototype gauge-field theory in curved space time. As we have noted, its more general form exhibits the full gauge-field theory character [3]. Due to these properties the N-body exponential metric is sometimes called the "magic result of Yılmaz" [6]. In contrast, the Schwarzschild metric a) is more complicated, b) it is only a 1-particle solution, c) the potential does not admit an arbitrary integration constant [7], d) it does not satisfactorily explain the N-body effects, e) it does not satisfy the harmonic (gauge) condition. EGR is a "test-particle" theory and is not an N-particle theory, nor is it a prototype gauge-field theory in curved space. To illustrate these points we shall first present the calculation of these effects in the new theory. As already noted, we shall do this by the N-body Hamiltonian derived from the Hamilton-Jacobi equation of the metric in the slow motion limit. It is

$$H = P_0 = \sum_A{}' \left\{ e^{-\phi} \sqrt{m^2 + e^{-2\phi} p^2} \right\}_A \tag{2.7}$$

which can also be written as $H = P_0 = \int \sqrt{-g}\tau_0^\nu dV_\nu$.

A similar Hamiltonian can be written for EGR but since it has only one body in the solution, it cannot lead to any N-body effects. (See Maxwell's statement, Reference [17]). The only way to make something interesting out of it is to add, by hand, some test particles and apply the Hamiltonian to them. A test body has no field, hence no active mass, since it is not in the solution, but it will mimic the limit in the N-body Hamiltonian when some particles have infinitesimal masses. As such, the test particles cannot interact among themselves because they are not in the solution. They do not represent active mass since they are put by hand. (The field equations do not recognize them.)

In references [3, 5] it is shown that the presence of t_μ^ν leads to the proof of the *Strong* Equivalence Principle (SEP) $m_i = m_p = m_a$.

3. CALCULATION OF THE FOUR CLASSICAL EFFECTS

We shall now calculate the first four effects with the above Hamiltonian of the N-body theory and show that they are fully explained by our N-body Hamiltonian. Reference will be made to test particles at appropriate places. Thereafter we shall also calculate the fifth test from the N-body Hamiltonian and show that this one cannot be calculated from a "test-body" theory. Therefore the experimental confrontation of the fifth effect (the 532″ per century planetary perturbative part of the perihelion advance of Mercury) will be considered as the *experimentum crucis* of the new theory. Our task is to calculate the following five effects:

1) Gravitational frequency shift
2) Bending of light near the Sun
3) Relativistic perihelion advance
4) Slowing of light near the Sun
5) N-body perihelion advance

3.1. Gravitational Frequency Shift

Consider the case of an atom, a very small body at rest ($p = 0$) at x, in the field of a very heavy body M, say, the Earth. The atom is assumed to have energy levels as standing wave configurations. Consider another atom, identical to the first but situated at x' where the frequency comparisons are made. The Hamiltonian of a stationary atom ($p = 0$) in the field of a heavy body is

$$H = me^{-\phi} \tag{3.1.1}$$

where m is the rest-mass (rest-energy) and ϕ is the field of the heavy body at x'. The same applies to the other atom with $H' = me^{-\phi'}$. Thus all the energy states and therefore also all energy differences between the states will appear shifted relative to the other atom by a factor $e^{-(\phi-\phi')}$. Since quantum mechanically $H = h\nu$, $H' = h\nu'$ we immediately find the gravitational frequency shift

$$\nu = \nu' e^{-(\phi-\phi')} \tag{3.1.2}$$

Note that in order to realize this configuration we need at least three bodies in the solution, although two of them are very small. Of course, other bodies are needed to perform the experiment, hence we actually need an N-body solution, which in principle should even include ourselves.

A corollary to the frequency shift just mentioned is the gravitational time dilation. Since an atom (hence an atomic clock) has a higher frequency at a higher altitude, it gains time relative to a reference level below it. From the above, the factor is $e^{-(\phi-\phi')}$. This factor is confirmed by an experiment performed by C. O. Alley and his associates involving two identical atomic clocks, one of which was flown a number of hours at a certain altitude. Upon return to the Earth the clock is found to gain an amount consistent with the theoretical result

$$\delta t = t' \left[e^{-(\phi-\phi')} - 1 \right] \tag{3.1.3}$$

There is also a special relativistic part $-\frac{1}{2}(v^2/c^2)t'$ which was also confirmed. Note that from a practical point of view this time gain effect may be the most important of all relativistic gravity effects because it is used in the operation of the multibillion-dollar Global Positioning System.

3.2. Gravitational Deflection of Light

Consider a massless particle (a photon or graviton) passing near a very heavy mass M at rest. The Hamiltonian of the massless particle ($m = 0$) is

$$H = e^{-2\phi} |\, p \,| \tag{3.2.1}$$

Since quantum mechanically $H = h\nu$, $p = h/\lambda$, $\upsilon = \lambda\nu = ce^{-2\phi}$, the space is here acting like a refractive medium with refractive index common to all wavelengths

$$n = e^{2\phi} \tag{3.2.2}$$

This leads to the observed deflection which, in the case of the Sun, is

$$\delta\theta = \frac{4M}{a} \tag{3.2.3}$$

where a is the closest distance of the light beam to the Sun.

Note that in this case we need at least three bodies in the solution, a photon (or graviton) plus the Sun and some more (equipment, measuring devices, etc.) to perform the experiment.

Note that if gravity is strong enough light can orbit the central body (in a quasi-stable orbit). In this case it is convenient to write the relevant Hamiltonian in spherical coordinates as

$$H = e^{-2\phi}\frac{L}{r} \tag{3.2.4}$$

where $L = rp$ is angular momentum.

Letting $\phi = M/r$ and setting the derivative of H with respect to r to zero, one has

$$r = 2M \tag{3.2.5}$$

Similarly, mass particles can have quasi-stable orbits. Using the more general Hamiltonian

$$H = e^{-\phi}\sqrt{m^2 + e^{-2\phi}\frac{L^2}{r^2}} \tag{3.2.6}$$

where m is the rest-mass. These and other interesting features of the orbits can be studied.

Note also that in this theory there are no event horizons, hence although gravitational collapse to dense compact objects may exist, the popular black hole concept does not apply. Such heavy collapsed objects without horizons are sometimes called "dark red holes" or "Yilmaz stars" [3]. They have qualitatively different properties than the usual concept. For example, radially directed light will always escape—red shifted.

3.3. Relativistic Correction to the Perihelion Advance

If the small body is still small but neither massless nor at rest, its Hamiltonian is more general. It is

$$H = e^{-\phi}\sqrt{m^2 + e^{-2\phi}p^2} \tag{3.3.1}$$

But if the velocity is small enough, and the field is weak enough, we can expand the Hamiltonian an order beyond the Newtonian limit and simplify it by the conserved Newtonian energy $E = p^2/2m - m\phi =$ constant, that is, by $p^2/m^2 = 2\phi$. (The constant does not contribute to the perihelion advance.) Upon expansion, the second order part is

$$\frac{1}{2}m\phi^2 - \frac{3}{2}m\phi\left(\frac{p^2}{m^2}\right) - \frac{1}{8}m\left(\frac{p^2}{m^2}\right)^2 \tag{3.3.2}$$

and upon simplification with $p^2/m^2 = 2\phi$ one finds the second order part to be $-3m\phi^2$, hence to second order the Hamiltonian is

$$H = \frac{p^2}{2m} - m(\phi + 3\phi^2) \tag{3.3.3}$$

From the Hamiltonian equations we find the equations of motion as

$$m\frac{d^2r}{dt^2} = -\left[1 + \frac{6\phi}{c^2}\right]\frac{mMr}{r^3} \tag{3.3.4}$$

Note that apart from the $6\phi/c^2$ term in the bracket, these are the Newtonian equations for a small body in the field of a heavy one. The extra term $6\phi/c^2$ gives the relativistic correction to the advance of the perihelion of Mercury as

$$\dot{\tilde{\omega}} = \frac{6\pi G(M+m)}{c^2(1-\varepsilon^2)a} \tag{3.3.5}$$

where a is the semimajor axis. For Mercury this formula gives 43″ per century. Note that here we need at least a two-body solution where $M + m$ is the Kepler Constant. In the case of Mercury m is negligible compared to M.

Note also that, in order for the relativistic effect to be calculated, one must first have a 2-body "bound system" (orbiting duo). That system itself requires at least an interactive 2-body solution. It is not the case that such solutions may exist in EGR but (due to nonlinearity, etc.) so far we have not been able to find them. It is the case that they (the kinds that are simultaneously required for the four classical tests) do not exist. (Note that we are not here concerned with a 2-body case alone, which can be treated as a 1-body case. We are concerned with at least a 10-body solution to treat the Solar system.) So in truth, in a "test-body" theory one cannot even start the calculation. This is discussed more fully in Section 3.5.

3.4. Gravitational Radar Time-Delay

In the new theory this is a corollary to the deflection of light as the same refractive index analogy covers it. We can calculate it with the same Hamiltonian

$$H = e^{-2\phi}|\,p\,| \tag{3.4.1}$$

Again a refractive index analogy

$$n = e^{2\phi} \tag{3.4.2}$$

holds, and the result of the calculation is

$$\delta t \simeq 4M \ln\left(\frac{l_1 + l_2 + a}{l_1 + l_2 - a}\right) \tag{3.4.3}$$

where l_1 and l_2 are the distances of the sending and receiving planets to the Sun and a is the distance between the two planets. In general relativity and in other theories there can be an additive term (bias) depending on the coordinates chosen (20 μs for the nonisotropic Schwarzschild metric). Although it can be eliminated by a special processing of the data, it is important to note that in the new theory it does not arise because r in M/r is identified as the true operational r satisfying the correct special relativity limit. Note, however, that we again need at least three bodies in the solution (and some more to perform the experiment).

These are the "Four Classical Tests" of a space-time theory of gravity. They are obtainable also via a test-body theory where only the central body is in the solution but the test-body is put by hand. A test-body theory is useful because it is simple and gives correct results in some important cases. The EGR as a test-body theory has gotten many correct results in this way. It is fortunate that this is so because the force of experimental confirmations of these results kept general relativity alive. Otherwise the space-time approach to gravity might have been abandoned and forgotten. The next treated case, namely the planetary perturbative part of the perihelion advance, requires a full N-body metric with at least N = 10 bodies, the Sun and nine planets.

3.5. Calculation of the N-Body Perturbative Part

This is the "crucial part" of the paper and we hope that we convey its essence and simplicity with the utmost clarity it deserves. Also, it is ancient knowledge. 150 years ago it was known to Leverrier and his colleagues. The surprising thing is that EGR is not able to cover it but YGR is.

In the third test above we have recovered the 43$''$ per century relativistic correction to the perihelion advance of Mercury. But the observed perihelion advance of Mercury is not 43$''$ per century, it is in fact 575$''$ per century. The observed result is the sum of two parts: a) the 532$''$ per century Newtonian N-body perturbative part, and b) the 43$''$ per century relativistic correction. Our theory is relativistic and has N-body interactive solutions, hence we can recover both these effects by using our Hamiltonian in its relativistic N-body form

$$H = P_0 = \sum_A{}' \left\{ e^{-\phi} \sqrt{m^2 + e^{-2\phi} p^2} \right\}_A \tag{3.5.1}$$

Repeating the previous procedure we find the general lowest order case to be

$$m\frac{d^2\mathbf{r}}{dt^2} = -\sum_A{}' \left[\mathbf{1} + \frac{6\phi}{c^2}\right] \frac{mm_A\mathbf{r}}{|r|^3} \tag{3.5.2}$$

or more explicitly,

$$\frac{d^2\mathbf{r}_k}{dt^2} = -\sum_{A\neq k} \frac{m_A\ \mathbf{r}_{AK}}{\mathbf{r}_{AK}^3}\left(1 + \frac{6}{c^2}\sum_{B\neq k}\frac{m_B}{r_{BK}}\right)$$

The crucial difference between (3.3.4) and (3.5.2) is that in (3.5.2) we have a *summation sign* $\sum_A'$ over the other eight planets. Let us now remove the $6\phi/c^2$ term causing the 43″ per century relativistic part. What is left is

$$m\frac{d^2\mathbf{r}}{dt^2} = -\sum_A{}' \frac{mm_A\mathbf{r}}{|r|^3} \tag{3.5.3}$$

and these are the Newtonian N-body equations of motion. Is there here a cause to produce advance of the perihelion for Mercury? Of course there is. The cause is the perturbations of planets on the motion of Mercury. This can be calculated and the result is the 532″ per century part of the total 575″. In Table I below, two calculations are shown, one by numerical integration and one with a semi-empirical formula (author, unpublished).

The overall total advance can be summarized in the case of the planet Mercury as

$$\dot{\omega} = 42.98 + 531.50\lambda \tag{3.5.4}$$

The semi-empirical formula used is

$$\dot{\omega}_N = 1216.1\sum_p \frac{m_\mu m_p}{(r_p - a)r_p(r_p + a)} \tag{3.5.5}$$

where m_μ is the mass of Mercury, $a = \frac{1}{2}$ and all quantities are in astronomical units. That the effect is due to planetary perturbations (mutual interactions) is clear from the products of the masses and the summation sign over the masses of the planets which goes over the other eight planets. EGR is not able to cover this effect simply because it does not have the necessary N-body solutions. Our N-body Hamiltonian above gives the total 575″ per century perihelion advance of Mercury seamlessly, that is, without separation into two parts. With our relativistic N-body Hamiltonian the computer calculates and prints out only one number which is closely the actual observed effect. Professor Carroll O. Alley calls this the "seamless calculation." Numerical integrations are done by Carl Hein. "Perihelion Advance of Mercury" Carl A. Hein and H. Yılmaz, Feb. 1990 (Unpublished)

Table I
Perturbative Contributions of Planets to Mercury's Newtonian Perihelion Advance

	Numerical	*Semi-empirical*
Venus	277.86	277.80
Earth	90.04	89.67
Mars	2.54	2.28
Jupiter	153.58	153.20
Saturn	7.30	7.39
Uranus	0.14	0.14
Neptune	0.04	0.04
Pluto	—	—
Subtotal	531.50	530.17

If we remove the $6\phi/c^2$ term, we get 532″ per century (actually about 531.61″). There was once thought to be a 2.3″ discrepancy but this was found by Narlikar and Rana [8] to be due to the precession of the ecliptic. The total uncertainty is about 0.1″ which translates into $|\lambda - 1| < 5 \times 10^{-3}$. This analysis can be applied to all other planets, producing nine demanding observed numbers. Note carefully that this can be done only if the theory has N-body interactive solutions like in the Newtonian theory, or like the new theory here under discussion. It cannot be done with a test-body theory. For each planet the result for the perihelion advance can be written as

$$\dot{\tilde{\omega}} = \lambda A + B \tag{3.5.6}$$

where for Mercury $A = 532''$, $B = 43''$ per century. Clearly the actual observed result $\dot{\tilde{\omega}} = 575''$ per century implies $\lambda = 1$.

It is usually implicitly assumed that Einstein's theory reduces to the Newtonian theory to the necessary order, hence would cover the planetary perturbative effect as a matter of course. This, however, is not the case. Einstein's theory reduces to the Newtonian theory only in first order, whereas this effect is second order even in the Newtonian theory (products of potentials appear in the equations of motion) as in

$$-\sigma_i \frac{d^2 x_\mu}{ds^2} = \sigma_p \partial_\mu \phi = \sigma_a \partial_\mu \phi = \underset{\Uparrow}{\nabla^2 \phi \partial_\mu \phi} \tag{3.5.7}$$

Therefore, Einstein's theory does not recover the planetary perturbative effects, because it does not satisfy a necessary second order correspondence to the Newtonian theory. That necessary second order correspondence is the existence of t_μ^ν with a coefficient $\lambda = 1$, and the new theory satisfies it. Therefore we must count this test as a *crucial test* in favor of the new theory.

4. THEORETICAL DETERMINATION OF λ

In EGR the four "test-body" tests are treated by assuming m to be the heavy central mass (which is in the solution) and considering a small test body (which is not in the solution but put by hand). This is done because EGR has no interactive N-body solutions. In the YGR there are no test bodies because all bodies are in the solution. Thus EGR violates the universal N-body symmetry of the gravitational interaction (one body is in the solution while the other is not), whereas the YGR satisfies the universal N-body symmetry. This argument leads to the $\lambda = 1$ theory without the need of any experiment.

Another, perhaps even stronger theoretical argument in favor of $\lambda = 1$ comes from a mathematical identity called the Freud identity [g]. This identity expresses the conservation of total energy momentum and reduces to the slow motion Newtonian limit as [3]

$$\partial_\nu\left(\sqrt{-g}\sigma u_\mu u^\nu\right) \equiv 0 \tag{4.1}$$

and our equations of motion are consistent with it [3, 9]. But it is known that in EGR the (geodesic) equations of motion are obtainable from its field equations under the assumption

$$\partial_\nu(\sqrt{-g}\sigma u^\nu) = 0 \tag{4.2}$$

which leads to the conservation of total rest-mass. But in the special theory of relativity, to which the theory must reduce as a correspondence limit, it is the total energy-momentum and not the total rest-mass that is conserved. These theoretical arguments again lead to the value $\lambda = 1$ and exactly so without any experiment.

5. PARAMETERIZING GRAVITY THEORIES

The parametric PPN method originated with A. S. Eddington, H. P. Robertson and others. The idea is to expand the metric into potentials with certain parameters and codify the theories with the parameter values to which it leads. Since observed effects correspond to certain values of these parameters, one can quickly judge a theory by simply looking at its parameter values instead of calculating the observed effects for each theory anew from scratch. For example, let

$$g_{00} = 1 - 2\alpha\phi + 2\beta\phi^2 \tag{5.1}$$

$$-g_{0i} = 4\phi_{0i} \tag{5.2}$$

$$-g_{ik} = \delta_{ik}(1 + 2\gamma\phi + 2\delta\phi^2) \tag{5.3}$$

In order for the theory to be viable one must have:

Gravitational frequency shift:	$\alpha = 1$
Gravitational light deflection:	$\alpha + \gamma = 2$
Relativistic perihelion shift:	$2\alpha(\alpha + \gamma) - \beta = 3$
Planetary-perturbative shift:	$\lambda = \delta = 1$

As to viability $\alpha = \beta = \gamma = \delta = 1$ we then have the partial list:

Table II
PPN Parameters

Theory	α	β	γ	δ	*SEP*	ϕ
Einstein:	1	1	1	$\frac{3}{4}$	No	$\frac{m}{r}$
Brans-Dicke:	1	1	$\frac{\omega+1}{\omega+2}$	$\frac{3}{4}$	No	$\frac{m}{r}$
Yilmaz:	1	1	1	1	Yes	$\sum_A \frac{m_A}{r_A} + C$

where $\lambda = 4\delta - 3$.

Note that the current practice of expanding the metric into N-body potentials is valid only in YGR since it is only in YGR that there are such N-body solutions to expand into.

Planetary perturbations are a most important part of celestial mechanics. So essential that from their presence and their precise magnitudes the planets Neptune and Pluto were first discovered (predicted) on paper before their discovery by observations. Let us ask the crucial question: "Are these planetary perturbations and, in particular, the planetary perturbative advances of the perihelia of the planets obtainable from Einstein's theory of general relativity?" Or putting it rather directly, "Could the planets Neptune and Pluto be so discovered if the only theory of gravity available at the time were EGR?"

The answer to this question is an emphatic *no*. We are compelled to say that EGR is only a test-body theory (one body in the solution, all other bodies being test bodies put by hand) and that it cannot meet the demands of the 5th test which requires N physical bodies with mutual interactions. We have seen above that such N-body solutions are possible only when $\lambda = 1$, hence the 5th test leading to this value of λ is here considered an *experimentum crucis* of the new theory of gravitation.

Remembering the oft-quoted statement of Einstein mentioned in the abstract, it is clear that with the present modification the right-hand side of the field equations now becomes as solid as their left-hand side. We believe Einstein would have been pleased to see this development.

Deep gratitudes are due to Professor Robert S. Cohen for inviting the author to contribute to this distinctive volume. I thank my colleagues Professor Carroll O. Alley and Dr. Yutaka Mizobuchi, and particularly Mr. Teruo Hiruma, President of Hamamatsu Photonics K.K., for their genuine interest and encouragement, and many valuable suggestions. Special thanks are due to Professor Michael D. Scadron for advice and discussions on the Hamiltonian method.

Hamamatsu Photonics K.K., Hamamatsu City, 435 Japan
Electro-Optics Technology Center, Tufts University,
Medford, MA 02155, USA

NOTES AND REFERENCES

[1] B. O. J. Tupper, Nu. Cim. **19B**, 135 (1974). Note that Tupper used spherical coordinates; unless otherwise stated we always use Cartesian coordinates.

[2] B. O. J. Tupper, Lett. Nu. Cim. **10**, No. 4, 627 (1974). It is in this short paper that Tupper made his important statement: "Since λ appears only in a second-order term, a suitable second-order experiment will be required to remove the final arbitrariness in the field equations."

[3] H. Yılmaz, Nu. Cim. **107B**, No. 4, 941 (1992). See also C. O. Alley, D. Leiter, Y. Mizobuchi and H. Yılmaz, "Energy Crisis in Astrophysics," astro-ph/9906458, 28 June 1999. For gravity wave solutions see Int. J. Theor. Phys. 20 (Dirac Issue), p. 899 (1982). See also "Quantum Mechanics and general relativity" Annals of the New York Academy of Sciences Vol. 480 p. 625–26 (1986).

[4] E. Schrödinger, "Space-Time Structure," Cambridge U. Press, p. 99 (1950).

[5] H. Yılmaz, "Perception and Philosophy of Science," Boston Studies in the Philosophy of Science, Vol. XIII, pp. 1-91, Eds. Robert S. Cohen and Marx Wartofsky, D. Reidel Publishing Company (1974). The talk was given on 28 October 1969 and manuscript submitted about a year later.

[6] Y. Itin, "Gravity on a Parallelizable Manifold, Exact Solutions," Los Alamos arXiv: gr-qc/9806110, 28 June 1998, p. 12.

[7] W. Pauli, "The Theory of Relativity," p. 192, Dover (1958). Note that Pauli criticized G. Mie's theory on this ground. We see here that general relativity has the same problem. The N-body solutions and integration constants were known as early as 1958 from our original paper, H. Yilmaz, Phys. Rev. **111**, 1417 (1958), but their connection to planetary perturbation was not realized until much later. The author realized it first in 1983 while reading a Scientific American article titled "Newton's Discovery of Gravity" by I. Bernard Cohen.

[8] J. V. Narlikar and N. C. Rana, Mon. Not. Astr. Soc. 213, 657 (1985) removed a small 2.31″ discrepancy by doing a more accurate calculation and identifying its origin in the precession of the ecliptic. For a method of calculation of the N-body perturbative effects, see D. Park, "Classical Dynamics and its Quantum Analogues," Springer-Verlag, p. 249 (1990) which is the analog of our semi-empirical formula. From the Narlikar-Rana paper one may infer that observationally $|\lambda - 1| < 5 \times 10^{-3}$.

[9] H. Yılmaz, "Correspondence Paradox in General Relativity," Lettere il Nu. Cim., Vol. 7, No. 9, 337 (1973). This was long before I knew of the Freud identity: P. Freud, Ann. Math. 40, 417 (1939). The title of Nu. Cim. article should have been, "Correspondence Difficulty in General Relativity."

[10] E. Schrödinger, Phys. Z, 19, 4 (1918).

[11] H. Bauer, Phys. Z, 19, 163 (1918).

[12] H. Weyl, "Space, Time, Matter," p. 270, Dover (1952).

[13] C. Burali-Forti, Academia di Science Fisische delle Scienze (Turin) Classe di Scienze Fis. Math. e Naturali (1922-23).

[14] A. Einstein, Sitzungsberichte Königlich P. Academie der Wiss. XXIV (1918).

[15] H. Yılmaz, "Toward a Comprehensible Physical Theory," Kluwer Acad. Publishers, S. Jeffers et al (Eds.), pp. 503-525 (1997).

[16] H. Yılmaz, "Conservation Theorems in Curved Space-Time," Physics Letters, Volume 92A, Number 8, p. 377 (1982).

[17] Kapek, M. in "The World View of Contemporary Physics," Kitchener, R. (ed.) SUNY Press, p. 96, 1988.

APPENDICES

The following four appendices are intended to balance the brevity of the text.

A. On the Philosophy of Science

This article was prompted by the invitation of Professor Robert S. Cohen to contribute to a distinctive volume on the philosophy of science, and in some sense may be considered as a sequel to my 28 October 1969 seminar at the Boston Studies in the Philosophy of Science [5]. My philosophy of science is evolutionary-adaptive in the sense that cognitive development from simple sense perceptions to the highest theories of physics is a single adaptive chain of developments. A distinctive feature of this approach as against other similar approaches is that the functional behavior of some perceptive organizations such as color vision can actually be derived from environmental distribution of stimuli plus material constraints plus evolutionary optimization with respect to survival. A most interesting aspect of the color vision theory is that the adaptation to broad-band illuminant changes leading to the law of color constancy is formally a Lorentz transformation (applied, of course, to perceptual variables instead of space-time). In other words, the human eye, in its long evolution, discovered the Lorentz transformations millions of years before Albert Einstein came up with the same formulas in the case of relativistic space-time transformations [5].

Can we therefore say that special relativity is an evolutionary-adaptive theory? It is at first strange to think of it that way but on closer thinking the answer has to be: why not? Thus we can think of special relativity as an adaptation of our ideas and measurements to the behavior of the material world under relative motion and no serious objection can be found against such an attitude. In the same spirit, can we say that Einstein General Relativity (EGR) is an adaptive theory with respect to the gravitational behavior of material bodies in the environment? Unfortunately we cannot answer this question affirmatively at present, because the equations of EGR lead to only an isolated 1-body solution, whereas there are very many gravitating bodies in the environment. In EGR one has to put these other bodies by hand, and this makes it a "test-body" theory in the sense that only the original one body is in the solution, and the other bodies, put by hand, are not in the solution. This destroys the universal interparticle symmetry of the gravitational interaction, hence despite some successful applications of the "test-body" concept, we cannot rest satisfied until the universal symmetry of the gravitational interaction is restored.

This and other objections mentioned in the text add up to the conclusion that the adaptation is not fully achieved and we must search for another theory which is adaptive. To this end, in 1958 the author initiated an N-body metric approach [7] and thereafter generalized it to a form which was discussed in my above-mentioned seminar (at the Boston Studies in the Philosophy of Science) and which appeared in Volume XIII in the Boston Studies in the Philosophy of Science series [5].

The present work contends that in the new theory (YGR) adaptation may now be considered achieved, because not only do the equations of YGR have the requisite N-body solutions, but the theory is compatible with all known phenomena concerning gravity as well. Furthermore, it exhibits a compelling gauge field theory analogy to the highly successful Maxwell gauge field theory of electromagnetism which is explicitly displayed in the text. This is highly relevant because the gauge field theories of Maxwell and of Yang-Mills type can be quantized and renormalized. This brings some sense of unity to our view of the forces of nature, because now all fields are of the nature of gauge fields.

For brevity, and for the purpose of effectively concentrating on the "Fifth Test," we have not dealt with the important experiments on the laser ranging of the Moon and the gravity radiation from the Binary Pulsar. For completeness we would like to state that both these experiments are consistent with the new theory. In the case of Moon ranging the result is that the Strong Equivalence Principle (SEP) is demonstrable and gravitational energy gravitates as in any other form of energy. As for the Binary Pulsar, it is shown (via the exact gravity-wave solutions of the theory) that the radiative energy is positive definite and the rate radiated is consistent with the observations. We plan to discuss these two experiments in a small volume the author is now in the process of writing.

As to how the other topics initiated in the above-mentioned paper fared, I am sorry to say that I did not pursue them with sufficient force apart from the theory of gravity presented in this article. The assumption was that if I succeeded making sufficient headway with gravity, researchers would automatically be motivated to look into my other works and, perhaps, make further progress in them. However, the events took a different turn. First of all, the work took longer than I expected. Although I did make sufficient headway in gravity, general relativists chose to attack it on frivolous grounds instead of looking at it as a viable alternative, which it is. Thus valuable time was spent on answering objections and writing articles covering the misunderstood points. This reminds one of Newton's well-known complaint: "A man either should not publish anything in his lifetime, or be slave to defend it for the rest of his life." By the way, it is not only the attacks that are a problem. There is also a tendency for someone to try to appropriate your results on equally frivolous or erroneous grounds, or even via professionally questionable routes.

Despite these obstacles the general approach caught the eyes of some prominent philosopher-scientists. Thus W. V. Orman Quine, in his book "Ontological Relativity" (Columbia University Press, 1969), referenced my work favorably and later stated that the work implies even the evolution of Kantian categories. S. S. Smith, Patriarch of Experimental Psychophysics, recognized my derivation of the fundamental psychophysical power law (for the first time) and helped to write my paper which appeared in "Perception and Psychophysics." Donald T. Campbell, then the president of American Psychological Association, prominently cited my work in his William James Lecture at Harvard University. In an invited talk at the SPIE meeting in 1990, the author for the first time derived the color analog of the Lorentz transformations for blue-yellow vision (namely, for the red-green blind) from von Kries' coefficient law. In that paper I stated that others independently arrived at the same conclusion but Dr. Michael H. Brill, a prominent color theorist and chairman of the session, corrected me informing us that the other derivation was not independent. I mention this not because it is important to me but because it is always important to set the record straight for the sake of further study.

In all these years the brightest spot of my interest in color vision was when I found that the relativity of color vision in its Galilean analog was first discovered by the great German poet Wolfgang von Goethe. In his book "Theory of Colours" (MIT Press, 1982, pages 36-37, Item 79), he describes an experiment almost equivalent to my "color box" demonstration shown in my above-mentioned paper [5]. He does not perform the reciprocal experiment but I believe this is implicit in his other experiments. It is unfortunate that his work was so badly treated by brutal physicists including the great Helmholtz, who enjoyed debunking Goethe with his thin long dark tube, removing the effect of the environmental illumination inducing the perceptual transformations.

B. On General Covariance

In all of our developments the author has always used Lorentzian (sometimes called Cartesian) coordinates of the special relativistic inertial base space (which is here called the background space), and this gives the impression that the theory under discussion may be a privileged frame theory. The question of what happens if we choose other coordinates (for example, spherical, ellipsoidal, rotating, etc.) of the background brings back the problem of nontensors occurring in these other coordinates of the background, which has been very confusing from the earliest times. This issue is now resolved since it was realized by the author (around 1993) that the problem is *not* with the choice of coordinates but, having made a choice, what kind of derivative operation is to be used to express the physical quantities and processes. If one uses the background-covariant (that is, the background-absolute) derivatives instead of the background-ordinary derivatives, the problem does not arise because then all tensors remain as tensors by definition (of the absolute derivative). This immediately also explains why there is no problem with the original choice of Lorentzian coordinates, because in that case the ordinary derivative is already the absolute derivative. The Lorentzian choice then becomes only a matter of personal taste or convenience. Thus although originally written in the Lorentz coordinates of an inertial background, the physical content remains the same independent of the choice of coordinates of the background. It amounts to simply saying that in all calculations one is to use the "background-absolute" derivatives instead of the "background-ordinary" derivatives.

Note that this is purely a mathematical matter and nothing is special about the choice of the background itself other than to say that we wish to see gravity be grafted onto the inertial space of special relativity, hence recover the latter as a correspondence limit in the absence of gravity. We can equally well graft gravity onto a Robertson-Walker background or onto a cosmological constant background or even onto a smoothed-out space of cosmic background radiation.

It turns out that something akin to this development was already initiated as early as 1940 by N. Rosen. He introduced beside $g_{\mu\nu}$ a second metric $\gamma_{\mu\nu}$, the latter representing the choice of coordinates other than Lorentzian (for example, polar, spheroidal, rotating, etc.) The question is how to calculate physical quantities if $g_{\mu\nu}$ is based on $\gamma_{\mu\nu}$ instead of the original $\eta_{\mu\nu}$, that is, how to eliminate the effect of $\gamma_{\mu\nu}$. Assuming $\gamma_{\mu\nu}$ be a coordinate transform of $\eta_{\mu\nu}$ one can show that

$$\left\{ \begin{matrix} \rho \\ \mu\nu \end{matrix} \right\} - \Gamma^{\rho}_{\mu\nu} = \Delta^{\rho}_{\mu\nu} \tag{B.1}$$

where $\left\{ \begin{smallmatrix} \rho \\ \mu\nu \end{smallmatrix} \right\}$, $\Gamma^{\rho}_{\mu\nu}$ and $\Delta^{\rho}_{\mu\nu}$ are Christoffel symbols with respect to $g_{\mu\nu}$, $\gamma_{\mu\nu}$ and $\eta_{\mu\nu}$ respectively. This operation takes away the effect of $\gamma_{\mu\nu}$ from $g_{\mu\nu}$ leaving behind only the effect of $\Delta^{\rho}_{\mu\nu}$, which is the effect of $g_{\mu\nu}$ with respect to the original $\eta_{\mu\nu}$. To implement his idea Rosen indeed correctly stated that henceforth the physical quantities were to be expressed in terms of $\Delta^{\rho}_{\mu\nu}$ instead of the ordinary symbols $\left\{ \begin{smallmatrix} \rho \\ \mu\nu \end{smallmatrix} \right\}$, but he did not mention that it is (or is equivalent to) the use of background-absolute derivatives of $\gamma_{\mu\nu}$. In fact the phrase "background-covariant derivative" was coined by us [15]. In his implementation Rosen divided out the background Jacobian as $\sqrt{-g}/\sqrt{-\gamma} = \sqrt{-\kappa}$ before differentiation which means that he considered $\sqrt{-\gamma}$ as constant under the background Γ differentiation. $\sqrt{-\kappa}$ then appears explicitly in most of the formulas, creating a sense of mystery and confusion. He did not need to do so because for any metric, under its own covariant (absolute) derivative the covariant derivative of the Jacobian is automatically zero. Apparently Rosen did not know this at the time he wrote his articles. In the statement of the author no such removal of $\sqrt{-\gamma}$ appears anywhere. However, it turns out that it is easier to implement the process on the computer in the Rosen form, hence we have it that way in our program.

We went through these seemingly tedious explanations because, not only are they important in showing the general covariance of the theory (any theory), they also show that Rosen's additional claim that the simultaneous presence of $\gamma_{\mu\nu}$ along with $g_{\mu\nu}$ is essential (the so-called bi-metric theory), and that the Riemannian space is fundamentally bi-metric, is not justifiable. The great insight of Gauss was that given $g_{\mu\nu}$, everything relevant can be intrinsically calculated. The bi-metric concept cannot be a guide to formulate distinct theories because the existence of $\gamma_{\mu\nu}$ does not make distinctions between theories. Rosen wrote the above formula as

$$\left\{ \begin{matrix} \rho \\ \mu\nu \end{matrix} \right\} = \Gamma^{\rho}_{\mu\nu} + \Delta^{\rho}_{\mu\nu} \tag{B.2}$$

in which $\gamma_{\mu\nu}$ appears like something to be incorporated instead of bypassed or eliminated. More specifically, such formulas form an additive group by virtue of the multiplicative group property of the metric mentioned in my Int. J. Theor. Phys. article, Reference [3], since Christoffel symbols are of the nature of logarithmic derivatives. To refer to any one of these as more fundamental than others makes the theory a privileged frame theory.

C. Historical Background

As is well known, Einstein adopted, as his gravitational field stress-energy, an expression z^{ν}_{μ} which he called the "stress-energy complex" of the gravitational field (he denoted it t^{ν}_{μ} but since we reserve the latter symbol for another quantity due to Yilmaz, we call Einstein's expression z^{ν}_{μ}). Einstein's expression turned out to have strange behaviors which cast doubt as to its acceptability as a field stress-energy for the gravitational field. For example, E. Schrodinger [10] calculated the z^{ν}_{μ} for the Schwarzschild metric in Cartesian coordinates and found all its components to be identically zero. On the other hand, H. Bauer [11] calculated it in flat space but in spherical coordinates (that is, with no gravity at all), and found that it will not vanish. It was later found by Einstein himself that z^{ν}_{μ} is neither localizable nor even symmetric. Because of these and other problems H. Weyl [12] called Einstein's choice of z_μ^ν as a gravitational field stress-energy "devoid of sense."

In the meantime further criticisms came from other respectable quarters. Levi-Civita, the founder of tensor calculus, pointed out that Einstein's theory does not satisfy the action equals reaction principle found in the Newtonian theory even in the limit. He attributed this to the issue of N-body solutions and tried to formulate a new set of field equations. On similar grounds, H. A. Lorentz proposed another set of field equations. Both these new field equations were later found to be objectionable and were discarded without solving their intended problems.

At least on the surface, all these issues appeared to be mostly theoretical. However, a well-known mathematician, C. Burali-Forti [13], brought the issue to the experimental arena. He pointed out that, as it stands, Einstein's theory cannot predict the 532″ per century planetary perturbative part of the perihelion advance of Mercury. No response was made to these statements at the time and, because of the lack of resolution of the N-body solutions, the issues were gradually forgotten.

As to Einstein himself, he never changed his mind, and despite severe criticisms and friendly warnings, he "stubbornly" held onto his original choice of z_μ^ν. He wrote [14]:

> "While general relativity met with agreement by most theoretical physicists and mathematicians, most colleagues of the subject raise objections to my formulation of the energy-momentum theorem. Since I am convinced to be correct with their formulation, I will explain in the following my view towards this question with necessary detail."

He then reproduces his original expression of z_μ^ν (including its asymmetry) which was so adamantly criticized by his illustrious colleagues. It is the same nontensor expression shown in our 1992 Il Nuovo Cimento article, Ref. [3]. (More on the early history of general relativity can be found in the series of five books, "Einstein Studies," edited by Don Howard and John Stachel and published by Birkhouser Publishing Company.)

We went over this piece of history because in recent years the author criticized general relativity on practically all of the above-mentioned grounds without knowing that they were already made decades earlier by others. This was met with great surprise and hostility from some quarters and efforts were made to portray these criticisms by the author and his associates as due to ignorance. In reality it is the opposite. As we have seen, the same criticisms were made from the earliest times by prominent scientists but due to lack of clear resolutions the problems were gradually forgotten. As it turns out the author only revived them.

There is, however, a difference. The difference is that the author, in addition, offers a resolution of the issues by pinpointing the sources of the difficulties and by providing a new framework where there are no such difficulties. He first points out that Einstein's z_μ^ν is a nontensor. If not eliminated by a general procedure it could make any theory coordinate dependent. But there are ways of eliminating the z_μ^ν, hence the coordinate independence is saved for any theory. He next postulates the existence of a true stress-energy tensor that enters the equations according to the paradigm mentioned. Then he shows that in this form there are indeed N-body interactive solutions, [3, 15, 16]. In the usual theory there are no N-body interactive solutions (metrics), so we cannot begin a legitimate calculation. Maxwell once said that "the concept of a single solitary particle is physically meaningless" as the third law (action=reaction) requires an interaction of at least two particles [17]. Thus a 1-body metric is never satisfactory in the slow motion Newtonian limit, by which, among other things, the 532″ per century planetary-perturbative part of the perihelion advance of Mercury can be correctly calculated.

D. Some Miscellaneous Remarks

In this Appendix we make some remarks which were omitted in the text for the sake of simplicity and rectify some misstatements in old articles which have long since been clarified. Corrections do not extend to obvious typographical errors.

a) The exponential metric (1.7) is a compact statement of the result of infinite number of local infinitesimal iterations from $g_{\mu\nu}$ to $g_{\mu\nu} + \delta g_{\mu\nu}$ although many global closed solutions exist. Local interpretation of the integral is compatible, among other things, with the requirement that ∂_x and ∂_y commute since infinitesimal iterations always commute.

b) In the Int. J. Theor. Phys. article, Reference [3], the impression is given that in the new theory the equations of motion follow from the field equations as in Einstein's theory. This is not the case. In the new theory the equations of motion of material particles arise from a different variation in the action as in all other field theories of physics. Also, in Reference [3] it was said that the two identities "clash." The word "confute" may be a better description, or you may invent your own word for the situation.

c) In the Boston Studies in the Philosophy of Science article, Reference [5], the formula (4.8) on page 29 should read $p = (f/n) \pm A^2 n(\delta f - n\delta v)/2$. Likewise with (4.7). Although the derivation is questionable, the formula itself is valid. This formula for the perception of pitch seems to offer a viable new theory of harmony in music. It also predicts a certain neural structure in the ear which may be possible to confirm by physiological investigation.

d) The gauge term $\partial_\alpha(\sqrt{-g}\, g^{\nu\rho}\partial_\rho\phi^\alpha_\mu)/\sqrt{-g}$ cannot arbitrarily be set to zero, because in the quantized theory it has the role of setting up the theory in various gauges (Feynman, Landau, Coulomb, axial, etc.) and deciding which terms are observable. The argument that it may be removed by adding an external counter term to the Lagrangian destroys the autonomy of the geometric approach.

In analogy to quantum electrodynamics, the gauge term can be parameterized as

$$\Box^2\phi^\nu_\mu + \frac{(\xi - 1)\partial_\alpha(\sqrt{-g}g^{\nu\rho}\partial_\rho\phi^\alpha_\mu))}{\sqrt{-g}} \Rightarrow \sigma u_\mu u^\nu \tag{D.1}$$

where $\xi = 1$ corresponds to the Feynman gauge, and $\xi = 0$ to the Landau gauge. We can see that in the Feynman gauge we have the simpler field equations [3, 5]

$$\Box^2\phi^\nu_\mu \Rightarrow \sigma u_\mu u^\nu \tag{D.2}$$

e) A most misleading argument in general relativity was that the field stress-energy t^ν_μ cannot be a tensor because it can be annihilated by a coordinate transformation. This argument is false because a counterexample is staring at us. It is the t^ν_μ of the exponential metric (1.9). What about the annihilation argument which still sounds valid? No. The original argument was valid because it was advanced when the theory was a "test-particle" theory which had no active mass, hence no Laplacian. Now we have a true N-particle theory with nonzero Laplacians at every mass point, and a nonzero Laplacian cannot be annihilated by a coordinate transformation. Therefore the process has to be a "compensation" in accordance with the strong principle of equivalence and not a coordinate transformation [3].

f) The slow motion limit (which is sometimes also called static) is often construed as saying that (since the solution is time-independent) nothing can move, nothing can happen. Those who make this kind of statement expose their ignorance about the Newtonian theory which is the correspondence limit of this theory. Now, as a solution to the Poisson equation, the Newtonian Potential is time-independent, so how does anything move in the Newtonian theory? Is it not obvious that the same procedure leading to motion in the Newtonian theory can be repeated here? While we leave this argument to them to work out themselves, we also point out that the theory has infinitesimal velocity dependent solutions into which the initial field can evolve. In other words, the time-indepedent initial field does not necessarily remain time-independent.

Finally, the whole philosophy, and also the whole physical basis of the new theory, can be paraphrased in three short statements extending special relativity and its measurement procedure locally to curved space-times:

1) Physical measurement is a local process;
2) Local signal velocity is a universal constant;
3) Laws of physics are local Lorentz covariant.

It is clear that the first two statements have to do with the local operational procedure of space-time measurements and define the local kinematics of the theory. However, without the third statement the theory would not be complete, because it has to do with the behavior (dynamics) of the material content of the space-time so defined. As is well known, further progress in this direction is the determination of the interaction terms via a principle of local gauge covariance. We hope to discuss this topic in a different communication.

Note added in proof: The Yılmaz theory is formulable as an axiomatic system if Bianchi and Freud identities are regarded as *axioms*, equations (1.3) and (1.4) as *postulates*, and the gauge conditions (1.5) as the *compatibility conditions* between Quantum Mechanics and the spacetime theory of gravity. The present exposition, however, has a rather definite goal; to focus attention on the simple yet nontrivial case of the N-body exponential metric (2.4) and its physical consequences *vis a vis* those of the 1-body Schwarz schild metric (2.1). We believe the case is strong in favor of the N-body exponential metric (2.4) of the Yılmaz theory.

YALÇIN KOÇ

IMPLICATIONS OF THE GEOMETRY OF QUANTUM MECHANICAL PERFECT CORRELATION FUNCTIONS CONCERNING "BELL'S THEOREM WITHOUT INEQUALITIES"

1. INTRODUCTION

By analyzing a system of three or more correlated spin-1/2 particles, Greenberger, Horne, Shimony and Zeilinger claim that "the EPR program contradicts quantum mechanics even for the cases of perfect correlations" (Greenberger 1990, p. 1131).[1] Greenberger et al. state that "this incompatibility with quantum mechanics is stronger than the one previously revealed for two-particle systems by Bell's inequality, where no contradiction arises at the level of perfect correlations" and conclude that their argument provides a proof of "Bell's theorem without resorting to an inequality" (Greenberger 1990, pp. 1131-1132).

Bell's arguments (Bell, 1964 and Bell, 1971) against local hidden variable theories proceed by means of inequalities; it is noted in (Koç, 1992) that in these arguments Bell does not consider the geometrical (or, algebraic) properties of the quantum mechanical correlation function (for a system of spin-1/2 particles in the singlet state). It is shown in (Koç, 1992) that, due to the geometry (or, algebraic properties) of the quantum mechanical correlation function, Bell's arguments in (Bell, 1964 and Bell, 1971) are inconclusive. In addition to this, it is asserted in (Koc, forthcoming) that Wigner's argument (Wigner, 1970) against local hidden variable theories is similarly inconclusive because of the geometrical (or, algebraic) properties of the quantum mechanical probability functions (for a system of spin-1/2 particles in the singlet state).

The significance of the geometrical (or, algebraic) properties of the quantum mechanical correlation functions, however, is completely disregarded by Greenberger et al. in 1990.

In section 2 in the present paper, we briefly summarize the main argument in (Greenberger, 1990). In section 3, by considering the geometrical (or, algebraic) properties of the quantum mechanical perfect correlation functions, we decompose the *perfect correlation function* in eqn. (10a) [see (Greenberger, 1990, p. 1134); also

G. Irzık & G. Güzeldere (eds.), Turkish Studies in the History and Philosophy of Science, 59-68.

eqns. (2) and (16) below] into the difference of two terms where one of the terms is a constant and the other is the product of a constant with a quantum mechanical *imperfect (or, statistical) correlation function* [see eqns. (16) and (17) below]; the decomposition, we should note, is purely geometrical.[2]

Greenberger et al. employ the two-valued functions $A_\lambda(\Phi_1)$, $B_\lambda(\Phi_2)$, $C_\lambda(\Phi_3)$ and $D_\lambda(\Phi_4)$ to 'demonstrate' the 'contradiction' (Greenberger, 1990, p. 1135); l denotes the "complete state of the four particles" (Greenberger, 1990, p. 1134). One must immediately note that since the only possible values, namely +1 and −1, of these functions correspond to the outcomes of the spin measurements of the four correlated particles, these functions are *not* additive; they are, however, multiplicative. Quantum mechanical correlation functions, on the other hand, are additive.[3]

Our analysis in section 3 shows that no possible combination of the functions $A_\lambda(\Phi_1)$, $B_\lambda(\Phi_2)$, $C_\lambda(\Phi_3)$ and $D_\lambda(\Phi_4)$ can produce the quantum mechanical imperfect (or, statistical) correlation function $E^\Psi(\hat{m}, \hat{n})$ [in eqns. (16) and (17) below] which we obtain by geometrically decomposing the quantum mechanical perfect correlation function $E^\Psi(\hat{a}, \hat{b}, \hat{a}, \hat{b})$ [in eqn. (16) below].

The quantum mechanical perfect correlation function $E^\Psi(\hat{a}, \hat{b}, \hat{c}, \hat{d})$ [in eqn.(2) below] is *bilinear*. On the other hand, the two-valued functions $A_\lambda(\Phi_1)$, $B_\lambda(\Phi_2)$, $C_\lambda(\Phi_3)$ and $D_\lambda(\Phi_4)$ are *not additive*; consequently, these functions are *not bilinear*. Although bilinearity is an algebraic property, it also reveals a geometrical feature. Thus, the geometry (or, algebra) of the quantum mechanical perfect correlation function $E^\Psi(\hat{a}, \hat{b}, \hat{c}, \hat{d})$ and the geometry (or, algebra) of the two-valued functions $A_\lambda(\Phi_1)$, $B_\lambda(\Phi_2)$, $C_\lambda(\Phi_3)$ and $D_\lambda(\Phi_4)$ are *incompatible* in view of bilinearity.

It therefore follows that Greenberger et al. obtain the 'contradiction' (Greenberger, 1990, p. 1135) by comparing two distinct types of functions whose geometries (or, algebras) are *already* incompatible. Since Greenberger et al. completely disregard the geometrical (or, the algebraic) properties of correlation functions, the incompatibility between the geometries (or, algebras) of the two distinct types of functions above is not manifest in their paper (Greenberger, 1990). The 'contradiction' (Greenberger, 1990, p. 1135) therefore is *not* significant and consequently the arguments in (Greenberger, 1990) do not constitute a proof of "Bell's theorem without inequalities".

2. "BELL'S THEOREM WITHOUT INEQUALITIES"

In view of Bell's argument of 1964 (Bell, 1971), Greenberger et al. state that "in the system of two spin-1/2 particles, contradictions develop only when one considers the quantum mechanical statistical predictions. This incompatibility is demonstrated by deriving an inequality from EPR's premises and then noting that the quantum mechanical statistical predictions do not satisfy this inequality." Then, they argue: "however, in the three-particle system, there is no point in deriving an inequality, or anything else for that matter, since the premises are inconsistent." (Greenberger, 1990, p. 1133).

To demonstrate the 'inconsistency' of EPR's premises [see (i)–(iv) in (Greenberger, 1990, p. 1132)], Greenberger et al. consider "a system of four spin-1/2 particles produced so that particles 1 and 2 move freely in the positive z-direction and particles 3 and 4 in the negative z-direction, as shown in Fig. 2" (Greenberger, 1990, pp. 1133-1134). If we express the orientations of the Stern-Gerlach analyzers for particles 1, 2, 3 and 4 respectively by the unit vectors $\hat{a}$, $\hat{b}$, $\hat{c}$ and $\hat{d}$ which are all in the xy-plane such that they make the angles Φ_1, Φ_2, Φ_3 and Φ_4 with the respective x-axes[4], then the quantum mechanical expectation value function E^{Ψ} of the product of the outcomes is:

$$E^{\Psi}(\hat{a}, \hat{b}, \hat{c}, \hat{d}) = -\cos(\Phi_1 + \Phi_2 - \Phi_3 - \Phi_4) \tag{1}$$

[eqn. (9), (Greenberger, 1990, p. 1134)]. Then, from eqn. (1), Greenberger et al. derive the following *perfect correlation functions*:

$$\text{If } \Phi_1 + \Phi_2 - \Phi_3 - \Phi_4 = 0,$$
$$\text{then, } E^{\Psi}(\hat{a}, \hat{b}, \hat{c}, \hat{d}) = -1 \tag{2}$$

And:

$$\text{If } \Phi_1 + \Phi_2 - \Phi_3 - \Phi_4 = \pi,$$
$$\text{then, } E^{\Psi}(\hat{a}, \hat{b}, \hat{c}, \hat{d}) = +1 \tag{3}$$

[eqns. (10a) and (10b) respectively, (Greenberger, 1990, p. 1134)].

After reconsidering the EPR premises, Greenberger et al. state that "the arguments in Sec. II can now be paralleled to infer the existence of four functions $A_\lambda(\Phi_1)$, $B_\lambda(\Phi_2)$, $C_\lambda(\Phi_3)$, $D_\lambda(\Phi_4)$ with values $+1$ or -1. These functions are the outcomes of spin measurements along the respective directions when the complete state of the four particles is λ." Greenberger et al. then "*restate* (10a) [eqn. (2) above] and (10b) [eqn. (3) above] in terms of the functions A, B, C, and D, the existence of which follows from the premises" [Greenberger, 1990, p. 1134] [Emphasis is mine]:

$$\text{If } \Phi_1 + \Phi_2 - \Phi_3 - \Phi_4 = 0,$$
$$\text{then, } A_\lambda(\Phi_1)B_\lambda(\Phi_2)C_\lambda(\Phi_3)D_\lambda(\Phi_4) = -1 \tag{4}$$

And:

$$\text{If } \Phi_1 + \Phi_2 - \Phi_3 - \Phi_4 = \pi,$$
$$\text{then, } A_\lambda(\Phi_1)B_\lambda(\Phi_2)C_\lambda(\Phi_3)D_\lambda(\Phi_4) = +1 \tag{5}$$

[eqns. (11a) and (11b) respectively, (Greenberger, 1990, p. 1134)]. Then, Greenberger et al. infer the following from eqn. (4) above:

$$A_\lambda(2\Phi) = A_\lambda(0) = \text{constant for all } \Phi \tag{6}$$

[eqn. (16), (Greenberger, 1990, p. 1135)]. By considering eqn. (5), Greenberger et al. obtain the following:

$$A_\lambda(\theta + \pi) = -A_\lambda(\theta) \tag{7}$$

[eqn. (18), (Greenberger, 1990, p. 1135)]. Thus, according to Greenberger et al., "the trouble becomes manifest and an actual contradiction emerges" (Greenberger, 1990, p. 1135) if we consider the case where $\Phi = \pi/2$ and $\theta = 0$:[5]

$$A_\lambda(\pi) = A_\lambda(0) \text{ and } A_\lambda(\pi) = -A_\lambda(0) \tag{8}$$

Greenberger et al. furthermore argue that "in the foregoing algebra, the argument of the function $B_\lambda(\Phi_2)$ was fixed throughout to be 0, which shows that the premises (i)-(iv) are also inconsistent when applied to a system of three spin-1/2 particles" (Greenberger, 1990, p. 1135).

3. GEOMETRY OF THE QUANTUM MECHANICAL PERFECT CORRELATION FUNCTION E^Ψ

Greenberger et al. maintain that "in the case of two spin-1/2 particles with total spin zero, the cosine of Eq. (3') [i.e., $E^\Psi(\Phi_1, \Phi_2) = -\cos(\Phi_1 - \Phi_2)$] plays a central role in proving that quantum mechanics contradicts the inequality. However, in the three-particle case, the specific form of the correlation plays no role in demonstrating a contradiction." (Greenberger, 1990, p. 1133).

Greenberger et al. consider neither the geometry nor the algebraic properties of the quantum mechanical correlation function in eqn. (1) and of the perfect correlation function in eqn. (2). Such a consideration, however, is crucial because of the following geometrical analysis.

Let us first consider the complex representations of the unit vectors $\hat{a}, \hat{b}, \hat{c}$ and $\hat{d}$ in the xy-plane:

$$\hat{a} = e^{i\Phi 1},\ \hat{b} = e^{i\Phi 2},\ \hat{c} = e^{i\Phi 3},\ \hat{d} = e^{i\Phi 4} \tag{9}$$

where the unit vectors $\hat{a}, \hat{b}, \hat{c}, \hat{d}$ make the angles $\Phi_1, \Phi_2, \Phi_3, \Phi_4$ with their respective x-axes. We rewrite eqn. (1) as follows:

$$E^\Psi(\hat{a}, \hat{b}, \hat{c}, \hat{d}) = E^\Psi(\hat{u}, \hat{v}) = -\hat{u} \,.\, \hat{v} \tag{10}$$

where, in the complex representation:

$$\hat{u} = e^{i(\Phi_1 + \Phi_2)} \text{ and } \hat{v} = e^{i(\Phi_3 + \Phi_4)} \tag{11}$$

$\hat{u}$ is obtained by rotating $\hat{a}$ an angle of Φ_2 in the positive (i.e., counter-clockwise) direction and $\hat{v}$ is obtained by rotating $\hat{c}$ an angle of Φ_4 in the positive direction. Both

$\hat{u}$ and $\hat{v}$ are of unit length and "." indicates the inner product of two real vectors. We should note that $\hat{u}$ and $\hat{v}$ are in the xy-plane.

The correlation function $E^{\Psi}(\hat{u}, \hat{v})$ in eqn. (10) above is *bilinear* and *symmetric*.[6] It therefore follows that the correlation function $E^{\Psi}(\hat{a}, \hat{b}, \hat{c}, \hat{d})$ in eqn. (1) is expressed in the *bilinear* and *symmetric* form $E^{\Psi}(\hat{u}, \hat{v})$ in eqn. (10), in a purely geometrical manner.

Let us now express in eqn. (11) as:

$$\hat{u} = \hat{m} + \hat{n} \tag{12}$$

such that, in the complex representation:

$$\hat{m} = e^{i(\Phi - \pi/3)}, \quad \hat{n} = e^{i(\Phi + \pi/3)} \quad \text{where } \Phi = \Phi_1 + \Phi_2 \tag{13}$$

$\hat{m}$ and $\hat{n}$ are also in the xy-plane. It should be noted that the vector $\hat{m} + \hat{n}$, which is defined in eqns. (12)-(13), is of unit length.[7] By means of eqns. (10)-(13), we can write the following:

$$E^{\Psi}(\hat{u}, \hat{v}) = E^{\Psi}(\hat{m} + \hat{n},\ \hat{m} + \hat{n}) = -(\hat{m} + \hat{n}) . (\hat{m} + \hat{n}) = -1 \tag{14}$$

One should also note that $E^{\Psi}(\hat{u}, \hat{v}) = E^{\Psi}(\hat{a}, \hat{b}, \hat{a}, \hat{b})$; therefore, eqn. (14) satisfies the condition $\Phi_1 + \Phi_2 - \Phi_3 - \Phi_4 = 0$ in eqn. (2) because, $\Phi_3 = \Phi_1$ and $\Phi_4 = \Phi_2$. Since E^{Ψ} in eqn. (14) is bilinear, we have:

$$E^{\Psi}(\hat{m} + \hat{n},\ \hat{m} + \hat{n}) = E^{\Psi}(\hat{m}, \hat{m}) + E^{\Psi}(\hat{m}, \hat{n}) + E^{\Psi}(\hat{m}, \hat{n}) + E^{\Psi}(\hat{n}, \hat{n}) \tag{15}$$

E^{Ψ} is furthermore symmetric. We therefore have the following quantum mechanical perfect correlation function from eqns. (10)-(15):

$$E^{\Psi}(\hat{a}, \hat{b}, \hat{a}, \hat{b}) = 2E^{\Psi}(\hat{m}, \hat{n}) - 2 = -1 \tag{16}$$

Eqn. (16) above displays a *purely geometrical decomposition* of the perfect correlation function $E^{\Psi}(\hat{a}, \hat{b}, \hat{a}, \hat{b})$; one of the constituents in the decomposition is the *imperfect (or, statistical) correlation function*:

$$E^{\Psi}(\hat{m}, \hat{n}) = E^{\Psi}(e^{i(\Phi - \pi/3)},\ e^{i(\Phi + \pi/3)}) = 1/2 \tag{17}$$

where $\Phi = \Phi_1 + \Phi_2$.

The functions $A_\lambda(\Phi_1)$, $B_\lambda(\Phi_2)$, $C_\lambda(\Phi_3)$ and $D_\lambda(\Phi_4)$ in eqn. (4) are two-valued; they are multiplicative but *not* additive. Therefore, *no* composition of these functions can produce the imperfect (or, statistical) correlation function $E^{\Psi}(\hat{m}, \hat{n})$ in eqn. (17). We therefore have a proof of the statement that the full geometry of the quantum mechanical perfect correlations *cannot* be reproduced in terms of the two-valued functions $A_\lambda(\Phi_1)$, $B_\lambda(\Phi_2)$, $C_\lambda(\Phi_3)$ and $D_\lambda(\Phi_4)$.

Eqn. (16) above expresses the fact that the imperfect (or, statistical) correlation function $E^{\Psi}(\hat{m}, \hat{n})$ in eqn. (17) is a *geometrical counterpart* of the perfect correlation function $E^{\Psi}(\hat{a}, \hat{b}, \hat{a}, \hat{b})$ in eqn. (16). That is, the perfect correlation function $E^{\Psi}(\hat{a}, \hat{b}, \hat{a}, \hat{b})$ is *geometrically* composed of counterparts one of which is the imperfect (or, statistical) correlation function $E^{\Psi}(\hat{m}, \hat{n})$ in eqn. (17). This is a very interesting geometrical (or, algebraic) peculiarity of quantum mechanics; no analogue of this geometrical feature, however, can be obtained in terms of the functions $A_\lambda(\Phi_1)$, $B_\lambda(\Phi_2)$, $C_\lambda(\Phi_3)$ and $D_\lambda(\Phi_4)$, in the allegedly complete state λ.

Since no possible combination of the two-valued functions $A_\lambda(\Phi_1)$, $B_\lambda(\Phi_2)$, $C_\lambda(\Phi_3)$ and $D_\lambda(\Phi_4)$ can produce an imperfect (or, statistical) correlation function, the geometry of the two-valued functions $A_\lambda(\Phi_1)$, $B_\lambda(\Phi_2)$, $C_\lambda(\Phi_3)$ and $D_\lambda(\Phi_4)$, for the allegedly complete state λ, is *inadequate* to reproduce the geometry (or, algebra) of the quantum mechanical perfect *correlation* functions.

Furthermore, the geometry (or, algebra) of the quantum mechanical perfect correlation function $E^{\Psi}(\hat{a}, \hat{b}, \hat{c}, \hat{d})$ is *incompatible* with the geometry (or, algebra) of the two-valued functions $A_\lambda(\Phi_1)$, $B_\lambda(\Phi_2)$, $C_\lambda(\Phi_3)$ and $D_\lambda(\Phi_4)$ because of the following reason. The geometry (or, algebra) of the function $E^{\Psi}(\hat{a}, \hat{b}, \hat{c}, \hat{d})$ satisfies *bilinearity*. The two-valued functions $A_\lambda(\Phi_1)$, $B_\lambda(\Phi_2)$, $C_\lambda(\Phi_3)$ and $D_\lambda(\Phi_4)$, on the other hand, are not *additive*; it therefore follows that the geometry (or, algebra) of these two-valued functions *cannot* satisfy *bilinearity*.

The geometry (or, algebra) of the two-valued functions $A_\lambda(\Phi_1)$, $B_\lambda(\Phi_2)$, $C_\lambda(\Phi_3)$ and $D_\lambda(\Phi_4)$ is, therefore, incompatible with the geometry (or, algebra) of the quantum mechanical perfect correlation functions, in view of bilinearity. Thus, the comparison of these two types of distinct functions is inconsequential because their respective geometries (or, algebras) are *already* incompatible. It then follows that the 'contradiction' [see (Greenberger, 1990, p. 1135)] is a product of these two incompatible geometries (or, algebras) and thus the 'contradiction' *itself* is insignificant for a proof of "Bell's theorem without inequalities".

4. CONCLUSIONS AND DISCUSSION

We showed in section 3 in the present paper that a quantum mechanical perfect correlation function is geometrically composed of an imperfect (or, statistical) correlation function. We furthermore argued that since the two-valued functions $A_\lambda(\Phi_1)$, $B_\lambda(\Phi_2)$, $C_\lambda(\Phi_3)$and $D_\lambda(\Phi_4)$ are not additive, these functions cannot reproduce the full geometry (or, algebra) of the quantum mechanical perfect correlation functions. This point is not manifest in (Greenberger, 1990) because Greenberger et al. completely disregard the geometrical (or, algebraic) properties of correlation functions.

In view of bilinearity, we showed that the geometry (or, algebra) of the quantum mechanical perfect correlation functions is *incompatible* with the geometry (or, algebra) of the two-valued functions $A_\lambda(\Phi_1)$, $B_\lambda(\Phi_2)$, $C_\lambda(\Phi_3)$ and $D_\lambda(\Phi_4)$. We thus argued that the 'contradiction' (Greenberger, 1990, p. 1135) is a product of two geometries (or, algebras) that are already incompatible and therefore, the 'contradiction' itself is inconsequential for a proof of "Bell's theorem without inequalities."

In section IV of their paper, Greenberger et al. also investigate a case which is without spin (Greenberger, 1990, pp. 1135-1136). Since a similar analysis of this case is possible and therefore similar objections can be raised, we do not consider it in detail in the present paper.

Greenberger et al. state: "Is there any point in designing yet another experiment along new lines, in order to *rerefute* the program of EPR? We think so, for two reasons. The first is sheer intellectual challenge. We would like to know what experiment would have been appropriate had history been different and had GHZ's demonstration been the first proof of Bell's theorem. The second is that the investigation of correlations among three or more particles can open a new, beautiful, and fruitful type of experimentation, of interest independently of EPR." (Greenberger, 1990, p. 1136), [Emphasis is mine.]. And accordingly, they discuss and suggest new experiments (Greenberger, 1990, p. 1138).

Let us first ask whether the EPR program has been previously refuted? Does, for example, the Aspect et al. experiment (Aspect, 1982) refute the EPR program?

It is obviously true that the Aspect et al. experiment tests a specific form of Bell's inequality. However, one should note, testing of the inequality is significant for the EPR program *only if* the EPR program entails the inequality. That is, the experimental testing of Bell's inequality is consequential *only if* there is a physical model of the EPR program such that Bell's inequality has a derivation in this model; otherwise, an experimental violation of the inequality does not imply a refutation of the EPR program.

Bell's model of the EPR program consists of the hypothetical local hidden variable theory in (Bell, 1964) and (Bell, 1971). In (Koç, 1992), however, it is shown that Bell's local hidden variable expectation value function [eqn. (1) on (Koç, 1992, p. 961)] is compatible with quantum mechanics *only if* it is bilinear, symmetric and rotationally invariant; and furthermore, if the local hidden variable expectation value function satisfies all of these geometrical (or, algebraic) properties, then Bell's inequality cannot be a mathematical consequence of Bell's own model (Koç, 1992, pp. 962-964). It therefore follows that the experimental violation of Bell's inequality is inconsequential with respect to the EPR program. The Aspect et al. experiment, by using time-varying analyzers, has contributed to experimental techniques in the domain of quantum mechanics; because of the explanation above, however, the Aspect et al. experiment does not have the consequence of refuting the EPR program.

Because Bell completely disregards the geometrical (or, algebraic) properties of expectation value functions in (Bell, 1964) and (Bell, 1971), his arguments totally miss the point we make above. We should note that the complete disregard of the geometrical (or, algebraic) properties of the quantum mechanical expectation value functions (and, of probability functions as well) is a generic feature of the different derivations of Bell's inequalities.

Thus, Greenberger et al. are confronted not with a 'rerefutation' but simply a 'refutation' of the EPR program. If the arguments of Greenberger et al. (Greenberger, 1990) are valid, then they would genuinely have the chance of refuting the program of EPR experimentally. In the present paper, however, we showed that the 'contradiction' (Greenberger, 1990, p. 1135) they obtain is a product of two incompatible geometries (or, algebras); therefore, the present form of their argument is inconsequential in view of "Bell's theorem without inequalities". Thus, it follows, the experiments Greenberger et al. suggest (Greenberger, 1990, p. 1138) are insignificant for the testing of the EPR program, independently of the number of particles involved in these experiments.[8]

We conclude by stating that an experimental test of eqn. (16) is significant for our understanding of the manifestation of quantum mechanical correlations in classical geometry.[9]

Bogazici University,
Department of Philosophy

5. NOTES

[1] Greenberger et al. specify the Einstein-Podolsky-Rosen propositions as: (i) perfect correlation, (ii) locality, (iii) reality and (iv) completeness and, state that the first of the premises is 'drawn' from quantum mechanics whereas the other three are 'quite plausible' (Greenberger, 1990, p. 1132). The alleged 'contradiction', then, is a consequence of the conjunction of (i)–(iv). Greenberger et al. also state that (i) is 'inconsistent' with the conjunction of (ii)–(iv) (Greenberger, 1990, p. 1134).

[2] We use the term 'imperfect (or, statistical) correlation' in the sense of Greenberger et al. (Greenberger, 1990, p. 1131) to indicate those correlations which are not perfect.

[3] See, for example, (Koç, 1992, p. 174).

[4] Greenberger et al. use $\hat{n}_1, \hat{n}_2, \hat{n}_3, \hat{n}_4$ instead of $\hat{a}, \hat{b}, \hat{c}, \hat{d}$ respectively.

[5] Because of simple logical curiosity, we cannot pass without commenting on the "emergence of an actual contradiction" suggested by Greenberger et al. In general logic, a contradiction indicates the impossible. The impossible, however, can never 'emerge' as 'actual'. Therefore, 'an actual contradiction' can never 'emerge'. Hence, we interpret "the emergence of an actual contradiction" only stylistically.

[6] Since the inner product is bilinear and symmetric, it immediately follows that the correlation function $E^{\Psi}(\hat{u}, \hat{v})$ in eqn. (10) is also bilinear and symmetric.

[7] We have: $\hat{m} . \hat{n} = \mathrm{Re}\,(e^{i(\Phi-\pi/3)} e^{i(\Phi+\pi/3)^*})$

$$= \cos(-2\pi/3)$$
$$= -1/2$$

Hence: $(\hat{m} + \hat{n}) . (\hat{m} + \hat{n}) = 2 + 2\, \hat{m} . \hat{n} = 1$

Therefore: $\|\hat{m} + \hat{n}\| = 1$

[8] Arguments similar to the ones in the present paper apply equally well to those of Bernstein et al.

[9] We can express in eqn. (16) as a four-place function as follows:

$$\begin{aligned} E^{\Psi}(\hat{m}, \hat{n}) &= E^{\Psi}(e^{i(\phi-\pi/3)}, e^{i\phi+\pi/3)}) \\ &= E^{\Psi}(e^{i\phi_1}, e^{i(\phi_2-\pi/3)}, e^{i\phi_1}, e^{i(\Phi_2+\pi/3)}) \\ &= E^{\Psi}(\hat{a}, \hat{g}, \hat{a}, \hat{h}) \end{aligned}$$

where $\hat{a} = e^{i\Phi_1}$, $\hat{g} = e^{i(\Phi_2-\pi/3)}$ and $\hat{h} = e^{i(\Phi_2+\pi/3)}$. It should be noted that the first $\hat{a}$ in $E^{\Psi}(\hat{a}, \hat{g}, \hat{a}, \hat{h})$ above denotes the orientation of the first Stern-Gerlach analyzer, $\hat{g}$ denotes the orientation of the second Stern-Gerlach analyzer, the second $\hat{a}$ denotes the orientation of the third Stern-Gerlach analyzer and $\hat{h}$ denotes the orientation of the fourth Stern-Gerlach analyzer. The correlation $E^{\Psi}(\hat{a}, \hat{g}, \hat{a}, \hat{h})$, then, can be experimentally tested.

6. REFERENCES

Aspect, A., J. Dalibard, G. Roger. "Experimental Tests of Bell's Inequalities Using Time-varying Analyzers", *Phys. Rev. Lett.* 47 (1982): 1804-1807.

Bell, J. S. "On the Einstein-Podolsky-Rosen Paradox", *Physics* 1 (1964): 195.

Bell, J. S. "Introduction to the Hidden Variable Question" in B. d'Espagnat, ed., *Foundations of Quantum Mechanics*. Academic Press, 1971.

Bernstein, H. J., D. M. Greenberger, M. A. Horne, A. Zeilinger, "Bell's Theorem without Inequalities for Two Spinless Particles", *Phys. Rev.* 47A (1993): 78-84.

Greenberger, D. M., M. A. Horne, A. Shimony, A. Zeilinger, "Bell's Theorem without Inequalities", *Amer. J. Phys.* 58 (1990): 1131-1143.

Koç, Y., "The Local Expectation Value Function and Bell's Inequalities", *Il Nuovo Cimento* 107B (1992): 961-971.

Koç, Y. , "Wigner's Inequality, Quantum Mechanical Probability Functions and Hidden Variable Theories" forthcoming *in Il Nuovo Cimento B.*

Wigner, E. P., "On Hidden Variables and Quantum Mechanical Probabilities", *Amer. J. Phys.* 38 (1970): 1005-1009.

PART II

EPISTEMOLOGICAL AND METHODOLOGICAL ISSUES IN SCIENCE

ÜMİT D. YALÇIN

QUINE'S ROBUST RELATIVISM

1. THE PROBLEM

It is fair to say that the main debate in the philosophy of science in the second half of the twentieth century has been a debate on relativism. More precisely, it has been a debate on whether the scientific quest culminates in a relativistic world-view. We all know the principal actors, pro and con: Hanson, Kuhn, Feyerabend, Goodman, Rorty and Putnam, amongst many others. Ironically, some in this list who have been critics of relativism, have been chastised, in turn, for harbouring theses with relativistic implications.

A historian of the period might remark that Quine, the dean of Anglo-American philosophy of science in the same period, has contributed very little to this debate. Quine's overall attitude seems to be that relativism is obviously an untenable position, and he will have nothing to do with it. When the occasion arises, he presents us with a sketchy refutation of relativism, and assures us that there are better alternatives (1960, 1975). And then, when all appears to be quiet on the relativism front, he suddenly announces that his "... view of science involves both relativistic and absolutistic strains" (1984, p. 294), and caps his short investigation of such strains with what appears to be a challenge:

> I was showing that scientific discourse radically unlike our own, structurally and ontologically, could claim equal evidence and that we are free to switch. Still we can treat of the world and its objects only within some scientific idiom, this or another; there are others, but none higher. Such, then, is my absolutism. Or does it ring relativistic after all? (p. 295)

Does this indicate a sudden shift in the way Quine views his philosophy? I don't believe so. For years, Quine has been aware of certain aspects of his philosophy that seem to have relativistic implications. Thus, already in *Word and Object*, he asks: "Have we now so far lowered our sights as to settle for a relativistic doctrine of truth– rating the statements of each theory as true for that theory, and brooking no higher criticism?" (1960, p. 24) But since he has always been convinced that relativism is a dead end, he has sought for alternatives. Whether he really believed that there were readily available alternatives, or whether he has been hoping that some saving alternative would materialize, is a matter for pure speculation.

G. Irzık & G. Güzeldere (eds.), Turkish Studies in the History and Philosophy of Science, 71-86.

In this essay, I propose to do two things: (a) evaluate the reasons Quine gives for dismissing relativism; and (b) determine whether Quine reaches a relativistic dead-end in his own philosophy of science.

To do either of these, I first have to clarify what 'relativism' is, for it is a vague and ambiguous doctrine, perhaps much more so than a number of other 'isms' in philosophy. This is indicated by deep disagreements in the literature over who should be called 'relativist'. These disputes run so deep that, in many cases, the disagreement does not seem to be merely about the facts of the case, so to speak, but also about the correct application of the term.

To avoid the philosophical quagmire such ambiguity might engender, I will formulate and give a positive account of two positions I shall call 'relational relativism' and 'conceptual pluralism'.[1] I believe that these two incorporate most, if not all, of the positions that have been covered by the blanket term 'relativism'.

2. TWO RELATIVISMS

2.1. Relational Relativism

There are many instances when we seem to be forced to 'discover' (*or* have to remind ourselves) that a certain predicate has to be relativized so that we can say what we wish to say without sounding paradoxical.[2] Usually, such relativizations are accompanied by a further stipulation that the additional subject position(s) is to range over a limited class of entities. Cultures, geographically specified groups, races, persons, historical periods, paradigms, spatial locations etc. have been considered suitable values. When we find ourselves considering such relativizations, we are knocking on the door of relational relativism.

The following seem to be typical attempts to formulate a relational relativism:

> The concept of absolute truth seems to be a concept of a two-term relation between statements (or perhaps propositions) on the one hand and facts (or states of affairs) on the other. But the concept of relative truth, as used by some relativists, seems to be a concept of a three-term relation between statements, the world, and a third term which is either persons, world views, or historical and cultural situations. (Meiland 1977, p. 571)
>
> Earlier in this century the special theory of relativity was sometimes taken as a model for relativism, though because of misunderstandings of the theory this often led only to confusion. Nevertheless, there is something to be said for the paradigm. On Einstein's view, such qualities as mass and velocity, once believed to be invariant or absolute, are now seen to be relative to inertial frameworks. To say that such qualities are relative is to say that they call for one more argument place or parameter than was formerly thought to be needed, and as a first approximation we may view relativism as a thesis that some concept ϕ requires relativization to some parameter p. (Swoyer 1982, p. 85)

So let us say that relational relativism is partly comprised of a claim that this or that predicate is to be relativized to some further parameter(s). To be a full-fledged thesis, more needs to be said about the logical properties of the relativized predicate. Minimally, one needs to choose between the two schemata below to further flesh-out one's relativism. Where 'P^n' is the original predicate, 'P^{n+1}' the relativized predicate, '$x_1, \ldots x_n$' the parameters the predicate originally was supposed to have, and 'y' is the "new" relativizing parameter, these are:

$$(\text{SRR})\ \forall x_1 \ldots \forall x_n \left(\exists y P^{n+1}(x_1, \ldots x_n, y)\ \&\ \exists y \neg P^{n+1}(x_1, \ldots x_n, y)\right)$$
$$(\text{WRR})\ \exists x_1 \ldots \exists x_n \left(\exists y P^{n+1}(x_1, \ldots x_n, y)\ \&\ \exists y \neg P^{n+1}(x_1, \ldots x_n, y)\right)$$ [3]

Let me illustrate with an example: Take the so-called relative identity thesis that derives from the work of Peter Geach (1962). According to this thesis, the predicate 'is identical with' is elliptical for the predicate 'is identical with under S', where 'S' marks a further parameter that ranges over sortal concepts. Having thus relativized identity, Geach also holds that for some entities *a* and *b*, *a* and *b* will be identical *F*s, but not identical *G*s. Notice that, as it stands, this is a weaker claim than an instance of SRR would make. An instance of SRR, as applied to identity, would assert that for any two entities, there is a sortal concept under which they are identical, and some other sortal concept under which they are not. As far as I can tell, Geach does not make it clear whether he adopts a relative identity thesis fashioned after WRR or SRR.

If you wish, we can think of the resultant positions as 'strong' or 'weak' relational relativism, so long as we keep in mind that these adjectives have also been used to mark quite different relativisms in the literature.

Whether a relational relativist *has* to embrace the stronger SRR is one of the important questions I will have to ignore in this essay. For reasons I will explain below, I will focus on this stronger thesis in the remainder of this essay. Suffice it to say, at this point, that the weaker relative identity thesis mentioned above is a *prima facie* example of a quite interesting relativism of the relational variety.

To turn to truth, structurally, relational relativism with respect to truth adds a further parameter to the predicate 'true'. I will call this two-place predicate 'true-in-X'.[4] The important thing to keep in mind is that according to relational relativism with respect to truth, 'true-in-X' is a primitive predicate and *not* to be understood in terms of 'true'. The occurrence of 'true' in 'true-in-X' does not have a semantic value on its own.

Relational relativism with respect to truth will be more fully formulated by the adoption of an additional component that corresponds, with suitable modifications, to SRR. That is, the full formulation of the thesis will incorporate an instance of the following schema (where 'S' ranges over *interpreted* sentences, and 'X' over whatever truth is supposed to be relative to):

$$(\text{SRRT})\ \forall S(\exists X(S \text{ is true-in-}X)\ \&\ \exists X^*(S \text{ is not true-in-}X^*))$$

If this is all there is to relativism, why are we not hailed by cries of 'relativism' in all those cases in which a predicate is relativized? There are well known examples of predicates such as 'tall' and 'old', which, according to some, are elliptical for 'tall for an *X*' or 'old for an *X*', where the relativizing parameters range over groups of individuals, races or what not.[5] But those who suggest such relativizations are *not* called relativists. What accounts for the tendency to react differently to different predicate relativizations?

I am not sure that I have a satisfactory answer to this question. It seems quite tempting to say that relativizing a predicate is considered relativistic when the relativization is to a subjective parameter; yet, I am not prepared to stake anything important on the correctness of this suggestion (would I not be a relativist if I said truth is relative to geographical locations, an apparently objective relativizing parameter?). Or, in a more pragmatic vein, it could be argued that whether a position will be called relativistic or not depends on how unexpected the relativization it introduces happens to be. Ironically, this approach would make the relativism of a given position itself relative to the expectations of a given reference class. Since I suspect that the application of the terms 'relativism' or 'relativistic' in the philosophical community is not guided by adherence to some strict explicit or implicit rule, I tend to favour this pragmatic account; but I would be hard-pressed to offer an argument for it.

Fortunately, nothing I will have to say subsequently depends on the availability of an answer to this question. It will be sufficient for my purposes to understand relational relativism as that sort of position which is called relativistic, and the relativism of which is grounded in the fact that the position involves the relativization of a key predicate.

2.2. Conceptual Relativism

I take conceptual relativism (perhaps better called conceptual *pluralism*) to be comprised of the following four theses:

(a) There are significantly different conceptual frameworks, (frames of discourse, conceptual schemes, languages, theories), A and B, in the sense that A does not have a concept that exactly matches the concept c in B (or vice versa).

(b) Apparent conflicts between the members of different conceptual frameworks (frames of discourse, conceptual schemes, languages, theories) such as A and B, are (sometimes? often? always?) to be resolved by noticing that concepts which appear to be the same are different concepts (that words or sentences which appear to be synonymous have different meanings).

(c) Neither conceptual framework is rationally preferable to the other.

(d) In some cases where we are confronted with a situation as in (a)-(c) above, we can find analogous words or concepts in the respective languages or conceptual frameworks.

The first three theses constitute the core of the position I call conceptual relativism, although, without taking time to support my claim, I submit that most conceptual relativists also accept (d).

Consider the perennial dispute about whether it is possible to love (with equal devotion etc.) two different people at the same time. Many of us might have witnessed or participated in arguments about the possibility of such a state of affairs, and might have come away feeling that there was a genuine disagreement between the opponents, and that one of them was wrong. If someone would argue that the 'opponents' were not disagreeing after all, because there were two different (yet perhaps analogous) concepts in question, that neither 'opponent' had[6] the concept of the other, and neither was being more or less rational than the other by utilizing her own particular concept, then this person would be a conceptual pluralist in the sense I define above.

The following serves as a good illustration of the position:

> Relativism as here presented does not assert that all statements are to be expanded to read, "According to framework *F, p*." On the contrary, the relativist asserts that all statements must be made *within* a framework, which need not be treated as part of the statement itself. All statements must be made in some language, but that does not mean that all statements must be prefaced by "in language L." (Devine 1984, pp. 407-408)
>
> ...Some critics of relativism have argued that a relativist cannot maintain the usual sense of "true," or the prescriptive force of moral judgments, while admitting that positions other than his own are, from some equally valid point of view, true or well-grounded. But the relativist need not make any such claim: **as he uses words, "true" or "well-grounded" engage the criteria characteristic of his framework. But he will also maintain that other frameworks are possible, [sic] that generate senses or uses of "true" or "well-grounded" analogous to his (he may himself use the word "orthodoxy" to describe them), and that no rational argument is available to show that his own (or his opponent's) position is true or preferable.** He may attempt to get his opponents to accept his framework, and to govern their lives accordingly. But he will not employ rational persuasion in this attempt: he will not argue with unbelief, but only preach to it. (Devine, p. 408; boldface added)[7]

The relational relativist does not take himself to be forwarding a thesis like conceptual relativism. Most relativists do not formulate their position in terms of assertions like "You and I do not use 'love' (or 'truth') in the same way", "By 'love' ('truth') you mean something different than I do", or "You do not have a concept that exactly matches my concept of love (truth)".

Of course, distinguishing relational relativism from conceptual relativism is one thing, knowing when to attribute one or the other position to a self-professed relativist is another. The problem is exacerbated by the fact that most relational relativists do not formulate their position explicitly in terms of the relativization of a predicate. Quine is no exception. In describing relativism, he uses expressions such as 'rating the statements of each theory as true for that theory', without saying much about how the locution 'true for' is to be interpreted. The kind of relativism he militates against can be either a version of SRRT, or conceptual relativism . Given the limitations of the present essay and for reasons of exegetical charity that will become clear towards the end of the essay, I will just assume that the position Quine has in mind is SRRT. The investigation of the alternative will have to wait for another occasion.

3. QUINE'S ANTI-RELATIVISTIC ARGUMENT

Despite his apparent scorn for relativism, all Quine has to offer against it is a three-sentence argument:

> Truth, says the cultural relativist, is culture-bound. But if it were, then he, within his own culture, ought to see his own culture-bound truth as absolute. He cannot proclaim cultural relativism without rising above it, and he cannot rise above it without giving it up. (Quine 1975, pp. 327-28)

Accordingly, we have a truth predicate relativized to cultures, and at this point, it might seem worthwhile to wonder: "Why cultures?" I don't know, but it does not really matter. For as we shall see, Quine's argument does not invoke any specific property of cultures as opposed to properties of other parameters to which various relativisms have been indexed. Like many anti-relativists he is not interested in such niceties: he focuses on an abstract relativism that relativizes truth to "whatever truth is relativized to", or to such nondescript place holders as "frameworks", "perspectives" or "cultures". I call the latter 'place holders' simply because, both for the relativist and his absolutist nemesis, they seem to serve the unique purpose of being concrete-sounding relativizing parameters about which nothing else has to be divulged. The relativist formulates his thesis, and the absolutist proceeds to refute it, both at the same level of lofty abstraction.

Notice also that Quine says neither exactly what the "culture-bound truth" of the relativist is supposed to be, nor why the latter has to be taken "as absolute". For the moment, let us call this culture-bound truth Φ, whatever it might be (or, whatever they might be, if there is more than one). Quine's whole argument seems to depend on assuming:

(ABS) The relativist has to accept that Φ is absolutely true

I am not convinced that granting ABS is sufficient for Quine's anti-relativistic argument to go through, but I will not pursue that point. Instead, I will question Quine's assumption of ABS. What in the world would happen if the relativist refuses to take anything to be absolutely true, insisting, as does SRRT, that every truth, including the thesis of relativism is relatively true?

Quine never says, but here are some recent attempts at showing the disastrous consequences of such a move:

(a) An attempt to unpack Quine's argument, and supply the missing steps in his argument comes from Harvey Siegel, who has adapted this argument against various relativisms:

> Substitute "framework" for "cultural" and "framework-bound" for "culture-bound" in this passage, and the argument wanted is at hand. The framework relativist, if correct, would be unable to recognize the equal status of alternative frameworks, for she would not be able to transcend her own. She would thus regard her own framework as absolute. But she does recognize alternative frameworks. So it cannot be the case that her framework constitutes the bound on her though that is essential to the framework relativist position. As Quine puts it, she cannot proclaim framework relativism without rising above it; and she cannot rise above it without giving it up. The recognition of the equal epistemological status of (and indeed of the independent existence of) alternative frameworks, which is necessary for framework relativism, necessitates as well the rejection of framework-boundedness, which is the central component of the framework relativist position. Thus, framework relativism cannot proclaim itself, or even recognize itself, without defeating itself. (Siegel, 1987, pp. 43-44)

The crucial point in Siegel's argument is reached when he asserts that "[t]he framework relativist, if correct, would be unable to recognize the equal status of alternative frameworks, for she would not be able to transcend her own.". What Siegel seems to say implies that 'transcending your framework' means, or entails, accepting a framework that does not recognize your framework as a framework. If so, it is not clear why the framework relativist cannot transcend her own framework.

Let R be the relativist's framework, and let A be some other framework, perhaps even some absolutist's framework. Let us assume that the relativist, by applying the criteria she uses to select frameworks, recognizes $\{f_1, \ldots f_n\}$ as frameworks. Let us also assume that not only R, but also A qualifies as a member of $\{f_1, \ldots f_n\}$. So, the relativist recognizes A as a genuine framework. Moreover, assume that A does not recognize R as a framework. Hence, the set of frameworks A recognizes is different from $\{f_1, \ldots f_n\}$. '*R* is a framework' is false-in-*A*. Since *R* recognizes *A*, *R* accepts that '*R* is a framework' is false-in-*A*. Thus, despite Siegel's claim to the contrary, here is a case where the relativist does *not* regard her own framework as absolute. Yet, no contradiction follows from this. The relativist accepts the following four propositions:

(1) '*R* is a framework' is false-in-*A*
(2) '*R* is a framework' is true-in-*R*
(3) '*A* is a framework' is true-in-*A*
(4) '*A* is a framework' is true-in-R

And still no contradiction follows even if we assume that the absolutist only recognizes absolutist frameworks, even though this would entail that '*R* is a framework' is absolutely false for the absolutist.

The diagnosis has to be that when Siegel starts saying "The framework relativist, if *correct*...", he already takes the framework relativist to be using an absolutist concept of correctness, rightness, or truth. But this is not the relativist concept or truth.

(b) The next example, which also targets the idea that relativism cannot be relatively true, is of very recent vintage:

> Suppose that relativism is merely relatively true, i.e. true in some perspectives and untrue in others. Consider the latter case, a perspective in which relativism is untrue. In such a perspective, call it *p*, not-relativism–that is, absolutism–is true. Now, absolutism is true only if there is some proposition that has the same truth value in all perspectives. That is in *p* there is some Φ such that [it is true in all perspectives that] Φ. However, it does not seem that *p* could contain such a proposition. Φ could not be the thesis of absolutism itself, since *ex hypothesi* there are perspectives in which it is untrue and relativism is true. Nor could Φ be the thesis of relativism, since *ex hypothesi* there are perspectives in which it is untrue. Nor do any other candidates for Φ look promising since–**given the assumption that there are perspectives in which relativism is true–we are guaranteed that the truth-value of every proposition Φ will vary across perspectives.** Hence, there is no proposition that is true in all perspectives; that is, for every proposition there are perspectives in which it is true and perspectives in which it is untrue. Then relativism is true in all perspectives, and this, I have already shown, entails that relativism is untrue. Thus it seems that relativism can be neither absolutely nor relatively true. The claim that everything is relative must be false. (Hales 1997, p. 36)

I will not focus on the second leg of the trilemma, for I am not sure exactly what it means to have an absolutist perspective in which relativism is absolutely true. So I grant that relativism cannot be the absolute truth that grounds an absolutist perspective. But the rest of the argument is not without problems. Let me take the last leg of Hales' trilemma first. Here, the claim

(1) The truth-value of every proposition Φ will vary across perspectives is supported by:
(2) Relativism is true for some perspective.

But (2) does not entail (1) when (1) is properly understood, i.e., read in a way that does not assume the absolute falsity of relativism. That is, (1) cannot be read as if 'The truth value of every proposition *F* will vary across perspectives' were something that were true or false simpliciter, without needing relativization to a perspective. Reading (1) in this manner would amount to assuming that (1) can be absolutely true or false. But this is something SRRT will deny. Hence, the proper reading of (1) will relativize (1) to the absolutist's perspective. But now, Hales' argument does not work.

To appreciate this, call one of the relativistic perspectives mentioned in (2), *R*, and the absolutist perspective *A*. Now, take the following instantiation of (2):

(3) Relativism is true-for-*R*.

and replacing SRRT for 'relativism' in (3), we get:

(4) '$\forall S\ (\exists X(S \text{ is true-in-}X)\ \&\ \exists X^*(\neg S \text{ is true-in-}X^*))$' is true-for-*R*

Now take a particular sentence, Φ, and let us assume that the relativist "mistakenly" insists that Φ is a sentence that can be absolutely true for the absolutist–how will Hales show the "mistake" of the relativist? He can say that given (4), and for some particular perspectives, C and D (not necessarily distinct from A and R):

(5) '(Φ is true-in-*C* & $\neg\Phi$ is true-in-*D*)' is true-for-*R*

This is fine, but does not amount to:

(6) '(Φ is true-in-*C* & $\neg\Phi$ is true-in-*D*)' is true-for-*A*

Yet, unless (6) is true, Hales has not shown that Φ's truth value varies across perspectives in *A*, for (5) is consistent with:

(7) '($\forall$X) (Φ is true-in-X)' is true-for-*A*

because conjoining (5) and (7) merely entails

(8) '$\neg\Phi$ is true-in-*D*' is true-for-*R* & 'Φ is true-in-*D*' is true-for-*A*

which is not contradictory. Hence, the relativist is after all not shown to be mistaken.

Consider now the first leg of the trilemma, and the claim that 'Φ could not be the thesis of absolutism'. The support for this claim comes from the assertion that '*ex hypothesi* there are perspectives in which it is untrue and relativism is true'. Well, there might be such perspectives among what the relativist counts as perspectives, but the issue is whether the thesis of absolutism is true in all perspectives *countenanced by absolutism*. And the problem for Hales' argument is that there is absolutely no argument given for the supposition that the perspectives in which absolutism is false are perspectives the absolutist recognizes as perspectives. To suppose that this is the case is to suppose that (according to the relativist), the absolutist sees eye-to eye with the relativist (and perhaps other perspectives) when it comes to the question as to what to count as a perspective. It is to take some sentence like:

(9) '$\{p_1 \ldots p_n\}$ are all and the only perspectives'

as being either absolutely true or absolutely false. But this is not something SRRT will concede. In fact, SRRT would insist that for any *p*, '*p* is a perspective' is true in some perspectives and false in some others. And it will be reasonable to think, barring an argument to the contrary, that the perspective of which SRRT is a part does *not* count as a viable perspective from the absolutist's perspective.

In conclusion, I submit that Quine's tacit assumption, that there has to be *something* that relativism has to take as being absolutely true, is not sustained by the current state of the debate on relativism. Relativism is not so easily refuted, if we take it seriously, and avoid interpreting it through absolutist spectacles (i.e., *not* forget the relativization of the truth-predicate). And given what comes next, perhaps Quine should be relieved by the apparent failure of his anti-relativistic argument.

4. CONSEQUENCES OF UNDERDETERMINATION: IS QUINE A RELATIVIST?

Unlike other philosophers who have argued that there is no such thing[8], Quine used to launch his philosophy of science by more or less assuming *the underdetermination of theories by evidence*. Here is one of his well-known, concise formulations:

> Physical theories can be at odds with each other and yet compatible with all possible data even in the broadest sense. In a word, they can be logically incompatible and empirically equivalent. This is a point on which I expect wide agreement, if only because the observational criteria of theoretical terms are commonly so flexible and fragmentary. (Quine 1970a, p. 179)

The problem that arises from underdetermination is initially an epistemological problem. We need to decide what to say about justification and knowledge when all possible evidence does not justify one theory above all others.[9] But the epistemology is intertwined with metaphysics. Some try to avoid the epistemological problem by adjusting their world-view elsewhere. Thus, such "fine-tuning" operations on notions of truth, meaning and reality lead to metaphysical systems far removed from the naive realism of common sense[10]: we arrive at esoteric systems such as instrumentalism, relativism, and the verificationist theory of meaning. This is exactly the kind of crossroads Quine seems to find himself when he says (of two theories): "Can we say that one, perhaps, is true, and the other therefore false, but that it is impossible in principle to know which? Or, taking a more positivistic line, should we say that truth reaches only to the observation conditionals at most, and, in Kronecker's words, that *alles übrige ist Menschenwerk*?" (Quine 1975, p. 327)

Those familiar with Quine's work will doubtless remember the various twists and turns this story takes in the last forty years. Of course, he is not satisfied with either option we have just seen him consider. Embracing scepticism would be a confession of failure for one who has upheld the scientific method as the one and only oracle, and adopting instrumentalism would be to confess that the scientific oracle is no less mysterious than the Delphic.

Quine has considered two other options at various points in his career:

The Ecumenical Solution

(E) The difficulty generated by two rival underdetermined theories can be avoided by re-writing one of the theories using terms alien to both theories, and thus render them both compatible. Then, we can declare the truth of both theories.

Quine has said that this solution was suggested to him by Davidson.[11]

Unfortunately, as I have argued elsewhere, the strategy does not work.[12] Here is why, in a nutshell: If underdetermination is a phenomenon about full-fledged, *interpreted* theories (as opposed to *uninterpreted*, merely syntactical *theory formulations*), and the incompatibility between them is genuine semantic incompatibility, no purely syntactical manipulation will make them semantically compatible. If you have two theories that disagree about the truth of, say, 'There are electrons', one cannot make them semantically consistent by replacing 'electron' with 'superpolyfragilistic' in one of the theories (re-writing one of the theories in Swahili will not help either).

Such remedies will help only if the incompatibility between the theories were only apparent, in which case the so-called underdetermination problem is also spurious.[13]

For his own reasons, Quine has also given up on the Davidsonian strategy. (1990, p. 100) We shall therefore lay it to rest.

The Sectarian Solution

(S) Somehow, the truth of the theory one holds is to be declared, while the rival is cast *as somewhat less than true.*

Let me immediately address the cryptic expression 'holding the rival theory as somewhat less than true'. I have to take recourse in such ambiguity, because Quine vacillates between various ways of characterizing the alethic status of the rival theory. To elaborate:

The sectarian solution is closely bound with Quine's sporadic tendency to declare the *immanence* of truth. Quine expresses this vision at a number of points, and in a number of ways: "Truth is immanent, and there is no higher. We must speak from within a theory, albeit any of various." (Quine 1981, pp. 21-22) The same goes, it appears, for the notion of factuality: "Factuality, or matterhood of the fact, is likewise immanent. We do not adjudicate between our aggregate system of the world and a rival system by appeal to a transcendent standard of truth or factuality." (Quine 1986, p. 367)

The transcendence of truth is best understood in light of an analogy Quine introduces elsewhere while discussing various grammatical categories: "A notion is immanent when defined for a particular language; transcendent when directed to languages generally." (Quine 1970b, p. 19) Hence, we arrive at the following:

(1) The predicate 'true' as used in a given theory stands for a concept which is uniquely a concept of that theory; other theories do not have that concept but, at best, only analogues of it.[14]

Quine further elaborates his position by adopting, at various points in his career, one or the other of the following theses:

(2a) The immanent truth predicate and the concept associated with it, though unique to each theory, are predicable of (applicable to) assertions of other theories (or suitable translations thereof).

(2b) The immanent truth predicate, and the concept associated with it are *not* predicable of (applicable to) assertions of other theories (or suitable translations thereof). The predicate only applies to assertions of the theory of which it is a part.

Hence we have at least two senses of the immanence of truth (and of sectarianism), and perhaps a third—for Quine often proposes a more detailed and stronger version of (2b)

(2c) The immanent truth predicate, and the concept associated with it are trivially *not* predicable of (applicable to) assertions of other theories (or suitable translations thereof) because the "assertions" of these theories are *not meaningful* (and hence trivially not translatable) from the perspective of the theory we are working with.

Let me first clarify one issue. Quine has often suggested that what he tries to express by these various theses is nothing but Tarski's semantic notion of truth. He emphasizes that Tarski has shown us that there is no generic truth predicate definable to apply to all languages: each language gets its own truth predicate. One cannot apply the truth-predicate defined for a given language to the sentences of some other language.

No doubt, there is some similarity between Tarski's notion of truth and what Quine calls his immanent notion of truth. Yet, Quine is mistaken in suggesting that the two are one and the same. Given Quine's and Tarski's customary use of 'theory' and 'language', the two cannot be the same. More than one theory can be formulated in a given language, and if each such theory needs its own distinct "Quinean" truth predicate, then none of these Quinean truth predicates can be identified with the Tarski-style truth predicate of the language in which they are couched. I mention this just to explain why I will not be talking of this Tarskian strand in Quine's thought.

One might object that I am ignoring Quine's tendency to identify theories and languages. Not so. I emphasized above that my claim is based on the "customary" use of 'theory' and 'language'. If Quine is really identifying the two notions, he is thereby clearly introducing a new way to use these terms. He needs to explain to us how the notion of truth developed by Tarski for languages (in the customary sense) applies to language (as identified with theories) in Quine's sense. Unfortunately, no explanation has been forthcoming in those contexts where Quine seems to identify the two.

To turn to the main issue at hand, the sectarian solution interpreted by means of (1) and (2a) would be to (a) declare the truth of one's own theory, and (b) the falsity of the rival theory. Call this *S1*. This strategy was briefly considered by Quine (1975, pp. 327-328) and rejected. As he puts it, "[m]ust we still embrace one theory and oppose the other, in an *irreducible existentialist act of irrational commitment*? It seems an odd place for irrational commitment, and I think we can do better." In any case, it is clear that (2a) would not commit Quine to a position that has relativistic tendencies for truth. Conceding that your rival's theory is as warranted as yours, but still calling it false may be irrational, but definitely not relativistic.

The sectarian solution interpreted by means of (1) and (2b) would amount to (a) declaring the truth of one's own theory, and (b) denying the applicability of 'true' or 'false' to assertions of the rival theory. Call this *S2*. Although *S2* presents Quine with an option that might be worth considering, he does not pursue it in much detail. He comes closest to adopting this stance when he likens the immanence of truth to the workings of the Tarski truth predicate. Perhaps he is sensitive to some of the concerns I voiced above regarding identifying theories and languages, but there is no need to speculate. For our purposes, it is more important to evaluate *S2*'s relativistic tendencies, if any.

In one sense, it is clear that *S2* is not a relativism of the relational variety, for it does not explicitly relativize truth.[15] At the same time, it is also clear that there is a very strong similarity between it and the position I called conceptual relativism in section 2. For *S2* asserts that:

(a) There are significantly different theories, *our theory* and *the rival theory*, in the sense that *the rival theory* does not have a concept that exactly matches the concept truth in *ours* (or vice versa).
(b) Apparent conflicts between the assertions of *our theory* and *the rival theory* are to be resolved by noticing that concepts, which appear to be the same, are different concepts (that words or sentences that appear to be synonymous have different meanings). In particular, *our theory* and *the rival theory* will not appear to be making inconsistent claims, because there will be no way of saying that *if our claims are true, the conflicting claims of the rival theory have to be false*.
(c) Neither theory is rationally preferable to the other, for both are equally warranted, equally simple, etc.
(d) *The rival theory* has a concept analogous to our predicate 'true' (if the Tarski model is to be adopted).

Before I say anything further about *S2*, let us compare it with the version of sectarianism Quine actually adopts, which I will call *S3*. Here is how Quine formulates it:

> The sectarian is no less capable than the ecumenist of appreciating the equal evidential claims of the two rival theories of the world. He can still be evenhanded with the cachet of warrantedness, if not of truth. Moreover he is as free as the ecumenist to oscillate between the two theories for the sake of added perspective from which to triangulate on problems. In his sectarian way he does deem the one theory true and the alien terms of the other theory meaningless, but only so long as he is entertaining the one theory rather than the other (1990, p, 100)

Thus, *S3* combines (1) and (2c) to (a) declare the truth of one's own theory, and (b) deny the applicability of 'true' or 'false' to assertions of the rival theory, and (c) deny the meaningfulness of at least some assertions of the rival theory. At the same time, *S3* treats the rival theory as equally warranted, and hence as a potential alternative to "our own". Moreover, this is not idle talk of a potential alternative, because we are free to adopt the rival in an attempt to gain an "added perspective from which to triangulate on problems".

So it seems like *S3* will fit the conceptual relativist mold as much as *S2*. The only difference will appear in the second clause of the conceptual relativist formulation, which will have to be modified to read:

(b*) Apparent conflicts between the assertions of our theory and the rival theory are to be resolved by noticing *that apparently meaningful utterances of the rival theory are in fact meaningless*. In particular, *our theory* and *the rival theory* will not appear to be making inconsistent claims, because there will be *trivially* no way of saying that *if our claims are true, the conflicting claims of the rival theory have to be false*.

I am not sure that *S3* is coherent. At first blush, it seems hard to understand how one can talk about adopting a "theory" one has characterized to be meaningless, and of gaining "an added perspective" from doing so. To underscore the difficulty, consider the diary of an imaginary Quinean scientist:

"Yesterday, I argued with my colleague and told him that he was wrong for embracing a putative theory rife with meaningless gibberish. He tried to convince me that I could not say this, because his theory is as warranted as mine. I did not dispute this fact; it is true that the two theories are equally warranted. But mine is the true one; this is entailed by my naturalism, by the immanence of truth. His is just meaningless.

Today he came back and told me that he would like to switch to my theory. He says that I convinced him yesterday that he was uttering meaningless sounds. But he was quite put off; I could tell this from his manner, the sad stoop of his shoulders. This really touched me, and I decided to surprise him. First I told him not to switch yet, and then I adopted his theory. Now we both hold the true theory, and we are in total agreement. We speak with one voice, as it should be; after all, this is what befits such good friends as my colleague and I. What of my old theory? What can I say ..."

Well, what can a Quinean say after such a switch of theories? That he was wrong yesterday, and that he falsely believed that the other theory was meaningful? But how can he say this if he knows that he can switch over again after lunch, and thereby re-establish both the meaningfulness and truth of his old theory? Can he seriously contend that meaningfulness (and truth) are determined, if not by convention, then by conversion? I suspect that the answer has to be in the negative.

To show *S3*'s coherence, one would have to argue that it is possible to regard a linguistic entity (a theory) as being both meaningless and capable of being understood (which seems to be a presupposition of being adopted) at the same time. One would also have to say something about what *adopting* a theory means. Is this notion to be cashed in terms of *belief* or *acceptance*? And if so, what is the exact import of these propositional attitude terms when wedded to Quine's behaviouristic semantics? Unfortunately (or perhaps fortunately), this is not the place to get entangled in these interesting yet complicated questions. Given our concerns, we can sidestep them and summarize the situation at hand as follows:

Quine is careful to avoid relational relativism by persistently refusing to explicitly relativize truth. Yet at the same time, he is slowly pushed to a position where he has to make a choice between the irrationalism of *S1*, and the conceptual relativism of *S2* and *S3* (if the latter makes sense). He is perhaps satisfied to avoid his own anti-relativistic argument (insofar as it is to be construed as targeting relational relativism; if the arguments in the present essay are correct, he need not have worried in the first place), but there is no doubt that he is still flirting with relativism in one of its guises. Yet, in typical Quinean style, he tries to eschew the substance of the problem by presenting it as a merely linguistic issue.

> The fantasy of irresolubly rival systems of the world is a thought experiment out beyond where linguistic usage has been crystallized by use. No wonder the cosmic question whether to call two such world systems true should simmer down, bathetically, to a question of words (1990, pp. 100-101)

Does this manoeuvre avoid the threat of relativism presented by *S2* and *S3*? I will be satisfied with answering this question conditionally. If Quine's arbiter is merely linguistic usage, and if by deferring to it he is able to avoid relativism, he is still faced with a substantial cost. For in that case the winner, after all, appears to be Kronecker. If the question of the truth of empirically equivalent rival theories is to be finally resolved by the "crystallization" of linguistic usage, and this means no more than our *deciding* what to *call* 'true', there is not, and there never was, much to truth in the first place.

East Carolina University, Philosophy Department

5. NOTES

[1] I borrow the term from Feyerabend who talks about a similar thesis (1987, p. 82). See also Meiland 1977.

[2] I will assume throughout this essay that relativizing an n-place predicate that actually denotes amounts to accepting that it denotes a relation between n+k entities, not n entities as we originally had supposed (where a unary relation is what is commonly known as a property). I also assume that this way of putting things does not commit me to realism about properties; I intend what I say to be compatible with the reducibility of properties. Finally, the locution 'relativizing a predicate' is a short way of saying 'accepting that the surface grammar of a predicate expression misleadingly presents it as an n-place predicate when, in fact, the predicate in question is actually an n+k-place predicate'.

[3] One can also contemplate variations where (for whatever reason) one would mix the quantifiers. I do not know how to motivate such cases, and leave it to the imagination of relativists to come up with such interesting relativisms.

[4] I am taking 'true' to be a monadic predicate that purports to denote a two-term relation between the world and a representation. In turn, 'true-in-X' is a two-place predicate denoting a three-term relation between the world, a representation and something else (whatever the relativist will take 'X' as ranging over).

[5] Geach called predicates such as 'big', 'tall' and 'good' attributive adjectives, proposing the following test to distinguish them from predicative adjectives:

> ...in a phrase 'an A B'...'A' is a predicative adjective if the predication 'is an A B' splits up logically into a pair of predications 'is a B' and 'is an A'; otherwise I shall say that 'A' is a (logically) attributive adjective. (1962, p. 32)

Geach's distinction, and the use he puts it to, suggest that he considers predicates like 'is tall' as always being elliptical for 'is tall for an X' (more simply, 'is a tall X') where the variable place is to be filled with an expression designating a class or reference group.

[6] One reason that I formulate language pluralism in terms of "having" or "lacking" a concept is because we have to distinguish this position from the simple case where two people who both happen to grasp ("have") the different senses of a word just happen to be using the word in its different senses. In other words, two people who disagree about whether banks have money in them, because they use different senses of 'bank' would not illustrate language pluralism as I understand it.

[7] Some construals of the incommensurability thesis associated with Kuhn and Feyerabend, and what Swoyer calls "weak relativism" (1982, p. 92) seem to accord with this characterization of language pluralism.

[8] For example, see Laudan 1991.

[9] Quine is quite explicit about this when, referring back to his own work on underdetermination, he remarks, "The truth of physical theory and the reality of microphysical particles, gross bodies, numbers, sets, are not impugned by what I have said about proxy functions and of wildly deviant and empirically equivalent theory formulations. Those remarks had to do *not* with what there is and what is true about the world, but only with *the evidence* for what there is and what is true about the world." (Quine 1984, p. 295. Italics added). See also Yalçın (forthcoming in Noûs).

[10] Two minor comments: first, I do not intend to mean that this departure from common-sense is to be deplored. When the genuine internal tensions and contradictions of the body of beliefs we loosely call common sense come to light, there is bound to be no commonsensically satisfying solution. Second, I am fully prepared to accept that the commonsense I am talking about is merely that of a certain dominant trend in Western Civilization (whatever that may be).

[11] There is also another "Davidsonian" strategy Quine sometimes talks about, which is (a) re-writing one of the theories using terms alien to both theories thus rendering them compatible, (2) joining these compatible rivals into one theory, and (3) declaring its unchallenged truth. See Quine 1990, p. 99. This variant also suffers from the problems I will shortly discuss in the body of the essay.

[12] "Solutions and Dissolutions of the Underdetermination Problem", in *Noûs*.
[13] What if underdetermination is cashed merely as a relation between theory formulations? In that case, the summary answer is that we end up with unmitigated instrumentalism, which misses Quine's self-professed "robust" realism by a wide margin. Please see Yalcin for details.
[14] This is a somewhat un-Quinean way of expressing a thesis, of which there may be no coherent Quinean formulation. Nothing discussed in the text turns on the pragmatic choice I make here.
[15] Whether such a relativization of truth is the only clear way to make sense of *S2* is an interesting question I cannot pursue in this context.

6. REFERENCES

Devine, P. E. "Relativism," *Monist* 67: 405-418,1984.

Feyerabend, P. K. "Notes on Relativism," in *Farewell to Reason*, 19-89. London and New York: Verso, 1987.

Geach, P. T. *Reference and Generality*. Ithaca: Cornell University Press, 1962.

Hales, S. D. "A Consistent Relativism," *Mind* 106: 33-52, 1997.

Laudan, L. "Empirical Equivalence and Underdetermination." *Journal of Philosophy* 88: 449-473, 1991.

Meiland, J. "Concepts of Relative Truth." *Monist* 60: 568-582, 1977.

Meiland, J. "On the Paradox of Cognitive Relativism." *Metaphilosophy* 11: 14-37, 1980.

Quine, W. V. *Word and Object*. Cambridge, MA: MIT Press, *1960*.

Quine, W. V. "On the Reasons for the Indeterminacy of Translation", *Journal of Philosophy* 67: 178-183, 1970a.

Quine, W. V. *Philosophy of Logic*. Englewood Cliffs, New Jersey: Prentice-Hall, 1970b.

Quine, W. V. "On Empirically Equivalent Systems of the World", *Erkenntnis* 9: 313-328, 1975.

Quine, W. V. "Things and Their Place in Theories", in *Theories and Things*. Cambridge, Massachusetts and London: Harvard University Press, 1-23, 1981.

Quine, W. V. "Relativism and Absolutism", *Monist* 67: 293-296, 1984.

Quine, W. V. "Reply to Robert Nozick," in *The Philosophy of W. V. Quine*, 364-367, 1986.

Quine, W. V. *Pursuit of Truth*. Cambridge: Harvard University Press, 1990.

Siegel, H. *Relativism Refuted*. Dordrecht: D. Reidel Publishing Company, 1987.

Swoyer, C. "True For," in *Relativism: Cognitive and Moral*, 81-108, 1982.

Yalçın, Ü. D. "Solutions and Dissolutions of the Underdetermination Problem," Noûs 35, No. 3, 394-418 (2001).

DAVID GRÜNBERG

CONFIRMATION OF THEORETICAL HYPOTHESES: BOOTSTRAPPING WITH A BAYESIAN FACE[1]

1. INTRODUCTION: THE PROBLEM OF TESTING THEORETICAL HYPOTHESES

The purpose of this paper is twofold. First, we shall propose an emended version of Glymour's bootstrapping, which we intend to be conducive to an evidential relevance relation rather than to a full-fledged confirmation theory. This emended version of bootstrapping avoids the counterexamples raised against Glymour's original, as well as those against several revised versions thereof. Second, we shall argue that the Bayesian method, taken alone, is not sufficient for testing *theoretical hypotheses*, but that, when combined with bootstrapping, can in principle yield such a method. We understand by a "theoretical hypothesis," a hypothesis whose vocabulary transcends that of the available evidence.

Indeed, the testing of theoretical hypotheses constitutes a problem that can be formulated—in Glymour's terms—as follows: "...how can evidence stated in one language confirm hypotheses stated in a language that outstrips the first?" (1980, 10). As mentioned above, we call such hypotheses "theoretical" (relative to the evidence).

Glymour (1980, 10) indicates that testing of such hypotheses is problematic. In particular, they cannot be confirmed by their instances, since the instances would be theoretical, and therefore could not constitute the evidence. Glymour (1980, 12-13) considers four methods, viz., elimination of theory, the deductive method (hypothetico-deductive method), the bootstrap method, and probabilistic strategies (in particular the Bayesian method) concerning the problem of testing theoretical hypotheses. He criticizes the first, second, and fourth of these methods, and propounds a version of the third one as an adequate solution to the problem of testing theoretical hypotheses.

The method of bootstrapping, however, has been subjected to heavy attacks, and confronted with numerous counterexamples. As a result of this, Earman (1992, 73-77), for example, has contended that there cannot be an adequate confirmation theory that is purely logico-structural.

Our view is that bootstrapping is a purely logico-structural procedure, but, as mentioned above, it is not a complete confirmation theory. However, we think that it can be combined with a confirmation theory, such as the Bayesian method, which is

G. Irzık & G. Güzeldere (eds.), Turkish Studies in the History and Philosophy of Science, 87-102.

not purely logico-structural, to yield an adequate method for testing theoretical hypotheses.

2. THE EMENDED EVIDENTIAL RELEVANCE RELATION

Glymour's method of bootstrap testing involves a ternary evidential relevance relation among a set E of evidence statements expressing that certain quantities possess observed or measured values, a hypothesis H (typically, a universal statement) being tested, and a theory T consisting of the set of consequences of certain hypotheses that sometimes, though not always, includes H itself. In the main, the non-logical vocabulary of H contains some terms not occurring in E, viz., the so-called theoretical terms. Whereas Glymour's method applies vacuously to testing of hypotheses devoid of theoretical terms, it seems to be indispensable for testing theoretical hypotheses.

Glymour's evidential relevance relation, "E is evidentially relevant to H with respect to T" is defined roughly by the following three conditions: (G1) E, H, T are jointly consistent, *Cons* (E, H, T) for short; and there is a set *Aux* of auxiliary hypotheses which are derived from T such that: (G2) there is a set V of values for the quantities occurring essentially in H which can be computed from E by means of *Aux*, and V constitutes a positive instance of H, in the sense of fulfilling the satisfaction* condition (i.e., that V entails the development of H for the individual constants occurring in E). (G3) there is a possible alternative evidence E' expressible in the vocabulary of E, and a set V' of values for the above-mentioned quantities that can be computed from E' by means of *Aux*, where the union of E' and *Aux* is consistent,[2] and V' is a negative instance of H, i.e., a positive instance of $\sim H$.[3]

As mentioned in the introduction, Glymour's original system of bootstrapping, as well as its several versions, has been subjected to heavy criticisms. We shall attempt to formulate an emended version that avoids them. We shall make the following three emendations. First, whereas Glymour treats quite differently qualitative and quantitative hypotheses, we introduce a unified framework consisting in a many-sorted first-order language for both kinds of hypotheses. Second, we replace Glymour's notion of computation of the values of quantities using a graph by straightforward deduction of these values (from *E and Aux*). Third, whereas Glymour's, as well as others', versions of the evidential relevance relation are three-place, ours is four-place, the fourth relatum being a subtheory T^{-H} of T representing an H- free portion of T.

We assume that E, H, T, and T^{-H} are formulated in a many-sorted first-order language L with identity. The language L refers to various physical and mathematical sorts. The physical function symbols are supposed to be defined in terms of primitive predicates. Consider a quantitative hypothesis H of the form

$$(X_1)\dots(X_n)R\,[q_1(X_1,\dots,X_n),\dots,q_n(X_1,\dots,X_n)] \quad (1)$$

where each $q_i (1 \leq i \leq n)$ is a function symbol expressing a physical magnitude, and R expresses a purely mathematical relation. (The variables $X_1,\dots,X_n$ range over physical objects.) By virtue of the definition of the function symbols, (1) is equivalent to

$$(X_1)\ldots(X_n)(\rho_1)\ldots(\rho_m)[[Q_1(X_1,\ldots,X_n,\rho_1)\&\ldots\&Q_m(X_1,\ldots,X_n,\rho_m)] \rightarrow R(\rho_1,\ldots,\rho_m)] \tag{2}$$

where each $Q_i(1 \leq i \leq n)$ is an $(n+1)$-ary predicate symbol. (The variables $\rho_1, \ldots, \rho_m$ range over real numbers.) We use the terms "quantity" and "values of quantity" in Glymour's (1980, 123-24) sense. Therefore, we shall call the formulas $Q_i(X_1, \ldots, X_n, \rho_i)$ *quantities*, and their respective instances resulting from the simultaneous substitution of individual constants for the variables, *values of these quantities*. On the basis of the notions of "quantity" and of "value of a quantity" we define in our unified framework, for any given set of quantities, a set V of values *with respect to evidence E* to be a set of values resulting from the simultaneous substitution of individual constants belonging to E for the variables occurring in these quantities. The set V of values computed in the bootstrap testing of a qualitative or quantitative hypothesis should be taken with respect to the given evidence E. In practice, we shall use the abbreviated formulation (1) instead of (2).[4]

We can now state our definition of the emended evidential relevance relation *Rlv*. Let us consider a non-empty set of sentences E, a sentence H, a (possibly empty) deductively closed theory T, and a subtheory T^{-H} of T belonging to a language L (as described above) such that: (i) If *not* $(T \vdash H)$, then $T^{-H} = T$, (ii) If $T \vdash H$, then *not* $(T^{-H} \vdash H)$ and $Cn(T^{-H} \cup \{H\}) = T$.[5] We say then evidence E is *evidentially relevant* to H with respect to theory T as relativized to subtheory T^{-H}, or $Rlv\,(E, H, T, T^{-H})$ for short, iff

(G1*) $Cons(E \cup \{H\} \cup T)$.

There exists a set *Aux* of logical consequences of theory T such that:

(G2*) There is a maximal set V of values with respect to E for at least all the quantities occurring essentially in H such that

(i) For all $v \in V, E \cup Aux \vdash v$
(ii) V is, with respect to E, a positive instance of the hypothesis H in the sense that it fulfills Glymour's satisfaction* condition.[6]

(G3*) There is a set E' whose non-logical vocabulary is included in that of E such that

(i) $Cons(E' \cup Aux)$
(ii) Condition (L): $Cons(E' \cup T^{-H})$;
and there is a set V' of possible values with respect to E for the same quantities referred to in (G2*) such that
(iii) For all $v' \in V', E' \cup Aux \vdash v'$
(iv) V' is, with respect to E, a negative instance of H, i.e., it is a positive instance of $\sim H$ with respect to E.

(G4*) If $\vdash H \leftrightarrow (H_1 \& H_2)$, the non-logical vocabulary of H_1 and H_2 are included in that of H, and *not* $\vdash H \leftrightarrow H_1$, *not* $\vdash H \leftrightarrow H_2$, then both of the quadruples E, H_1, T, T^{-H} and E, H_2, T, T^{-H} satisfy separately conditions (G1*) - (G3*).[7]

We obtain the corresponding conditions (NG1*)-(NG4*) for the definition of a *negative evidential relevance relation Nrlv* by making the following modifications in the definition of *Rlv*: (NG1*) results from replacing the consistency condition in (G1*) by $Cons(E \cup T^{-H})$ · (NG2*) results from (G2*) by substituting in (iii) "*negative instance*" for "positive instance." (NG3*) results from (G3*) by substituting in (iv) "*positive* instance" for "negative instance," and by adding $Cons(E' \cup \{H\} \cup T)$. (NG4*) results from (G4*) by substituting (NG1*)-(NG3*) respectively for (G1*)-(G3*).

We can classify the *types* of evidential relevance (positive or negative) holding among a quadruple E, H, T, T^{-H} in the following way: 1. E is *bootstrap relevant* to H with respect to T and T^{-H} in case $H \in T$, and E is *nonbootstrap* relevant to H with respect to T and T^{-H} in case $H \notin T$.[8] 2. E is *strictly bootstrap relevant* to H with respect to T and T^{-H} in case each set of auxiliaries *Aux* has a member which cannot be deduced from T^{-H} alone; and E is *nonstrictly bootstrap relevant* to H with respect to T and T^{-H} in case $H \in T$, and there is a set of auxiliaries *Aux* such that every member of *Aux* can be deduced from T^{-H} alone.

3. CHRISTENSEN'S PAIRS OF COUNTEREXAMPLES

We assume that the evidential relevance relation *Rlv* defined in section 2 is the formal counterpart of an intuitive evidential relevance relation expressed in an interpreted language. Our emended relation is four-place in contradistinction to Glymour's original three-place relation. The need for a four-place relation is shown by Christensen's pairs of examples indicating that evidential relevance is not completely determined by the three relata E, H, and T alone. Indeed, Christensen exhibits two examples both referring to the same relata E, H, and T, but with different axiomatizations such that, in the first one, E is not intuitively relevant to H with respect to T, whereas, in the second one, E is intuitively relevant to H with respect to T.

Let us consider Christensen's (1983, 478-79) pair of examples:

$$E\colon \{Ra,\ Fa\}$$
$$H\colon (x)(Rx \rightarrow Bx)$$
$$T\colon Cn(\{H, H'\}) = Cn(\{H,\ H''\})$$

where H' is $(x)(Rx \rightarrow Fx)$, and H'' is $(x)[Rx \rightarrow (Bx \leftrightarrow Fx)]$. Interpreting the non-logical constants R as "is a raven", B as "is black", and F as "has a particular type of feather", we see that the evidence E is not intuitively relevant to H with respect to T when T is axiomatized by the set $\{H,\ H'\}$, whereas E *is* intuitively relevant to H with respect to T when T is axiomatized by the set $\{H,\ H''\}$. Therefore, Glymour's formal evidential relation agrees with intuitive relevance in one axiomatization of the theory, but disagrees with another. Hence, the following equivalence condition is violated: Given two axiomatizations of one and the same theory, if E *is* evidentially relevant to H with respect to one axiomatization, E is also evidentially relevant to H with respect to the other axiomatization.[9]

Christensen's pair of examples, by referring explicitly (besides to the theory T) to particular axiomatizations of T, give rise, in fact, to the introduction of a fourth relatum into the evidential relevance relation, which is, in our opinion, the notion of an H-free subtheory T^{-H} of T. Indeed, we see that T^{-H} in the first example of the pair is equal to $Cn(\{H'\})$, whereas in the second is equal to $Cn(\{H''\})$. Clearly, these two subtheories are different, and it is precisely this difference that explains the unequal behavior of the members of Christensen's pair of examples.

Furthermore, the set $E' \cup T^{-H}$ is inconsistent in the first example of the pair, and consistent in the second one, so that our legitimacy condition (L) is violated in the first one, and satisfied in the second. Therefore, regarding Christensen's pair of examples, there is perfect agreement between our formal relation *Rlv* and the corresponding intuitive relation. The same agreement obtains also between *Rlv* and the corresponding intuitive relation involved in the other pairs of examples given in Christensen 1990.

4. IMMUNITY OF *RLV* FROM NARROW MINDEDNESS AND GULLIBILITY

A formal evidential relevance relation R (of any arity) may give rise to two different kinds of inadequacy. First, it might be that E is intuitively relevant to H with respect to T (and possibly a fourth relatum) but R does not hold. Then R is said to be "too strong" or "narrow minded." Second, R holds but E is not intuitively relevant to H with respect to T (and possibly a fourth relatum). Then, R is said to be "too weak" or "gullible."[10] There are thus two kinds of counterexamples for relation R, viz., those which exhibit narrow mindedness, and those which exhibit gullibility of R.

In the following two subsections, we shall consider counterexamples for Glymour's original evidential relevance relation, and several variants thereof, showing that they are all avoided by the emended relation *Rlv*.

4.1. Immunity of Rlv from Narrow Mindedness

Glymour's (1975, 1980) original system of bootstrapping is free of narrow mindedness. But, later versions are not so, due to the introduction of restrictive rules for the purpose of avoiding gullibility. These are Edidin's (1981, 295-6; 1983, 47-53) restriction *not* ($Aux \vdash H$), van Fraassen's (1983, 30, 42, n.17) clause (2-1) (2), Glymour's (1983b, 6) clause (iv) of *Schema I*, Glymour's (1983a, 627-28) restriction (R) or clause (4) of his definition of bootstrapping, Earman and Glymour's (1988, 261-62) condition *Cons* ($E' \cup Aux$) in clause (G3), and clause (R) or (R'), and finally Culler's (1995, 572) clause (iii) of (v-R).

Indeed, all cases in which the set *Aux* contains the hypothesis H being tested, or more generally *Aux* entails H, are precluded in these versions. Cases in which

$H \in Aux$ have been called "macho bootstrapping" (Earman and Glymour 1988, 262), and we shall call all cases in which $Aux \vdash H$, "quasi-macho bootstrapping." Not only macho, but also quasi-macho bootstrapping entails that $E' \cup Aux$ is inconsistent. Edidin's (1981, 1983), and Earman and Glymour's (1988) accounts imply that they reject even quasi-macho bootstrapping. In our formulation, the difficulty of macho or quasi-macho bootstrapping can be explained by pointing out that where $E' \cup Aux$ is inconsistent, any arbitrary set of possible values would follow therefrom, so that any hypothesis could be confuted. However, we shall show that examples that are in Glymour's formulation cases of macho or quasi-macho bootstrapping, can be reconstructed in our system as satisfying the condition $Cons(E' \cup Aux)$ and thus *not* $(Aux \vdash H)$.

For example, consider the bootstrap test of Mach's so-called definition of mass. Given

$$E: \{a(\mathrm{P}_1, \mathrm{t}_1) = \mathrm{r}_{11},\ a(\mathrm{P}_2, \mathrm{t}_1) = \mathrm{r}_{21},\ a(\mathrm{P}_1, \mathrm{t}_2) = \mathrm{r}_{12},\ a(\mathrm{P}_2, \mathrm{t}_2) = \mathrm{r}_{22}\}$$
$$H: (x)(t)[m(x) = -a(\mathrm{P}_1, t)/a(x, t)],$$
$$T: Cn(\{H\})$$
$$T^{-H}: Cn\varnothing$$

where the r_{ij}s are real numbers such that $\mathrm{r}_{11}/\mathrm{r}_{21} = \mathrm{r}_{12}/\mathrm{r}_{22}$, m is the mass function taken to be constant over time, a the acceleration function, P_1 a particle selected by convention to have the unit mass, P_2 the other particle interacting P_1, and t_1, t_2 different time moments,[11] take Aux: $\{m(\mathrm{P}_2) = -a(\mathrm{P}_1, \mathrm{t}_1)/a(\mathrm{P}_2, \mathrm{t}_1)\}$ and E': $\{a(\mathrm{P}_1, \mathrm{t}_1) = \mathrm{r}'_{11},\ a(\mathrm{P}_2, \mathrm{t}_1) = \mathrm{r}'_{21},\ a(\mathrm{P}_1, \mathrm{t}_2) = \mathrm{r}'_{12},\ a(\mathrm{P}_2, \mathrm{t}_2) = \mathrm{r}'_{22}\}$ where $\mathrm{r}'_{11}/\mathrm{r}'_{21} \neq \mathrm{r}'_{12}/\mathrm{r}'_{22}$. Then conditions ($\mathrm{G}_1$*) - (G4*) are satisfied so that $Rlv(E, H, T, T^{-H})$ holds.

We see that Aux consists of a universal instantiation of H, but in Glymour's versions, where the computation of the values of the quantities involves a graph, $Aux = \{H\}$ so that the example turns out to be a case of macho bootstrapping, and, therefore, is precluded by Glymour's new versions of bootstrapping. However, in this example, E is indeed intuitively relevant to H with respect to T and T^{-H}; but as shown above Rlv holds among E, H, T and T^{-H}. Of course, here Aux is derived from $Cn(\{H\})$ so that this example, in our terminology, is a case of strict bootstrapping, though not a macho or a quasi-macho one.

This example, as well as those given in Zytkow 1986, 102-104, viz., the testing of the law of conservation of momentum and Ohm's law, are cases of *ineliminable* strict bootstrapping, in the sense that they cannot be reduced either to a case of direct empirical test, or one of nonstrict bootstrapping. Such kinds of reductions are not exactly parallel to the reduction made for the testing of the ideal gas law, as exhibited in Edidin 1981, 294-95, Edidin 1983, 50-51, and van Fraassen 1983, 42, n.17. Furthermore, it goes against the claim in Earman and Glymour 1988, 262-63, viz., the claim that the reduction applies to Zytkow's examples, as well as to similar ones in general. Indeed, the testing of H (Mach's alleged definition of "mass") cannot be reduced to the direct empirical test of H': $(x)(t)[-a(\mathrm{P}_1, t)/a(x, t) = -a(\mathrm{P}_1, t')/a(x, t')]$, because $m(x)$ does not mean merely the common value of the ratios $-a(\mathrm{P}_1, t)/a(x, t)$ and $-a(\mathrm{P}_1, t')/a(x, t')$. On the other hand, consider the alleged reduction of the testing of H to the concurrent nonstrict bootstrap testing of

$$\{(x)(t)(t')[m(x, t) = m(x, t')], (x)(t)[m(x, t) = -a(\mathrm{P}_1, t)/a(x, t)]\}$$

with respect to the theory consisting of the logical closure of this set and the relevant subtheories, i.e., the closure of the second member of the set for the test of the first member, and the closure of the second for the first. Now the second member of the set is not analytic, since it does not express the definition of the function m (assumed, in this context, to depend also on time). But, then, its test is a case of strict bootstrapping so that our example is really a case of ineliminable strict bootstrapping.

Let us now consider two examples of nonstrict bootstrap relevance both of which are cases of intuitive relevance, but are disallowed by various versions of bootstrapping.

First example: As stated by Glymour (1983a, 629) his definition of bootstrapping, by virtue of restriction (R), rules out not only cases of macho bootstrapping, but also some cases—in our terminology—of nonstrict bootstrapping (which are cases of intuitive relevance). Such cases involve "the confirmation of a universally quantified conjunction" (*ibid.*, 629). Earman and Glymour's (1988, 263) example is a case in point. We shall show that this example (whose intuitive legitimacy will be exemplified by the interpretation given below) is formally legitimated by our system. Consider

$$E: Oa$$
$$H: (x)Gx$$
$$T: Cn(\{(x)[Gx \,\&\, (Gx \leftrightarrow Ox)]\})$$
$$T^{-H}: Cn(\{(x)(Gx \leftrightarrow Ox)\})$$

where x ranges over ordinary material objects, Gx means that object x is ultimately constituted of quarks, Ox means that x has a large number of specified observational features, and a denotes a particular observed material object. If we take, then,

$$Aux: \{(x)(Gx \leftrightarrow Ox)\}$$
$$E': \sim Oa$$

it is easy to see that all conditions of Rlv are satisfied.

Second Example: Consider Christensen's (1990, 656) example. Given

$$E: \{Sa, Pa\}$$
$$H: (x)(Ex \leftrightarrow Zx)$$
$$H': (x)(Sx \leftrightarrow Zx)$$
$$H'': (x)(Px \leftrightarrow Ex)$$
$$T: Cn(\{H, H', H''\})$$
$$T^{-H}: Cn(\{H', H''\})$$

take

$$Aux: \{H', \ H''\}$$
$$E': \{Sa, \ \sim Pa\}.$$

Although in the interpretation Christensen assigns to *E, Z, S*, and *P*, evidence *E* is intuitively relevant to *H* with respect to *T* and T^{-H}, it is precluded by Glymour's (1983a) and Earman and Glymour's (1988) versions of bootstrapping. However, it is easy to show that *Rlv* holds so that our system is not narrow minded with respect to this example either. Note that the second member of the pair given in Christensen's (1983) example, as well as the second members of the other pairs of examples given in Christensen 1990, can be treated in the same way.

4.2. Immunity of Rlv from Gullibility

Edidin (1981, 1983), Christensen (1983, 1990), Grimes (1987), and Earman (1992) have given counterexamples showing the gullibility of Glymour's (1975, 1980, 1983a,b), van Fraassen's (1983), Zytkow's (1986), and Earman and Glymour's (1988) systems of bootstrapping, although various restrictive rules, viz., *Cons* $(E' \cup (Aux - \{H\})$ given in Glymour 1980, 131, and the ones mentioned in subsection 4.1, for preventing this defect have been successively introduced by these systems, but at the cost of creating or increasing narrow mindedness. Indeed, gullibility is considered to be a more serious deficiency than narrow mindedness.[12] Mitchell's (1995) system also fails to avoid gullibility. For instance, in Grimes' (1987) counterexample, which will be discussed below, the truth, and thus the confirmation, of the auxiliary does not really depend, for structural reasons, upon the truth of the hypothesis being tested, despite Mitchell's contention to the contrary.[13] It follows that Mitchell's structural rule (1995, 246) is not really violated by Grimes' counterexample, and consequently, Mitchell's system cannot avoid it.

We can show now that our evidential relevance relation *Rlv* is not subject to any of the above-mentioned counterexamples which have hitherto been given, because they all violate condition (*L*). For example, Edidin (1983, 46) has given a counterexample (which is a simpler version of the one given in Edidin 1981, 295) exhibiting the gullibility of Glymour's (1975) system. This counterexample has been blocked by Glymour's (1980, 131) restrictive rule mentioned above. Nevertheless, Christensen (1990, 650-51) has shown that the computation used in Edidin's counterexample can be modified in such a way that all of these restrictive conditions are satisfied. Thus, the counterexample is not really ruled out by any of the systems of bootstrapping proposed so far. However, in our emended system condition (*L*) is violated in both versions of the counterexample, so that Edidin's famous counterexample is really barred only within our system. Christensen (1983, 473, 474-76, 478-79) has given counterexamples which show the gullibility of Glymour's (1980) system strengthened with the addition of Edidin's restriction $not(Aux \vdash H)$.

On the other hand, Christensen (1990, 648-656) has given further counterexamples satisfying all the four restrictions and thus also shown the gullibility of Glymour's

(1983a) and Earman and Glymour's (1988) systems. Finally, Earman (1992, 240, n.13) has given a counterexample to van Fraassen's (1983) system. Now we see that all these counterexamples violate condition (*L*) so that they are avoided by our emended system.

Whereas all of the above-mentioned counterexamples were cases of strict bootstrapping, the one advanced in Grimes (1987, 103-104) is a case of non-bootstrapping. Given

E: (any sentence)
H: (any hypothesis)
T: $Cn(\{I,\ (E \rightarrow I)\,\&\,(\sim E \rightarrow I^*)\})$, if E is true (first case), and
$Cn(\{I,\ (\sim E \rightarrow I)\,\&\,(E \rightarrow I^*)\})$, if E is false (second case),[14]

where T_1 denotes $Cn(\{I,\ (E \rightarrow I)\,\&\,(\sim E \rightarrow I^*)\})$, and T_2 denotes $Cn(\{I,\ (\sim E \rightarrow I)\,\&\,(E \rightarrow I^*)\})$ such that $T_1 = CnT_1$ and $T_2 = CnT_2$. In this example I is the conjunction of the members of a positive instance, I^* is the conjunction of the members of a negative instance, of H with respect to E, *not* $(H \vdash I)$, *not* $(I \vdash H)$ (i.e., *H and I* are mutually independent)[15], $Cons(\{E\&H\&I\})$, and $Cons(\{\sim E\&H\&I\})$. Take then *Aux to* be $T_1 - \{I\}$ in the first case, $T_2 - \{I\}$ in the second, and E' a sentence such that $E' \vdash\sim E$ in the first case, and $E' \vdash E$ in the second. Then, we see that condition (*L*) is violated, since the sets $E' \cup T_1^{-H}$ and $E' \cup T_2^{-H}$ are respectively equivalent to $E' \cup Cn(\{E\&I\})$ and to $E' \cup Cn(\{\sim E\&I\})$ both of which are inconsistent.[16] Thus Grimes' counterexample is avoided by our system, but not by the ones involving the restrictions mentioned above, since they are all satisfied by Grimes' example. As already indicated above, neither is this counterexample barred by Mitchell's structural rule, since, considering that *I* and *H* are mutually independent, we see that the truth (hence the confirmation) of *Aux* (in both cases) does not depend on the truth of *H*, but only on that of the conjunction of the members of a positive instance *I* of *H*.

5. COMBINATION OF BOOTSTRAPPING WITH THE BAYESIAN METHOD

The Bayesian method of incremental confirmation can be described as follows:

(1) Hypothesis H is *incrementally confirmed* by evidence E with respect to background knowledge K iff $\Pr(H|E\,\&\,K) > \Pr(H\,\&\,K)$,
(2) Given that $0 < \Pr(H|K) < 1$, $\Pr(H|E\,\&\,K) > \Pr(H\,\&\,K)$ iff $Pr(E|H\,\&\,K) > \Pr(E|\sim H\,\&\,K)$.[17]

Here Pr (.) stands for the absolute probability function, Pr(.|.) for the conditional probability function, $\Pr(H|K)$ is the *prior* probability of H, $\Pr(H|E\,\&\,K)$ is the *posterior* probability of H, and $\Pr(E|H\,\&\,K)$ is the *likelihood* of H.

Given a universal hypothesis H of the form $(x)(Fx \rightarrow Gx)$, one can incrementally confirm it by means of an instance-statement of the form $Fa\ \&\ Ga$ (cf. Earman 1992, 71-72). Indeed, we can expect that

$$\Pr(Fa\ \&\ Ga|H\ \&\ K) > \Pr(Fa\ \&\ Ga|\sim H\,\&\,K),$$

so that the hypothesis is incrementally confirmed (where K expresses the background knowledge, and that background knowledge is a realistic one).

In case the hypothesis H is non-theoretical, Fa & Ga can be used as evidence for the Bayesian confirmation of H. But, if H is theoretical relative to the available evidence, i.e., at least one of F and G is theoretical, the statement Fa & Ga can no more constitute the available evidence. Assume, for example, that F is theoretical and G non-theoretical. Then, the available evidence reduces to Ga. But, then, given that K expresses a realistic background, it is plausible to accept that $\Pr(Ga|H \& K)$ and $\Pr(Ga| \sim H \& K)$ are almost equal, so that H is neither incrementally confirmed nor disconfirmed.

Let us illustrate the hypothesis H by taking Fx to mean "x is a cloud that is electrically charged during a certain period of time," and Gx to mean "x gives rise to lightning." Then, H means "All electrically charged clouds give rise to lightning." (Here F is theoretical and G is non-theoretical.) We have seen that the evidence Ga alone is insufficient to incrementally confirm the hypothesis H. So, how can we incrementally confirm H? It seems that the Bayesian incremental confirmation of H requires the conjunction of the statement Fa to the evidence Ga in one way or another. This can result from the introduction of Fa in the background knowledge K. However, the background knowledge must be a kind of secure knowledge which can be taken for granted. But, the epistemic status of Fa that contains the theoretical term (or "quantity") F is quite different. Note that we mean here by a "theoretical term" not merely one which does not occur in the available evidence, but also one whose value cannot be ascertained without recourse to some law or other of the theory that contains essentially the term in question.[18] It follows that the truth value of Fa depends on the truth value of a theory which itself is in need of being tested.

One purpose of bootstrapping is to compute the values of such theoretical quantities. Therefore, our proposal is to divide the process of confirmation of a theoretical hypothesis into two steps. In the first (bootstrap) step, we deduce theoretical values, such as the statement Fa from the evidence and the theory, and in the second (Bayesian) step, we incrementally confirm the hypothesis on the basis of an instance-statement that contains not only the available non-theoretical evidence, but also the computed values of theoretical quantities.

Let us apply this procedure to the above-mentioned clouds-lightning example.

Step 1: The bootstrap step: Given

E: $\{Q_1(a, t_1), Q_2(a, t_1), Q_1(a, t_2), Q_2(a, t_2)\}$
H_1: $(x)(t)[Q_1(x, t) \rightarrow (Fx \leftrightarrow Q_2(x, t)\ ]$[19] (so that F is theoretical)
T: $Cn(\{H_1\})$
T^{-H}: $Cn\varnothing$

where $Q_1(x, t)$ means "x is brought close to a standard body (which in fact is electrically charged) at time t", $Q_2(x, t)$ means "x is pushed or attracted by the same standard body at time t", and as said above, Fx means "x is a cloud that is electrically charged during a period of time so that t_1 and t_2 are time moments belonging to this period." Take now

Aux: $\{Q_1(a, t_1) \to [Fa \leftrightarrow Q_2(a, t_1)]\}$
E': $\{Q_1(a, t_1), Q_1(a, t_2), Q_2(a, t_1), \sim Q_2(a, t_2)\}$.

Then, we see that $Rlv(E, H, T, T^{-H})$ holds, so that the computed value Fa is a suitable one for the next Bayesian step.

Step 2: The Bayesian step. Once Fa is established, in the sense that it has been successfully computed, it can be used for determining the incremental confirmation of hypothesis H, viz., $(x)(Fx \to Gx)$. Fa can either be conjoined to the available evidence as a theoretical evidence item, or else, if one refuses to admit theoretical evidence, it can be incorporated into the background knowledge, since it is already established. Thus, in either case, the posterior probability of H takes the following form:

$$\Pr(H|Ga \,\&\, Fa \,\&\, K).$$

This is formally identical to the posterior probability of the raven hypothesis. (See Earman 1992, 69-73.) Hence, one may expect—for realistic background knowledge—that the hypothesis will be incrementally confirmed.

One might object to the use of the bootstrapping test by taking as background knowledge $K = Q_1(a, t_1)\&Q_2(a, t_1)\&H_1$ (where H_1 is as stated above). Then, since K entails Fa, $\Pr(H|Ga\&K) = \Pr(H|Ga\&Fa\&K)$ so that the bootstrapping step might be dispensed with. However, such a procedure consists in nothing but incorporating the bootstrapping step into the Bayesian one. Therefore, the objection fails.

Lastly, one might question the epistemic status of *Aux* itself, i.e., the problem of confirming *Aux*. If *Aux* is not supposed to be confirmed, then the confirmation of the theoretical hypothesis in question, via the combined method of bootstrapping and Bayesianism, would be qualified only as *relative* confirmation. On the other hand, in order to speak of the *absolute* confirmation of the theoretical hypothesis, every bit of the theory from which *Aux* is derivable should be confirmed in the same way, i.e., by means of the combination of bootstrapping and Bayesianism.

6. THE RATIONALITY AND OBJECTIVITY OF THE COMBINED METHOD

As pointed out by Earman (1992, 145-153) one way of showing the rationality and objectivity of the Bayesian method is by having recourse to the Gaifman and Snir theorem. Consider a hypothesis H whose vocabulary is included in that of the evidence (hence a non-theoretical one). An infinite set of evidence sentences $E_1, \ldots, E_n, \ldots$, called the *evidence matrix*, is said to *separate* the set of models of the language just in case for any two models w_1, w_2 there is an E_i such that w_1 is a model of E_i and w_2 is a model of $\sim E_i$. Then, the theorem can be expressed roughly by means of the following two propositions:

(i) In any given model of the language, the posterior probability of an hypothesis H, given a sequence of n elements of the evidence matrix separating the set of models of the language, taken negated or unnegated, when n goes to infinity, tends to 1 if H is true in the model, and to 0, if H is false in the model. (Each element of the evidence matrix is taken unnegated in case it is true in the model, and negated if it is false in the model.)

This proposition expresses the "washing out of priors." (See Earman 1992, 141f.) Indeed, it expresses that, for each person, whatever the prior probability assigned by that person to the hypothesis is, the posterior probability will take in the limit the same value, which is therefore an objective one.

(ii) In any given model of the language, the difference of posterior probabilities (as defined above) assigned by any two persons, as n goes to infinity, tends to 0, provided that the persons assign the probability 0 to exactly the same sentences.

This expresses the "merger-of-opinion" (Earman 1992, 147). Indeed, the proposition expresses that the posterior probabilities of the different persons, whatever the prior probabilities are, will tend in the limit to equal values. Therefore, again, the common value is objective.

In case the vocabulary of the hypothesis H transcends that of the evidence, the evidence matrix can no more separate the set of models of the new language, since now the models refer to theoretical terms which do not occur in the evidence. (See Earman, 1992, 151.) But, separability is a necessary condition for the applicability of the Gaifman and Snir theorem. (See Earman 1992, 151, n. 19.) Hence, the convergence results of the Gaifman and Snir theorem are not applicable for theoretical hypotheses.

To extend these results to theoretical hypotheses, Earman (1992, 151) introduces the two equivalent notions of weak observational distinguishability (*wod*) and strong observational distinguishability (*sod*). The "more usual" notion of *sod* can be defined as follows: Two alternative theories T_1, T_2 (or theoretical hypotheses) are *sod* if there is an observation sentence O such that T_1 semantically entails O, and T_2 semantically entails $\sim O$ (Earman 1992, 151). Then, given a partition of theories $\{T_1, \ldots, T_n\}$ which are pairwise *sod*, it is possible to select a single theory (as the true one) on the basis of our knowledge of the truth value of the entailed observation sentence. This procedure is "simple eliminative induction." However, as remarked by Earman (1992, 152), simple eliminative induction cannot generally supplant the Bayesian apparatus, in case the entailed observation sentences are complicated (e.g., when they are multiple quantified). In such cases, one must apply the Bayesian method for determining the truth value of these sentences. Earman (1992, 167) calls such a combination of the Bayesian method with simple eliminative induction a "sophisticated eliminative induction."

However, sophisticated eliminative induction is applicable only if the alternative theories are observationally distinguishable. But, there are alternative theories which are not observationally distinguishable. (This is the case of theories which are underdetermined by observation.) In such a case "the conditions needed to make sophisticated eliminative induction work" are missing (Earman 1992, 167).

The bootstrapping method, in combination with the Bayesian one, is applicable both in case of observationally distinguishable, and of observationally indistinguishable theories. The application consists in computing the values of theoretical quantities, establishing thus an evidence matrix that contains also theoretical elements. Such an evidence matrix separates the set of models of the language so that the convergence results of the Gaifman and Snir theorem can be obtained.

Note that in contradistinction to sophisticated eliminative induction, the combination of bootstrapping with the Bayesian method does not presuppose a partition of theories in order to select the true one. Indeed, the Gaifman and Snir theorem can be applied to a single theoretical hypothesis, just as it applies to an observational one.

To illustrate, consider the following two alternative theories which are not observationally distinguishable, so that sophisticated eliminative induction does not apply, even if completed to a partition:

$$\begin{aligned}
&E\colon\ Fa\\
&E'\colon\ \sim Fa\\
&\mathrm{H}_1\colon\ \exists x\, Gx \,\&\, (x)(Fx \leftrightarrow Gx)\\
&\mathrm{H}_2\colon\ \sim \exists x\, Gx \,\&\, (x)(Fx \leftrightarrow Gx)\\
&\mathrm{T}_1\colon\ Cn(\{\mathrm{H}_1\})\\
&\mathrm{T}_2\colon\ Cn(\{\mathrm{H}_2\})\\
&\mathrm{T}_1^{-\mathrm{H}_1} = \mathrm{T}_2^{-\mathrm{H}_1} = Cn\varnothing\\
&Aux\colon\ (x)(Fx \leftrightarrow Gx)
\end{aligned}$$

Since T_1 does not have contingent observational consequences, T_1 and T_2 are non-*sod*. However, we can apply bootstrapping to compute a value of the theoretical predicate G to show that E is positively relevant to H_1 (with respect to T_1 and $\mathrm{T}_1^{-\mathrm{H}_1}$) and negatively relevant to T_2 (with respect to T_2 and $\mathrm{T}_2^{-\mathrm{H}_1}$). However, the confirmation procedure, and the justification of the confirmation procedure do not end up here. On the object-level of confirmation, having computed, say, Ga, H_1 should be probabilified via the Bayesian method. On the meta-level of justification of the confirmation procedure, reiterated computation of the values of the theoretical quantity will lead to the construction of a theoretical evidence matrix that makes the Gaifman and Snir theorem applicable.

In general, by means of bootstrapping, we can ascertain the values of theoretical quantities so that a theoretical evidence matrix can be constructed. This evidence matrix, being theoretical, can separate the set of models of a theoretical language, so that it can be used within the Gaifman and Snir theorem. Indeed, just as the elements of an observational evidence matrix are successively checked, the elements of a theoretical matrix are successively computed by bootstrapping (cf. Earman 1992, 146-147).

7. CONCLUSION

We conclude with a recapitulation, first, of the emended version of bootstrapping, and, second, of the ways of using bootstrapping, in combination with the Bayesian method, for testing theoretical hypotheses.

As to the former, the emended evidential relevance relation has three merits. First, it realizes the unification of the methods of bootstrap testing of quantitative and qualitative hypotheses. Second, by replacing the graph method of computation with straightforward deduction, it becomes possible to reconstruct cases of macho or quasi-macho bootstrapping as cases of non-macho strict bootstrapping so that the emended system becomes free of narrow mindedness. Third, by introducing a fourth relatum into evidential relevance relation, we secure the full determinacy of the relation, and thus avoid the underdeterminacy involved in the three-place relation as expounded in Christensen's pairs of examples. Furthermore, the fourth relatum allows the formulation of the legitimacy condition (L) whose satisfaction secures the avoidance of gullibility.

As to the latter, there are three points at issue. First, given a single theory (or theoretical hypothesis), the computed values of the theoretical quantities can be used—either as part of the background knowledge or of evidence—in the Bayes' theorem for incremental (as well as for absolute) confirmation of the hypothesis in question. Second, in case two theories (or theoretical hypotheses) are not observationally distinguishable, bootstrapping (together with Bayesianism) will play the role that simple or sophisticated eliminative induction does, either on the level of confirmation or on the level of the justification of the confirmation procedure via the Gaifman and Snir theorem. Lastly, given a single theory (or theoretical hypothesis) again, the values of theoretical quantities computed via successive application of bootstrapping can be used to extend the evidence matrix so that it will separate all possible models of the theoretical language. Then the Gaifman and Snir theorem will be applicable for justifying, i.e., for showing the rationality and objectivity of, the combination of bootstrapping with the Bayesian method.

8. NOTES

[1] I would like to thank John Earman, Gürol Irzik, Erdinç Sayan and David Davenport for helpful suggestions.

[2] As it stands, this condition first appears in Earman and Glymour 1988, 261 as part of their condition (G3).

[3] We adopt these conditions from Earman and Glymour 1988, 261. Cf. also Glymour 1975, 414, and Glymour 1980, 130-31.

[4] Concerning quantitative languages, Edidin (1981, 293) has pointed out the following: "Since (in at least some formalizations) equations are atomic formulae, there will be hypotheses that are confirmed by a body of evidence relative to a theory... *qua* first-order formula... but not... *qua* quantitative hypotheses...." Edidin proposes to remove this discrepancy between the two versions of bootstrapping by requiring that in a combined form of bootstrap testing "a computation for a given sentence yield numerical values for all the quantities in it that take numerical values, as well as truth values for the atomic formulae" (1981, 300). We think that Edidin's proposal, as it stands, is ad hoc. By transforming any quantitative hypothesis into a first-order language with identity without primitive function symbols we obtain a natural solution to the difficulty expounded by Edidin. However, in practice, to make use of (1) instead of (2) is to follow what Edidin has suggested. Thus our treatment of the problem can be best construed as a rational justification of Edidin's proposal.

[5] 'Cn' denotes the logical consequence operation and is defined as follows: $Cn(S) = \{K: S \vdash K\}$, for any set S of sentences. (K is a variable ranging over sentences.)

[6] See Glymour 1975, 414, condition (ii), and Glymour 1980, 130.

[7] (G1*) - (G3*) correspond to (G1)-(G3) in Earman and Glymour 1988, whereas (G4*) corresponds to clause (v) of the Bootstrap Condition in Glymour 1980, 132.

[8] A nonbootstrap evidential relevance relation is developed in Culler 1995. See especially *ibid.*, 572 where the relation is defined by condition (ν-R).

[9] This follows from the general equivalence condition stated in Earman and Glymour 1988, 262, n.2.
[10] We follow Glymour 1983a, 629 in our use of "narrow minded" and "gullible."
[11] Note that *H* is not analytic, even though it is called "Mach's definition of mass." (See, e.g., Sneed 1979, 59-60, 135.)
[12] Glymour (1983a, 629) writes: "It is better for a formal confirmation theory to be narrow minded than for it to be gullible. . . . "
[13] See Mitchell 1995, 252.
[14] Since Grimes (1987, 103), in both cases, takes *I* to be true, *I* should belong to theory *T*. Note that Culler (1995, 571-72, n.4) also considers *I* as an additional member of *T*.
[15] See Grimes 1987, 104, and 107, n.3.
[16] Note that Grimes' counterexample turns into a counterexample of nonstrict bootstrapping by the addition of *H* to theory *T* (in both cases) which again violates (*L*).
[17] Cf. Earman 1992, 70.
[18] See Glymour 1980, 107.
[19] This is a bilateral reduction sentence of the form mentioned in Carnap 1936, 440.

9. REFERENCES

Carnap, R. (1936), "Testability and Meaning", *Philosophy of Science* **3**: 440-41

Christensen, D. (1983), "Glymour on Evidential Relevance", *Philosophy of Science* **50**: 471-481.

Christensen, D. (1990), "The Irrelevance of Bootstrapping", *Philosophy of Science* **57**: 644-662.

Culler, M. (1995), "Beyond Bootstrapping: A New Account of Evidential Relevance", *Philosophy of Science* **62**: 561-579.

Earman, J. (1992), *Bayes or Bust ? A Critical Examination of Bayesian Confirmation Theory*. Cambridge, MA: The MIT Press.

Earman, J. and Glymour, C. (1988), "What Revisions Does Bootstrapping Need? A Reply", *Philosophy of Science* **55**: 260-264.

Edidin, A. (1981), "Glymour on Confirmation", *Philosophy of Science* **48**: 292-307.

Edidin, A. (1983), "Bootstrapping Without Bootstraps," in J. Earman, (ed.), *Testing Scientific Theories*. Minneapolis: University of Minnesota Press, pp. 43-54.

Edidin, A. (1988), "From Relative Confirmation to Real Confirmation", *Philosophy of Science* **55**: 265-271.

Glymour, C. (1975), "Relevant Evidence", *The Journal of Philosophy* **72**: 403-426.

Glymour, C. (1980), *Theory and Evidence*. Princeton: Princeton University Press.

Glymour, C. (1983a), "Discussion: Revisions of Bootstrap Testing", *Philosophy of Science* **50**: 626-629.

Glymour, C. (1983b), "On Testing and Evidence," in J. Earman, (ed.), *Testing Scientific Theories*. Minneapolis: University of Minnesota Press, pp. 3-26.

Grimes, T. R. (1987), "The Promiscuity of Bootstrapping", *Philosophical Studies* **51**: 101-107.

Mitchell, S. (1995), "Toward a Defensible Bootstrapping", *Philosophy of Science* **62**: 241-260.

Sneed, J. D. (1979), *The Logical Structure of Mathematical Physics*. Dordrecht: Reidel Publishing Company.

van Fraassen, B. (1983), "Theory Comparison and Relevant Evidence", in J. Earman, (ed.), *Testing Scientific Theories*. Minneapolis: University of Minnesota Press, pp. 27-42.

Zytkow, J. M. (1986), "What Revisions Does Bootstrap Testing Need?", *Philosophy of Science* **53**: 101-109.

ERDİNÇ SAYAN

IDEALIZATIONS AND APPROXIMATIONS IN SCIENCE, AND THE BAYESIAN THEORY OF CONFIRMATION

1. INTRODUCTION

Idealizations enter into scientific analysis or explanation in at least two ways. An idealization may be embodied within the law or theory itself. For example, insofar as Newton's second law is conceived as applying only to point-masses, that law contains an idealization as part of its content. Sometimes idealizations take the form of assumptions conjoined to a theory from outside. For instance, the assumption that the universe contains only two bodies is an idealization that may be employed in some contexts as input to Newton's law of gravitation and to the second law of motion (as when deriving Kepler's laws from Newton's theory). Scientists must resort to idealizations and approximations for several reasons. There may be a lack of: (i) necessary data with required accuracy, (ii) mathematical-analytical or computational power, (iii) necessary auxiliary theories. Without idealizing and simplifying assumptions, such as assumptions of linearity, of negligible masses, perfect vacuums, frictionless planes, isolated thermodynamic systems, perfectly elastic bodies, perfectly uniform electric fields, ideally rational economic man, perfectly competitive market, and the like, working out the observable implications of theories is often impracticable. Computational facilitation afforded by idealizations and simplifications makes them vital elements of scientific activity in both natural and social sciences.

Despite the pervasive use of idealizations and approximations in science, their role has been ignored or misunderstood by philosophers. Idealizing and simplifying assumptions are strictly speaking *false* statements; hence they amount to distortions of reality. Still, interestingly, they are routinely employed in explanations of phenomena or when testing a scientific theory for truth. The role of idealizing and simplifying assumptions is critically important for theories of confirmation to account for. Yet, well-known theories of confirmation, such as the h-d (hypothetico-deductive), "bootstrapping," and Bayesian approaches, appear largely oblivious to the relevance of idealizations and approximations to confirmation. In testing contexts, the interaction between a theory and the idealizations under which it operates is an interaction of truth with falsehood—the (putative) truth of the theory with the falsehood of the idealizations and approximations. This facet of theory testing deserves more

G. Irzık & G. Güzeldere (eds.), Turkish Studies in the History and Philosophy of Science, 103-112.

attention than it has received from philosophers of science. In this paper, I shall be concerned with how the Bayesian account of confirmation can square with idealizations and approximations. But first, I want to expose some of the complications that idealizations and approximations create for the h-d model of confirmation of scientific theories.[1]

2. THE H-D METHOD'S MISTREATMENT OF IDEALIZATIONS AND APPROXIMATIONS

On the h-d methodology, a prediction *P* is derived from the conjunction of theory *T* with *I*, where *I* is a conjunction containing initial condition values and auxiliary hypotheses. *P* is then checked against nature to see if it is true or not, and a confirmation or disconfirmation decision is made accordingly. The h-d confirmatory and disconfirmatory argument schemata are as follows:

(HD)

CONFIRMATION:		DISCONFIRMATION:	
$T \& I \vdash P$	true	$T \& I \vdash P$	true
P	true	$\sim P$	true
$\therefore T \& I$	may be true	$\therefore \sim T \vee \sim I$	true.

'$\vdash$' above stands for logical entailment. The crucial point here is that *I* typically contains a set of idealization statements and approximations. But, since idealizations and approximations are already known to be false statements, *I* is known to be false. It is this circumstance that vitiates the h-d method. For even if theory *T* is true, we can hardly expect its conjunction with the false *I* to yield a true prediction *P*. This is because the falsehoods contained in *I* will have distorting effects during the derivation of *P* from *T*—unless the false elements in *I* fortuitously cancel each other out completely to cause zero distortion on the computed value of *P*. Since the chances of such total error cancellation are usually not very high, shall we require in order to decide that theory *T* may be *true* that its prediction *P* be *false*? Conversely, if *P* turns out to be *true*, shall we take that as a sign that *T* is *false*? (Such judgments, of course, would amount to a complete reversal of the h-d method.) But what if there really was a fortuitous error cancellation, the low odds notwithstanding?

Now, if a true *P* was obtained as a result of error compensation, two possibilities come to mind. One possibility is that *T* is in fact true, and the biases and errors introduced by the idealizations and approximations contained within *I* cancelled out during the computational derivation of *P*, with the result that their net distortion on *P* was nil. The second possibility is that *T* is actually false, and the errors contained *both* in *T* and in *I* conspired to cancel out during the computations, leading to a true *P* by sheer luck.[2] All these considerations make it clear that neither the truth nor falsehood of *P* can provide us with any indication whether *T* is true or false. A true *P* is supportive of neither the truth nor falsehood of *T*, as it is compatible with both. Similarly with a false *P*.

We can put the problem in more truth-functional terms. Even though the disconfirmation schema is deductively valid, the truth of its conclusion $\sim T \vee \sim I$ does not supply any hint as to the truth or falsehood of T. This is because the falsehood of I renders $\sim T \vee \sim I$ true regardless of the truth-value of T. As for the confirmation schema, the conclusion "T & I may be true" flies in the face of one the conjuncts, viz. I, being known to be false. In short, the original h-d model proves to be too crude to handle the intricate problems that arise when the derivation of a prediction from a theory is mediated by idealizations and approximations.[3]

3. BAYESIANIZATION OF IDEALIZATIONS AND APPROXIMATIONS

According to the standard Bayesian conception of testing, the incremental confirmation or disconfirmation of an hypothesis by a given piece of evidence requires comparison of the prior probability of the hypothesis with the posterior probability of that hypothesis on that evidence, where those probabilities are understood as subjective probabilities, or degrees of belief, conforming to the probability axioms. Let the hypothesis be theory T, the evidence be P, and let B be the background knowledge relative to which the theory is being tested. The Bayesian criteria for confirmation and disconfirmation are given as:

$$\begin{aligned} &\text{P confirms } T \text{ iff } \Pr(T|P \,\&\, B) > \Pr(T|B) \\ &P \text{ disconfirms } T \text{ iff } \Pr(T|P \,\&\, B) < \Pr(T|B). \end{aligned} \tag{1}$$

The posterior probability of the theory, $\Pr(T|P \,\&\, B)$, is related to its prior probability, $\Pr(T|B)$, by Bayes's theorem as follows:

$$\Pr(T|P \,\&\, B) = [\Pr(P|T \,\&\, B)/\Pr(P|B)]\ \Pr(T|B) \quad (\text{provided } \Pr(P|B) \neq 0). \tag{2}$$

From (1) and (2), another set of necessary and sufficient conditions are obtained for incremental confirmation and disconfirmation of T by P relative to B:

$$\begin{aligned} &P \text{ confirms } T \text{ iff } \Pr(P|T \,\&\, B) > \Pr(P|B) \\ &P \text{ disconfirms } T \text{ iff } \Pr(P|T \,\&\, B) < \Pr(P|B). \end{aligned} \tag{3}$$

How does the Bayesian model sketched above accommodate the role of idealizations and approximations in the confirmation and disconfirmation of T? Where do the idealizing assumptions and approximations figure in the Bayesian framework? In standard Bayesian treatments of scientific reasoning, they simply don't; they are entirely left out of the picture. But, as we shall see, idealizations and approximations pose challenges that the Bayesian model of confirmation must cope with, as must any theory of confirmation.

To see the problems involved, let us focus on the expression $\Pr(P|T \,\&\, B)$, called "the likelihood of P." $\Pr(P|T \,\&\, B)$ stands for the probability that we assign to observation P on the basis of the assumed truth of theory T and of our background beliefs B. Now, as we have said, in most testing situations, if we didn't employ any idealizations or approximations, analytic and computational complications would simply make it impossible for us to derive P as a prediction from T. And without the ability to derive P from T, we (more precisely, a perfectly rational agent) would fail to assign to $\Pr(P|T \& B)$ the value it deserves. For when we are unable to demonstrate that P is deducible from T, if only through mediation of some idealizations and approximations, our rational estimate of the value of $\Pr(P|T \& B)$ couldn't be much different from that of $\Pr(P|B)$. In that case, it follows by (3) that no incremental confirmation or disconfirmation of T by P would be gained. To illustrate this point, suppose that we are wondering if P, the observed period of a certain pendulum, provides any incremental confirmation for T, the Newtonian laws of motion. Also suppose, for the sake of the example, that the Newtonian laws have been newly conceived and that they have not been tested by too many observations yet. Without the ability to derive the period of the pendulum from the Newtonian laws, albeit by employing some idealizations and simplifications,[4] we wouldn't have any antecedent idea what the period might be, given Newtonian laws, other than what our general background knowledge or experience about the world would lead us to expect about the value of the period. That is to say, without such ability, we could only assign to the likelihood $\Pr(P|T \& B)$ of the period a value which is about the same as the value we assign to its prior likelihood (or "expectedness"), $\Pr(P|B)$. Consequently, the period of the pendulum would not count as confirmatory evidence for the Newtonian laws, although, under the circumstances we imagined, it should.

Therefore when P, say, the period of a pendulum, has been derived from T, say, the Newtonian theory, with the help of some idealizations and approximations I, this fact needs to be duly represented in the Bayesian scheme. For I is clearly an important part of the confirmation or disconfirmation of T by P. The question is, how is this to be done?

Let us start with the suggestion that I should appear as a conjunct in the condition clause of the likelihood of P. Thus the likelihood of P becomes $\Pr(P|T \& I \& B)$. The idea might be to reflect the fact that prediction P was derived not from T alone but from T in conjunction with I. This suggestion amounts to making I part of the background knowledge, in other words, expanding the background knowledge into $I \& B$. With this new background knowledge, (2) becomes:

$$\Pr(T|P \& I \& B) = [\Pr(P|T \& I \& B)/\Pr(P|I \& B)] \Pr(T|I \& B). \tag{4}$$

This, however, wouldn't do. As we have said, idealizations are false statements given what we know about the world, which is to say that I is inconsistent with B. And this means that every probability expression in (4) becomes degenerate. Such a result, of course, would be unappealing to Bayesians.[5]

Another suggestion might be to incorporate the information that P was derived from T in conjunction with I, into the background knowledge. That is, to make $T \& I \vdash P$ part of the background knowledge. With this modification, the likelihood of P becomes:

$$\Pr(P|T \& (T \& I \vdash P) \& B), \tag{5}$$

and the criteria in (3) now need to be rewritten as:

$$P \text{ confirms } T \text{ iff } \Pr(P|T \,\&\, (T \,\&\, I \vdash P) \,\&\, B) > \Pr(P|(T \,\&\, I \vdash P) \,\&\, B)$$
$$P \text{ disconfirms } T \text{ iff } \Pr(P|T \,\&\, (T \,\&\, I \vdash P) \,\&\, B) < \Pr(P|(T \,\&\, I \vdash P) \,\&\, B). \quad (6)$$

Now, what value can we assign to the probability expression (5)? The condition clause of (5) demands that we assume that T is true and that T together with I entails P. On the other hand, we know from B that I is false. So the probability we are considering in (5) is the probability of a *true* P (since P is an observation actually made) following deductively from an assumedly *true* T conjoined with a *false* I. A little reflection shows that this can happen only if the errors or falsehoods within I cancel out during the computational derivation of P from T & I. For, if the errors introduced by I do not cancel out one another, we can't obtain a true prediction from a true T conjoined to I, since those errors would cause distortion (from truth) in our prediction. Hence, the likelihood of P as expressed in (5) is, in fact, the probability that the errors caused by I would completely nullify one another during the process of derivation of P from T & I, given all that we know (viz. B) and under the assumption that T is true.

There are a couple of problems with (5). First, can we really assign any probability value to the occurrence of a complete cancellation of errors during the derivation of P? Remember that the reason we resort to idealizations in the first place is because we lack the ability to quantitatively handle all the variables involved in the case. When idealizing assumptions are imperative computationally, a quantitative grasp of the interaction of errors they create is likely to be beyond our reach. In such cases, the best we have got are hunches to the effect that the distortion of our prediction resulting from our idealizations and approximations must be "insignificant" or "negligible," but we are far from being able to turn such hunches into determinate probability distributions for the range of errors. In response, Bayesians could argue that even though it may be infeasible for real people to come up with the requisite probability assignments, e.g. to total error cancellation, those values nevertheless *exist* for a perfectly rational agent (one whose degrees of belief satisfy the axioms of the probability calculus) who is as competent as can be in error probability "guesstimation." Bayesians could suggest that we take the Bayesian criteria of confirmation and disconfirmation to be intended for such idealized agents, we imperfectly rational/irrational mortals being only approximations to them to various degrees.[6]

A more decisive objection to (5), and hence to (6), would be the following. In (5) and (6), the actual observation (represented by the P on the left-hand side of '|') is exactly what is predicted (represented by the P on the right-hand side of '|'). Therefore, (6) restricts the cases of theory testing to those where the prediction from the theory exactly matches reality. But such cases are rare, because, as we have said, a complete cancellation of errors, which must take place if the prediction is to closely match reality, seldom happens. What predominantly happens in actual testing situations is that the theory in conjunction with I entails P_T, which differs from the actual observation P. The confirmation or disconfirmation decision is then made on the basis of the size of the discrepancy between the observation P and the prediction P_T: If the scientist judges the discrepancy to be "sufficiently small," she may consider the theory to be confirmed, and if the discrepancy looks "too large" to her, she may regard it as disconfirmed. So the following version of the likelihood of P would pay respect to the importance of the discrepancy between P and the prediction P_T in theory testing:

$$\Pr(P|T \,\&\, (T \,\&\, I \vdash P_T) \,\&\, B). \tag{7}$$

Accordingly, the Bayesian condition for confirmation can be stated as:

$$P \text{ confirms } T \text{ iff } \Pr(P|T \,\&\, (T \,\&\, I \vdash P_T) \,\&\, B) > \Pr(P|(T \,\&\, I \vdash P_T) \,\&\, B). \tag{8}$$

Similarly for the disconfirmation condition.

Does (8) now finally capture in a Bayesian format the role of idealizations and approximations in testing of theories? I think we can begin to be optimistic. Let us first point out that we can give an alternative formulation to the Bayesian condition for confirmation which is equivalent to that in (8) and is perhaps intuitively more accessible:

$$P \text{ confirms } T \text{ iff } \Pr(P|T \,\&\, (T \,\&\, I \vdash P_T) \,\&\, B) > \Pr(P| \sim T \,\&\, (T \,\&\, I \vdash P_T) \,\&\, B). \tag{9}$$

In (9), '$\sim T$' stands for the negation of the theory.[7] Thus according to (9), P confirms T just in case the fact that $T\&I$ entails P_T renders P more probable on the supposition that T is true than on the supposition that T is false.[8]

An alternative to treating P as the evidence is to take the evidence to be ΔP, where 'ΔP' stands for the magnitude of the discrepancy between P and P_T; that is, ΔP expresses how far off the mark P_T has turned out to be. This would give us the following versions of the likelihood and the confirmation condition, respectively:

$$\Pr(\Delta P|T \,\&\, (T \,\&\, I \vdash P_T) \,\&\, B) \tag{10}$$

$$P \text{ confirms } T \text{ iff } \Pr(\Delta P|T \,\&\, (T \,\&\, I \vdash P_T) \,\&\, B) > \Pr(\Delta P| \sim T \,\&\, (T \,\&\, I \vdash P_T) \,\&\, B). \tag{11}$$

Put in words: A theory is confirmed by an observation just in case the truth of the theory makes the discrepancy between the prediction and the observation more probable antecedently (i.e. before the observation is made), than does the falsehood of the theory, given all that we know.[9]

Conditions (9) and (11) explicitly take into account the essential use made of idealizations and approximations as well as the fact that theoretically-based predictions that utilize them will not, in general, fit the data. (9) and (11) do not require that we know how to compute the magnitude of the observed error ΔP. They only require us to be able to compare the probability of this error on the assumption that the theory is true with the probability of the same error on the assumption that the theory is false. How feasible, in actuality, is it to make such probability comparisons? As we have said earlier, when the computations needed to extract a prediction from the theory are complex, as is so often the case, we cannot quantitatively monitor the behavior of errors in I. Thus, our assignment of a value to (7) or (10) will generally have to be based on hunches and intuitive judgments, and the most we are able to do will be to guesstimate a rough probability for a given amount of discrepancy between our prediction and observation. But using guesstimates and intuitive determinations is an essential part of the Bayesian methodology, as when appraising the prior probability of the hypothesis or the expectedness of the prediction.

I think our revised version of the Bayesian strategy of theory testing as expressed in (9) or (11) is not at all unworkable–even for real agents. In many circumstances, we find the explanation of a phenomenon by an hypothesis more credible than an explanation of that phenomenon by the negation of that hypothesis. In those cases, we think the phenomenon would have a higher antecedent probability to occur, were the hypothesis true, than were it false.

Let us illustrate this point with the following (admittedly not scientifically sophisticated) example. Suppose I am shooting at a target in a shooting gallery with my new gun. After a round of shooting, the distribution of the bullet holes on the target board suggests to me the hypothesis or "theory" that there is something wrong with the aiming mechanism of my gun. What leads me to think so is that the average distance of the bullet holes from the bull's-eye seems larger than it would be if my gun were not defective. Here, the normal-gun hypothesis is T, the defective-gun hypothesis is $\sim T$, the bull's-eye is the predicted location of my hits, P_T, and the observed average deviation of the bullet holes from the bull's-eye is ΔP. Taking into consideration all the factors which I believe to be normally responsible for my small deviations from the bull's-eye, it seems to me that the discrepancy between my prediction and my actual observation is so large that it cannot convincingly be explained except by a manufacturing defect in the gun. (My neglecting some of the said factors enabled me to derive P_T from T. For example, I assumed that the gun didn't kick and that the bullets followed a linear path after my firings. Without such "idealizations and approximations," it would be impossible for me to derive P_T from T.) Hence the truth of my defective-gun hypothesis, $\sim T$, accounts for the data better than does the truth of the normal-gun hypothesis, T. This is what makes me think that the data confirm the defective-gun hypothesis. These judgments have a translation in the language of subjective probability: The truth of my defective-gun hypothesis would make my antecedent degree of belief in (or expectation of) the data, ΔP, stronger, under the circumstances, than the truth of the normal-gun hypothesis would. In probabilistic terms, $\Pr(\Delta P | T \& (T \& I \vdash P_T) \& B)$ is smaller in this example than $\Pr(\Delta P | \sim T \& (T \& I \vdash P_T) \& B)$.

A final emendation is called for. As Laymon points out, when deriving a prediction from the theory, the scientist sometimes simplifies the theory itself by using idealizing assumptions in order to reduce the computational difficulties. But as a consequence of such simplification, the theory becomes transformed into something else. A familiar example of this kind of transformation of the theory to be tested is found in a standard derivation of the period of a simple (i.e. idealized) pendulum. Newtonian mechanics gives the equation of motion for a simple pendulum as $ml(d^2\theta/dt^2) = -mg\sin\theta$, where m is the mass of the pendulum bob, l is the length of the suspension cord, θ is the angular displacement of the bob from the vertical, and g is the gravitational acceleration. Let this equation stand for the theory T we want to test. The solution of the equation is facilitated if we make the simplifying assumption $\sin\theta = \theta$, which is approximately correct for small angles θ when the angle is measured in radians. We use this assumption to modify the theory into $ml(d^2\theta/dt^2) = -mg\theta$. The solution of this simpler equation yields $p = 2\pi(l/g)^{1/2}$ for the period p, which can then be compared with observed values of the period (for small angles of oscillation). These data are taken by the scientist to be testing the original theory T, although it was the transformed theory $ml(d^2\theta/dt^2) = -mg\theta$, or T', that yielded the prediction $p = 2\pi(l/g)^{1/2}$.[10] Relaxing our confirmation criterion to allow for such cases of theory testing, we obtain the following most general expression of the Bayesian incremental-confirmation condition:

$$P \text{ confirms } T \text{ iff } \Pr(P|T \,\&\, (T' \,\&\, I \vdash P_T) \,\&\, B) > \Pr(P| \sim T \,\&\, (T' \,\&\, I \vdash P_T) \,\&\, B), \qquad (12)$$

or alternatively:

$$P \text{ confirms } T \text{ iff } \Pr(\Delta P|T \,\&\, (T' \,\&\, I \vdash P_T) \,\&\, B) > \Pr(\Delta P| \sim T \,\&\, (T' \,\&\, I \vdash P_T) \,\&\, B), \qquad (13)$$

where T' may or may not be the same as T.

Middle East Technical University

4. NOTES

1 For an illuminating discussion of the serious problems that the use of idealizations present for the h-d approach to confirmation, see Laymon 1985, pp.147-155. My exposition of some of these problems in what follows largely draws from Laymon 1985.

2 Let me register one caution here. The talk above of *cancellation* of errors actually makes sense only if T is a fairly complex theory capable of dealing with a number of factors or variables. To see this, consider an experimental setting in which I am trying to predict the trajectory of an electron emitted by an electron gun. Let us ignore the quantum-mechanical complications for the sake of simplicity and assume that the Newtonian theory can be used for the purpose. So the inputs I need are the initial values of the position and the velocity vector of the electron plus the net force acting on the electron as a function of time. The net force is the vector sum of all the forces acting on the particle. Suppose that only two forces are present in our setting: the gravitational force and another force on the electron caused by a magnetic field. As it happens, the magnetic pull on our electron is equal to the gravitational force on it and opposite in direction throughout the period of time under consideration. Suppose further that, to simplify the prediction process, I employ an approximation or simplification in accordance with which I ignore both the gravitational and the magnetic fields. Thus I assume that there is no force acting on the particle after it leaves the electron gun. The (linear) trajectory I predict for the electron comes out correct, because the gravitational and the magnetic forces on the electron annul each other, resulting in a zero net force, which happens to be

what I assumed. So here we have a true theory, viz. Newtonian mechanics, yielding a true prediction thanks to an adventitious cancellation of the two erroneous assumptions I made.

Sometimes, however, a true law or theory may yield truth when conjoined with falsehood, even though there occurs nothing we can describe as cancellation of falsehoods. This kind of situation can arise with laws that are of relatively simple and nonquantitative nature, for example. Thus consider the simple true law "All copper wires conduct electricity," and the statement "This is a copper wire." Suppose the latter is false, because the wire in question is made of aluminum. The two statements entail the true conclusion "This wire conducts electricity," but not via any cancellation of errors; for only one error is involved here, viz. the supposition that the wire is made of copper. When a true theory conjoined with falsehoods entails a P which turns out to be true, this circumstance can be described in more general terms as "accidental" or "coincidental" entailment of truth by falsehood. Cancellation of errors is one way by which that can happen. For simplicity of expression, I shall continue to talk of "cancellation of errors," rather than of the more general phenomenon of "accidental entailment of truth by falsehood."

[3] I believe the kind of problems Laymon displays for the h-d model (as well as for the modified versions of that model that employ the notion of "approximate truth" in place of "truth" in the (HD) schemata) also haunt Glymour's bootstrapping theory of confirmation as found in Glymour 1980, Ch.5. "Computations" in Glymour's sense must typically employ idealizations and approximations, and we often have no way of tracing computationally the behavior of the errors caused by them. This makes the bootstrapping procedure of theory testing more complex than Glymour seems to realize, as he talks very little about errors, idealizations, and approximations. For Laymon's critique of the bootstrapping account, see Laymon 1983.

[4] Such idealizations and simplifications may include the assumptions that the bob of the pendulum is a point-mass, that its suspension string is weightless, that the angle of swing is small, etc. These assumptions are resorted to in introductory physics textbooks to facilitate the otherwise difficult derivation of the period of the pendulum from Newton's laws.

[5] Laymon makes similar points in an attempt to criticize the Bayesian theory of confirmation. In a recent brief treatment of the issue, he notes the following:

> Bayesian accounts fare no better [than the h-d accounts]. Distinguishing... between the underlying theory and the idealizations needed for an actual calculation, Bayes' theorem takes the form,
>
> $$\mathrm{P}(t\,\&\,i|e) = \frac{\mathrm{P}(e|t\,\&\,i)\mathrm{P}(t\,\&\,i)}{\mathrm{P}(e)}$$
>
> where e represents the empirical evidence[, t is the theory, and i contains the idealizations and approximations used]. Because the idealizations are false, $\mathrm{P}(t\,\&\,i) = 0$. Therefore, the evidence cannot affect the probability of (t&i). This is just the Bayesian analogue of the problem caused by the use of idealizations for hypothetico-deductivism. Consider also the typical case in science where the theory-produced prediction is false because of the distortion introduced by the idealizations. In such a case $\mathrm{P}(e|t\,\&\,i)$ will be zero. So once again, there will be no change in the probability of (t&i). Trying to avoid such difficulties by separating i from t and conjoining it with e [i.e., the suggestion we have considered above] leads to similar disappointments for Bayesians. (Laymon 1998)

Now, there is no denying that Laymon's rendering of the posterior and prior probabilities will be a nonstarter. But no wise Bayesian would take observation e (P in our notation) to be testing the *conjunction* of the theory with the idealizations, as the falsehood of the idealizations is certain. The Bayesian could focus on the posterior and prior probabilities of t rather than those of t&i. And, as we shall see, the prospects of that working out are brighter than Laymon seems to think.

[6] Cf. Howson and Urbach 1993, pp. 420-423.

[7] We can show that the confirmation condition in (3), viz. $\Pr(P|T\,\&\,B) > \Pr(P|B)$, holds if and only if $\Pr(P|T\,\&\,B) > \Pr(P|{\sim}T\,\&\,B)$ holds (provided $\Pr(T|B) \neq 1$). First, let us note that $\Pr(P|B) = [\Pr(P|T\,\&\,B)\Pr(T|B)] + [\Pr(P|{\sim}T\,\&\,B)\Pr({\sim}T|B)]$. Substituting this in the confirmation condition in (3), we have, $\Pr(P|T\,\&\,B) > [\Pr(P|T\,\&\,B)\Pr(T|B)] + [\Pr(P|{\sim}T\,\&\,B)\ \Pr({\sim}T|B)]$. Rearranging, we obtain $\Pr(P|T\,\&\,B)[1 - \Pr(T|B)] > \Pr(P|{\sim}T\,\&\,B)\Pr({\sim}T|B)$. Since $\Pr({\sim}T|B) = 1 - \Pr(T|B)$, cancellation yields, assuming $\Pr(T|B) \neq 1$, $\Pr(P|T\,\&\,B) > \Pr(P|{\sim}T\,\&\,B)$. The proof that (8) holds if and only if (9) holds proceeds along the same lines.

[8] T does not have to logically entail P for the confirmation condition (9) (and the other versions of the confirmation condition we shall give below) to work. Even if $T\&I$ implies P_T in a weaker sense than logical entailment, (9) still applies.

[9] The occurrence of '$T \& I \vdash P_T$' in the condition clauses of the probability expressions in (7)-(11) above might be found questionable. Since $T \& I \vdash P_T$ is a logico-mathematical truth, or some kind of tautology, isn't it a redundant element in those condition clauses? The question here is whether Bayesian agents are to be construed as logically omniscient, so that all logical entailments are always part of their background knowledge. One line of response to the "problem of old evidence," a challenge raised by Glymour (1980, pp. 85-93) against the Bayesian approach to confirmation, has led to a controversy over whether logical entailments can be taken as new knowledge and hence as evidence by Bayesian rational agents (see, e.g., Earman 1992, Ch.5; Garber 1983; Howson 1991; Zynda 1995), which would mean that those agents are not logically omniscient. Since we are treating $T \& I \vdash P_T$ not as evidence but as part of the background knowledge, our formulations of the confirmation criteria are not affected by that controversy. If the rational agents are supposed to be logico-mathematically omniscient, then $T \& I \vdash P_T$ is and has always been part of B. (Then why didn't the rational agent know all along that T was confirmed? Well, because the evidence P wasn't available to her all along.) If the rational agents are not supposed to be logico-mathematically omniscient, then the discovery that the entailment $T \& I \vdash P_T$ holds must figure as a new additional piece of information to B. Which way the rational agent ought to be construed does not matter as far as we are concerned; either way, $T \& I \vdash P_T$ is assured to be part of the background knowledge.

Let us note incidentally that our proposed versions of the confirmation criterion in (9) and in (11) are both vulnerable to the problem of old evidence. For if P is a long-known fact at the time of the introduction of the theory T, then the information $(T \& I \vdash P_T) \& B$ contained in our condition clauses is sufficient to render the likelihoods and expectednesses of both P and ΔP too close to 1.

[10] The example is taken from Laymon 1985, pp. 151-153. Notice that testing of a theory by modifying the very theory to be tested would be problematic for the standard h-d and the bootstrapping approaches.

5. REFERENCES

Earman, John, 1992, *Bayes or Bust?* (Cambridge and London: The MIT Press).

Garber, Daniel, 1983, "Old Evidence and Logical Omniscience in Bayesian Confirmation Theory," in John Earman (ed.), *Testing Scientific Theories* (Minneapolis: University of Minnesota Press), pp. 99-131.

Glymour, Clark, 1980, *Theory and Evidence* (Princeton: Princeton University Press).

Howson, Colin, 1991, "The 'Old Evidence' Problem," *British Journal for the Philosophy of Science* **42**, 547-555.

Howson, Colin, and Urbach, Peter, 1993, *Scientific Reasoning: The Bayesian Approach*, 2nd ed. (Chicago and La Salle: Open Court).

Laymon, Ronald, 1983, "Newton's Demonstration of Universal Gravitation and Philosophical Theories of Confirmation," in John Earman (ed.), *Testing Scientific Theories* (Minneapolis: University of Minnesota Press), pp. 179-199.

Laymon, Ronald, 1985, "Idealizations and the Testing of Theories by Experimentation," in Peter Achinstein and Owen Hannaway (eds.), *Observation, Experiment, and Hypothesis in Modern Physical Science* (Cambridge: The MIT Press), pp. 147-173.

Laymon, Ronald, 1998, "Idealizations," in *Routledge Encyclopedia of Philosophy*, Version 1.0 (London and New York: Routledge).

Zynda, Lyle, 1995, "Old Evidence and New Theories," *Philosophical Studies* **77**, 67-95.

A.M.C. ŞENGÖR

REPEATED INDEPENDENT DISCOVERY AND 'OBJECTIVE EVIDENCE' IN SCIENCE: AN EXAMPLE FROM GEOLOGY[1]

> Out yonder there was this huge world, which exists independently of us human beings and which stands before us like a great, eternal riddle, at least partially accessible to our inspection and thinking. The contemplation of this world beckoned like a liberation, and I soon noticed that many a man whom I had learned to esteem and to admire had found inner freedom and security in devoted occupation with it. The mental grasp of this extrapersonal world within the frame of the given possibilities swam as the highest aim half consciously and half unconsciously before my mind's eye. Similarly motivated men of the present and of the past, as well as the insights which they had achieved, were the friends which could not be lost. The road to this paradise was not as comfortable and as alluring as the road to religious paradise; but it has proved itself as trustworthy, and I have never regretted having chosen it.
>
> Albert EINSTEIN, 1949

1. INTRODUCTION

The purpose of this paper is to show, from the viewpoint of a geologist interested in the history and philosophy of his subject and on an example from the history of geology, that evidence matters in science and that without evidence there can be no science. In this paper I summarize the history of a discovery repeated four times, independently of one another, at different times and in different places, within different theoretical contexts, during the twentieth century. It is the history of the discovery and rediscovery again and again of mélanges, a chaotically mixed group of rocks indicating an environment of intense shear deformation[2]. This summary is taken from a larger paper on the same topic, in which the details of the geological arguments are given (Şengör, 2003), which are largely omitted here. The discoverers of mélanges have worked in cultural environments different from one another at different times in history and, at least one, in a completely different theoretical context from the rest. Two of the heroes of the story even come from two different non-western cultures: One is Turkish and the other is Chinese.

The history related below is embedded within the context of European science. I here make no attempt to compare European science with what has been called

G. Irzık & G. Güzeldere (eds.), Turkish Studies in the History and Philosophy of Science, 113-136.

science in other cultures (mostly by students who grew up within a European social and scientific context!). The sole purpose of the present contribution is to show that a dialogue between man and Nature outside him is possible and constitutes the essence of science.

2. ON THE POSSIBILITY OF OBJECTIVE EVIDENCE

Before I summarize the historical data, I wish to say a few words about objectivity, as it is the possibility of objective evidence that is denied by those who think that science is just a social construct. In its shortest and most comprehensive definition, objectivity is independence of individual whim (Popper, 1935, p. 16; 1980, p. 44; 1994, p. 18). In view of the human incapability of gathering perfect data concerning the world outside us, because of a variety of factors such as the imperfection of our sense organs to reproduce a faithful picture of the stimulants, or the in-built or acquired biases, distorting whatever message our sense organs give us, the claim that we can acquire knowledge independent of our imperfections must, at first, be thought surprising. The first step in understanding how objectivity is achieved is to realize that no one person can possibly be objective. The knowledge an individual acquires is beset with all sorts of distortions. These distortions, however, are different for different persons owing to the different kinds and/or degrees of their imperfections, and the differences in their biases, in processing information. When such a group of 'imperfect individuals' with diverse biases regards an object or evaluates a statement, there is always a large margin of disagreement as to what is being regarded or being evaluated. However, there is usually also a significant *area of overlap*. Science, indeed all rational life, takes its departure from this very *area of overlap*. Contestants disagreeing on the margins of the area of overlap seek to remedy their imperfections with a view to enlarging the area of the overlap *on the basis of what is contained in it*.

Microscope, for instance, is built on principles constituting an area of overlap amongst many people. But it helps to *enlarge* another area of overlap on the nature of microscopic objects, which, otherwise, would have been only much more indirectly perceived and most likely differently conceived by different individuals. Similarly, on the basis of the common understanding of physics, forming an area of overlap, even the fiercest critics of scientific objectivity do not step out of windows on the top floor of the Empire State Building in New York or go on expensive safaris with the hope of enjoying views of live dinosaurs.

This appreciation of the nature of objectivity (Popper 1957, p. 155-156; 1966, esp. p. 217-219; also see 1935, pp. 16-19; 1980, p. 44-48, esp. note *1 on p. 44; 1983, p. 48; 1994, pp. 18-21 and esp. note *1) emphasizes the indispensability of the evidence and the importance of the presence of a scientific *community, amongst the members of which an area of overlap of agreement on data can be achieved*. There can indeed be a one-man science, if that man had perfect sense organs and were moreover perfectly rational: but even then it would progress very much more slowly than if it were in a community. However, there can be no science of monads unable to communicate with their surroundings.

3. ROLE OF EVIDENCE IN TESTING HYPOTHESES BORN OF CONJECTURE

It was Karl Popper's great insight that most science must forever remain conjectural *and yet all science is capable of progress* (Popper, 1935, 1980, 1994). Popper was able to show this by deducing from Hume's Law (Grice, 1970; cf. Şengör, 2001) that although universal statements can never be verified by any number of favorable instances, they can be falsified by a single contrary instance. This demonstrates the importance of evidence in evaluating opinions on Nature, which are called scientific hypotheses. Once a hypothesis is falsified on the basis of observation reports, the new data bank includes all the observations the falsified hypothesis had been able to explain, plus those new observations that falsified the hypothesis. Therefore any new hypothesis must be able to account for all the older observations *plus* the new ones that defeated the older hypothesis. This means that the new hypothesis has a broader database; it is able to explain more than its predecessor[3]. This is the extent of scientific progress we can hope for. In only individual and trivial cases can we get more: for example, in some cases, we can recover ancient objects in their entirety, thus obtaining a non-hypothetical, 'complete' knowledge about what they are. But all interesting scientific issues are doomed to remain conjectural.

Yet this conjectural knowledge is capable of progressing (not just changing) in the sense of being able to explain more and more of the universe in which we live. This progress has two components: explanatory hypotheses and observation reports. In schools and universities students are taught the methods of observation. However, there can be no corresponding instruction on how to generate hypotheses as Einstein also pointed out in a well-known passage (Einstein, 1981, 110-111). Any way to come up with an explanatory account about a set of observations is legitimate, so long as it can be inconsistent with observation, i.e. testable. This is to say that the hypothesis ought to be able to come into contact with the world outside us. Evidence, i.e. observations, provide that contact and the scientific community ensures the objectivity of the observations.

It is of course possible that not only individuals, but also entire communities may have ulterior motives in supporting or suppressing this or that view (cf. Martin et al., 1986). If it were not so, Lysenko's genetics could not have survived in the Soviet Union as long as it did (see esp. Graham, 1993, ch. 6). Neither could the Church suppress science for such a long time as it did in the early middle ages (Eicken, 1887). What makes it impossible even for large communities (whole nations, empires) to impose their whim on the opinion of the entire human society is the inevitable chance discovery of, and infiltration of reports about, contrary instances.

Some philosophers of science have tried to establish a symmetry between falsification and verification. They pointed out that falsification is in as much need of a 'final decision' that something must be true as is verification. On this basis they have claimed that Popper's asserted asymmetry between falsification and verification was illusory (e.g. Lakatos, 1970; Chalmers, 1990). I think this is mistaken and results from what essentially amounts to denying the possibility of communication. But this is not the place to get into a discussion on this.

Others have claimed that psychology and sociology of research are probably equally as decisive as its logic in gaining theories acceptance or rejection (Kuhn, 1970, 1990, 1992; Feyerabend, 1988, 1991; see Laudan et al. 1986, for a summary of these and some other anti-Popper views; also see Gavroglu, 1994, on the importance of discourse and O'Hear, 1995, on some further criticism of Popper's views). I find all these criticisms expressing certain truths, but they do not undermine Popper's primary thesis. As I pointed out elsewhere (see Şengör, 2001, endnote 8), that nearly the whole of the early Middle Ages in Europe thought nonsense about the geography of the earth was a socially-conditioned phenomenon brought about by the tragedies of the last days of Western Rome including the rise of Christianity, although this does not detract one iota from the vast superiority of Ptolemy's *Geographike Uphegesis* of the 2nd century AD over those later thoughts. The superiority of the *Geographike Uphegesis* is established simply by its better agreement with observation (which does not, however, make it 'right'). The Arabs generated a much better map than Ptolemy in the 9th century AD by criticizing it on the basis of their own observations (Sezgin, 1987, 1993, 2000, esp. ch. I, section D). That this remained unknown to the west until about the 13th century was also a socially-determined circumstance, but this does not make the Arab geography any less superior in the late Middle Ages. As soon as some sensible Portuguese got news of it, they put it to very good use. Below, when I speak of the superiority of some scientific views with respect to others, I refer exclusively to the degree they correspond with observation and not to the number of their adherents or to their geographic spread. I thus find Kuhn's statement that 'As in political revolutions, so in paradigm choice—there is no standard higher than the assent of the relevant community' (1970, p. 94) totally untrue (also see Kuhn's unconvincing Machette Lecture, where he tries to support his case with inapposite similes and metaphors instead of documenting it on a real example with compelling detail: Kuhn, 1977; see also Notturno, 1999, p. 46). By Kuhn's criterion (that has five sub-criteria), none of the great pioneers of science could possibly have done their pioneering job, as it was by definition done against the prevailing opinions among their peers. They chose their paradigms not on the basis of a popularity contest, but on the basis of the compatibility with the relevant observations. When the plate tectonics 'revolution' (to use Kuhn's inapposite terminology) was underway, the majority of the community was hostile to it and yet its creators were not deterred by this hostility, because they had data that none of the older theories could explain. Plate tectonics explained all those plus the older observations. As the new data and interpretations were repeatedly published and discussed, the community gradually came round.

Although many psychological factors enter into observation, theory building and testing, ultimately it is observations that decide the fate of theories, notwithstanding all the 'discourse' and all the communities that may be involved. In the following paragraphs I relate the repeated discovery of mélanges with a view to showing that the discoverers all shared an area of overlap created by their observations. I emphasize how the discoverers struggled with what to them seemed extraordinary, strange and unique instances and how evidence led them, step by step, to identical conclusions independent of each other. In this step by step approach, they were all led by their own conjectures and their ability to face refutations brought about by the evidence. In the end, they and the community agreed that they were all looking at the same sort of thing that had not been commonly known before. Thus, common knowledge was increased. This increase had nothing to do with social context or cultural milieu or historical contingency except in the truism that without previous geological discoveries, mélanges could not have been discovered.

4. DISCOVERY OF MÉLANGES

Initial discovery: Edward Greenly (1895-1919)

Mélanges were recognized later than other kinds of structural rocks (Şengör and Sakınç, 2001), because the understanding of their nature grew out of the appreciation of the significance of what in the Scottish Highlands had been called a crush breccia in the late nineteenth century, a broken-up rock formed by crushing under an advancing thrust sheet. Edward Greenly (1861-1951) was the one who invented the concept (and the term) of mélange, while he was mapping in Anglesey, Wales (Greenly, 1919a, b). He had come to Anglesey after working for the Geological Survey of Great Britain in 1895. He had been a member of the legendary Scottish Highlands team (Strahan, 1919; Greenly, 1928-1932; Oldroyd, 1990) and was thus closely familiar with the concept of structural rocks associated with thrusting (Lapworth, 1883, p. 121; 1885, pp. 558-559). The high state of disruption of the rocks, which Greenly called the Mona Complex (Greenly, 1919a, p. 39), had been known already in the last quarter of the 19th century (see the review in Greenly, 1919a, pp. 1-13). Matley (1913) had called them 'crush-conglomerates' and 'crush breccias' using the terminology developed in the Highlands. Greenly compared them with what he thought were similarly produced crush-conglomerates from the Isle of Man (1919a, p. 65). He knew that he was looking at structural rocks, of the kind that had become so familiar to the workers of the Survey in the Scottish Northwest Highlands, but it was the much larger scale on which they occurred that induced Greenly to give them a new name. Greenly distinguished blocks, slivers, fragments floating in a highly sheared 'matrix'.

Greenly's description of mélanges depicts a tectonically disrupted and internally strained phyllite-sandstone sequence, to the extent that the original sedimentary geometries can no longer be recognized. He emphasized that although several members within a mélange could be mapped separately—if they are large enough to be shown on the chosen mapping scale—where this is not possible, the entire mélange should be mapped as a single unit (Greenly, 1919a, p. 66, note 1). Greenly's inferences have stood the test of time almost intact. Shackleton (1954) has shown that the sequence Greenly had inferred in the Mona Complex was upside down, but his inferences concerning the mélange and the mechanism of its formation remain unassailed. Shackleton (1969) believed that Greenly's mélange must have formed as a submarine slide breccia. This stemmed from his inability, in pre-plate tectonic days, of conceiving how such a chaotic mixture of rocks could be produced tectonically and then get overlain by sedimentary rocks. Shackelton correctly recognized that all kinds of environments were represented in Greenly's 'general mélange' and that all were brecciated (Shackleton, 1969, pp. 9-10). Where extreme deformation did not destroy original boundaries, the breccia contours were seen to be angular. The sizes of the blocks ranged from microscopic to several km. Shackleton could not causally associate the disruption and the mixing to any set of recognized structures affecting the mélange body in its entirety (*ibid.*, p. 10; also see Wood, 1974). He thus concluded that only a submarine slump could explain its origin. This was the common opinion before plate tectonics (with the exception of Bailey and McCallien, 1950a, 1950b, 1953, 1961, 1963: see below; also see Sir Edward Bailey's opinion on the tectonic origin of the Anglesey mélanges in Wood, 1974, p. 335), but once surfaces of displacement could be imagined along which movement could be measured by thousands, or indeed tens of thousands of kilometers (along subduction zones), all the characteristics that had earlier stumped the students of mélanges could be easily explained.

The important thing here for the topic of theory independence of observations and inferences is the following: Greenly was a geologist of the heyday of the discovery of great nappes, i.e. of structures of large horizontal displacements. The larger the amount of horizontal displacement discovered, the greater was the splash a geologist would make. He thus interpreted the Anglesey mélanges as essentially giant crush breccias formed by colossal thrusting. But his fundamental inference, *on the basis of the internal evidence of the mélange*, was that the mélange was a body of rock formed by a high state of disruption and mixing, in a shear environment, of pre-existing rock bodies. The thrust interpretation was a theoretical garment he placed on his fundamental observations and inference.

Robert Shackleton grew up at a time of reaction against immense horizontal motions. He was more of a field geologist (and an excellent one) than a theoretician. When he mapped Anglesey, he observed the same rocks as Greenly and, apart from changing his stratigraphy, he made the same fundamental inference about the mélange. However, he did not like Greenly's theoretical garment for it, so he changed thrusting to gravity sliding. Note that this change altered nothing on how the mélange was mapped and how it was interpreted using its own internal evidence.

First rediscovery: Levi Noble (1941)

'Chaos structure' is a term introduced by the United States Geological Survey geologist Levi Fatzinger Noble (1882-1965) in 1941 to describe the great confusion of rock types encountered in the Black Mountains, in the southern part of the Amargosa Range to the northeast of the Death Valley in eastern California. In describing it Noble used terms very similar, in some instances identical with those employed by Greenly. The depiction of the chaos structure is nothing more than the re-invention of the concept of mélange in complete ignorance of Greenly's work. Noble described its chief characteristics in the following words:

(1) The arrangement of the blocks is confused and disordered—chaotic.
(2) The blocks, though mostly too small to map, are vastly larger than those in anything that could be called a breccia; most of them are more than 200 feet in length, some are as much as a quarter of a mile, and a few are more than half a mile in length.

(3) They are tightly packed together, not separated by much finer-grained material.
(4) Each block is bounded by surfaces of movement—in other words, each is a fault block.
(5) Each block is minutely fractured throughout, yet the original bedding in each block of sedimentary rock is clearly discernible and is sharply truncated at the boundary of the block. Commonly the bedding, even of incompetent beds, is not greatly distorted.

> None of the geologic terms in common use appear exactly to fit this mosaic of large tightly packed individual blocks of different ages occupying a definite zone above a major thrust fault. *The feature suggests a fault breccia on a cyclopean scale, yet it is not a fault breccia in the orthodox sense.* Although it is a thrust plate, it is shattered over large areas to a degree that appears to be unique. *But more important than these considerations is the fact that the great areal extent of the feature makes it impossible to map separately the geologic units of the mosaic and makes it necessary therefore to treat the assemblage, despite its heterogeneity, much as a geological formation be treated. Like a formation, then, it requires a name that will indicate both its type locality and its character.* "Amargosa does the one, for the Black Mountains, in which Virgin Spring lies, form part of the Amargosa Range; and "chaos" as the preceding paragraphs have tried to show, does the other. (Noble, 1941, pp. 963-965, italics mine).

Note that entirely independently of Greenly, Noble also made the analogy of the chaos structure with fault breccias, but noted that the scale was 'cyclopean.' He too noted that its individual rock types could not be mapped separately and the chaos had to be mapped as a unit. Noble clearly noticed that the whole thing was related to shearing on an immense scale and, following the fashion of his day, assumed that the generative fault was a thrust.

Later Wright and Troxel (1969, 1984) showed that the thrust interpretation was wrong and that the faults creating the chaos were normal dip-slip in nature, bringing younger rocks over older. They noted that mappable bounding faults in the chaos always omitted section and assumed that the chaos was a product of coalescing listric normal faults, as Gilbert had depicted nearly a century earlier in the same place (Gilbert, 1875, his fig. 12). Wernicke and Burchfiel (1982, p. 109) pointed out, however, that the same effect can be produced along planar normal faults of large displacement. Their model is more attractive as it allows much greater magnitudes of displacement to lead to chaos formation. Wright and Troxel (1984, plate I, cross-section A–A′) appear to have adopted this suggestion.

Now let us look at the chaos from the viewpoint of our central theme of independence of observations from theories: Noble was unaware of Greenly's work. Yet, when confronted with a similar rock type, he invented the same interpretation to explain them. We do not of course know how many hypotheses Noble had initially formulated and then discarded to land finally on the chaos interpretation. But we can safely assume that he did not at once create the chaos interpretation. He must have first tried to map the individual members within the chaos as ordinary stratigraphy (this is what any geologist would first try). Failing in that, he must then have tried to identify fault-bounded packages (if stratigraphy fails it is usually because of later deformation; so the geologist tries to define what sort of deformation it is. In Noble's area, fault repetition would have been the obvious first choice). Not being able to find fault packages with coherent internal stratigraphy, Noble must have despaired and tried different fault arrangements. He probably found that whatever fault arrangement he assumed he always was able to find shear zones compatible with it. He finally must have realized that the entire rock body was riddled with shear zones and broken up along them. What he discovered was essentially the same thing as Greenly's mélange, and, as both Noble and Greenly wrote in English, it is remarkable that they often chose the very same words to describe homologous elements in their respective study areas.

From the perspective of our present-day understanding, we know that there are differences between mélange and chaos. The two main differences between the mélanges that form along subduction zones and the chaos structure forming along large normal faults are the amount of displacement along the generative fault zones and the evolution of the ambient pressure/temperature regimes as the rocks evolve. Along subduction zones entire palaeogeographic realms disappear and their representatives are now only encountered as exotic blocks within the mélange. Erosion is the only agent that can destroy the record of former environments in an extensional environment and therefore, in principle, chaos has no 'exotic' blocks. Chaos structure evolves in a progressively unroofing environment where ambient pressures and temperatures continuously decrease. Mélanges, especially those along subduction zones, have more complicated paths of evolution because of the vicissitudes of the tectonic regimes reigning in accretionary wedges and subduction channels.

But when Greenly and Noble were mapping, subduction was not known. Neither did they know anything about high pressure/low temperature metamorphism. From the viewpoint of the level of knowledge then, what they found was essentially the same thing and structurally it remains so today.

Second rediscovery: W. J. McCallien, Oğuz Erol and Sir Edward Bailey (1947-1950)

The word mélange did not become popular after Greenly. Even he himself did not use it later when summarizing his Anglesey work (e.g. Greenly, 1922). The word and the concept were revived, when William J. 'Mac' McCallien (1902-1981) and Oğuz Erol (1926—), McCallien's doctoral student in the University of Ankara, were surprised by the jumble of limestones and pillow lavas occurring in a matrix of sheared schists, serpentinites and, in places, even mafic volcanics near Ankara in Central Turkey in the late forties of the twentieth century. As I described the evolution of McCallien's and Erol's thought on the basis of their hypotheses invented to account for their observations in Şengör (2003) in detail, I do not repeat it here. Their train of reasoning was very similar to those of Greenly and Noble and, independently of both, they generated a terminology which is substantially the same as that employed by Greenly and Noble independently of one another! Let us see what Erol, the less experienced of the Ankara team, wrote about their discovery in his thesis:

'The first characteristic that catches the eye in this blocky series, of which we have tried above to outline the rock types, is the presence of large limestone blocks and intermediary material that surrounds them. The sedimentary parts of this intermediary material were sheared, mixed up and the igneous parts were pushed in between various rock types. The "pillow lavas" within the igneous parts are clear evidence for submarine eruptions (Bailey, 1936, p. 1721). The green tuffs must also be of volcanic origin. Thus we found it appropriate to call this mixed belt containing: blocks of possible Permo-Carboniferus (perhaps also Triassic) age, a flysch series whose age is provisionally assigned to the Mesozoic, greywackes thought to be still older, pillow lavas, and other eruptive material, the Elmadağı Blocky Series (= Boulder Bed Series). We do not wish to say much about the origin of these mixed series, which is still a topic of debate. But, it is clear that the blocks within the Elmadağı series are not a result of the "northerly tilting like fish scales of the Elmadağı Massif and its Mesozoic cover during the Cainozoic" as believed by Chaput (1931, p. 83[4]). The generation of the blocks must have occurred in an environment in which the pillow lavas, indications of older (probably Mesozoic) submarine eruptions, could spread, because the blocks and the intermediary material only rarely are in contact with fracture zones. Some of the normal faults in the area cut the Tertiary and are thus younger.

In brief, I wish to stress especially that this series is a "mixed series" consisting of the blocks and of the intermediate material. Although we do not have clear evidence for its age, we provisionally treat it as Mesozoic. The real age can only be established after finding evidence for the age of the intermediary material.

This "Boulder Bed Series," in which other rock types predominate over the green eruptives has been distinguished from the usual "Mesozoic Mixed Series," in which the serpentinites are dominant and which occurs in the southern part of our area.' (Erol, 1949, pp. 21 f.; 1956, p. 18).

This was precisely what Greenly had described, but when Erol committed the above to writing, neither he nor his teacher McCallien had been aware of Greenly's (or Noble's) work. Professor Erol (personal communication, 15th December, 1994) remembers that even before he started his doctoral thesis work (1945-1946) they had started using the terms 'matrix' and 'block' to describe what they were seeing on the basis of an honor's thesis mapping Erol had earlier done. Eventually they even distinguished in the field a sedimentary matrix consisting mainly of slates with subordinate sandstones from an igneous matrix formed dominantly from oxidized mafic volcanics and mafic tuffs. They further often talked about the matrix being 'sheared' forming a 'plastic medium' in which the blocks were 'churned'.

McCallien kept asking Erol to establish a stratigraphy in this mess and Erol kept coming back to him with the plea that it did not seem possible. Finally both agreed that they were looking, in the whole area, at a jumbled mixture of blocks floating in the two matrices. It was in the Çaldağ region that Erol also noted that the blocks and the foliation defined planes dipping to the north. McCallien hypothesized that this probably indicated a north to south tectonic transport. Erol further thought that the non-metamorphic clastic rocks, wedged into the jumbled rocks he was mapping in the Kıbrısyayla district in the Elmadağı region resembled the regulary bedded Liassic clastics in the Yakacık area and correlated them. All this indicated that the area had been highly tectonised and, apparently, during the Alpine orogeny.

Here too, we see a couple of frustrated geologists in an area that refused to yield to an ordinary, layer-cake stratigraphy. Neither McCallien nor Erol had known any other kind of geology and neither was an experienced theoretician. They tried hard to make the area conform to their prejudices, but to no avail. Finally, they tuned down their observations to ever smaller scales and at the end realized that they were looking at a highly broken-up and mixed rock package and the mixing was accompanied by much shearing. They had no idea how the shearing was accomplished, but they inferred that it had to have happened in an environment of deep sea.

Later, McCallien asked his old friend, the renowned British tectonician Sir Edward Bailey (1881-1965) whether he would want to go and tour this extraordinary area to see whether a reasonable interpretation could be formulated. Bailey accepted and their one month's travels around Ankara and Alaca districts in 1950 showed Bailey what his colleague and his student had seen earlier and it was presumably Bailey who remembered that a similar association had been described from Anglesey in 1919 by Greenly. They thus decided to use Greenly's terminology and to call the rock association the Ankara Mélange (Bailey and McCallien, 1950a, b; 1953).

Bailey's contribution was to tell McCallien that a similar thing had also been found elsewhere and had been interpreted in terms of thrusting. Here Bailey invoked the same interpretation. However, the defining characteristics of the mélange, namely its disrupted character, the multifarious nature of its blocks, the various geological environments and ages represented by the rocks and fossils of its blocks and its matrices and the pervasively sheared character of the matrix, that it was probably associated with some sort of thrust tectonism (and that in the Ankara region this thrusting had been from north to south) had been all worked out by McCallien and Erol before Bailey's arrival. One could easily see this by comparing Erol's unpublished doctoral thesis (Erol, 1949) and the famous Bailey and McCallien paper (1953).

Third rediscovery: Kenneth J. Hsü (1963-1968)

Kenneth Jinghwa Hsü (1929—), a native of Yangtze in the Chinese province of Zhejiang, tells how he had reinvented the mélange concept in his account of the history of thought on the Franciscan Complex in California (Hsü, 1985, esp. pp. 57 f.). It was a Swiss geologist, Héli Badoux (1911-2001), who told Hsü that what he thought he had invented had been invented already by Bailey and McCallien in 1950 (evidently Badoux had been unaware of Greenly and Noble) and that it had been called mélange (Hsü, 1985, p. 57). But Hsü's story is incomplete in his 1985 paper. I once had a conversation with Ken in his home in Zurich about independent discoveries and the importance of the evidence in science. Ken later came to Pasadena in California to stay with me for a few days in June 2001, while I was a Moore Distinguished Scholar at Caltech. During that visit, he told me that our previous conversation in Zurich had made him think back and that he had remembered that he had forgotten an episode in his own mélange adventure when he wrote his 1985 paper. He told me about it and I asked him to put it on paper. Below I give in full Ken's letter to me relating the missing bits in his mélange discovery (written on 11th June, 2002):

> Hi Celal
>
> I discovered the mélange phenomenon in 1963, as I wrote previously in my publications, when I first went to the beach to cool down, after several days of utter frustration. Then the evidence was clear. My knowledge of rock mechanics with [*David*] Griggs and [*John*] Handin was the key to my understanding. John had just published some papers on the differential ductility of sandstones and shales. The broken formations are a typical manifestation of the different rock-mechanics behaviors. It is not big a step to postulate their mixing with exotic components to produce mélanges.
>
> I always thought that I would like to have a label. I translated the descriptive term exotic boulder beds into Spanish and used that in my first Shell report of 1963. After I went to Riverside [*University of California at Riverside*] in 1964, my West Coast friends all expressed an interest to see the structures in the field. I finally arranged in the early summer of 1966 a trip for John Crowell, then at UCLA, Cliff Hopkins, then at UC Santa Barbara and Bob Garrison. Crowell asked me if he could bring Héli Badoux along. I had come to know him in 1964 when I was detained by the French immigration, because I had tried to leave the country without a visa (having entered France through mountain-trails). Badoux had to talk to the border police for more than half an hour, and perhaps had to give a small bribe before I was rescued. I had, therefore, no problem to welcome Badoux as my guest.
>
> Our first day was spent on the coast near Morro Bay. There was some red rocks which no American geologist knew what they were. I had had the rock thin-sectioned, but could not determine petrographically their mineralogy. I had then received an X-ray report: the rocks consisted exclusively of antigorite. Funny, isn't it? Badoux, with his experience in the tropics, took one look, and said that it was a serpentinite. There were pelagic cherts and limestones, and pillow lavas around, the serpentine thus completed the third member of Steinmann's Trinity.
>
> For some reason, I went to bed early that evening. John, Cliff, and Bob stayed up late and discussed with Badoux all evening about what they saw. Next morning, Bob Garrison told me at breakfast that he had witnessed at an historical occasion when the term mélange was invented for the Franciscan.
>
> John Crowell then told me of the evening's discussions, and advised me to adopt the term. I remember that he said, a good merchandise should have a good label. The trade mark mélange was just what I needed to sell my ideas to the public.
>
> Only then, did I recall my trip with Ernst Kündig [*1901-1981*] in 1959 or 1960. Kündig had worked many years for Shell to look for oil in "eugeosynclinal terranes." [*see Kuendig, 1959*]. When he was about to retire as Shell's Chief Geologist, he received, as his golden handshake, a fulfillment of his lifelong dream to come to California to see the Franciscan, and to get a sample of a glaucophane schist for his friend and mentor Professor Scheumann. Nobody in Shell knew anything about the Franciscan. I was one of the few who had come from the West Coast. They gave me a week or ten days to contact the people at Menlo Park and to find out where to find good outcrops. I went to reconnaitre in April, and the Kündigs arrived in May. We flew then to LA, rented a car and drove up to the Morro Bay country.
>
> Those were the days when everybody was afraid to go to the beaches. We made only roadside stops. Kündig saw immediately the similarity of the Franciscan rocks to those in the Ankara Mélange, pointing out to me the exotic blocks of radiolarian chert in the meadows on both sides of the highway.
>
> After several days on the coast, we ended up at Big Sur. I remember the evening at a fish restaurant, when Kündig and I argued all evening. He was talking about the eugeosyncline, and was wondering about the missing miogeosyncline. I was presenting my idea that the Franciscan is thrust over the flysch of the Great Valley sequence from the west. (I now think that the Franciscan is underthrust the foreland basin sediments). I thought then that I could do some palaeo-current measurements to verify my hypothesis. Kündig may or may not have used the word mélange for the Franciscan, but he must have talked about the Ankara Mélange in Anatolia. I do not remember whether or not he mentioned Bailey and McCallien. I myself did not read Bailey and McCallien until after I left Shell in 1963. In 1964, there was an AGI summer Institute to the Apennines. John

Maxwell and his Italian friends took us through the argille scagliose [*'scaly shales'*] everywhere. Those were the heydays of the gravity-sliding theory [*see, for example, Maxwell, 1959; Page, 1963*]. The argille scagliose and the Ankara mélange were assigned the same origin of having slid down slope in a deep sea basin. I have always been impressed by the phenomenon of penetrative shearing. It is not pure shear, but simple shear. It is not 50 or 75% like deformed oolite, the shear strain must have been many times more than unity. The shear cannot be caused by compression, the shear must have been caused by the displacement of an overlying (or underlying) slab. The only mechanism which could do the job, in my opinion of those days, had to be an allochthon of very large dimensions. Such an allochthon, after Hubbert and Rubey, 1959, could only have been a gravity-sliding body. I compared, therefore, the Franciscan with the argille scagliose, and compared the origin of Franciscan with the bottom moraine of glaciers. I was troubled, of course, because moraines are not thick deposits, but the Franciscan was supposedly "25,000 feet thick."

I first presented my idea on the Fransiscan structures at the Wegmann Symposium in the spring of 1966 [*Hsü, 1967*]. My structural cross-section clearly shows that I understood the pervasively sheared nature of the mélanges. It is noteworthy that the Italian literature on argile scagliose was referred to, but Bailey and McCallien's Ankara Mélange article was not cited by me in the article published in 1967. The manuscript was probably written before the 1966 trip with Badoux. The Bailey and McCallien articles were cited in my 1968 (Principles of Mélanges in GSA [*Hsü, 1968*]) and 1969 (with Ohrbom) articles [*Hsü and Ohrbom, 1969*]. I believe that I did not read the Greenly memoir until after I took the trip with Badoux, either in Riverside during the 1966/67 Winter Semester, or more likely in the Spring of 1967 when I was writing the manuscript for GSA, at about the same time when I also read Bailey and McCallien. In the early 1970's, I went with Robert Schackleton to Anglesey. Robert, I believe, was Bailey's student, or his assistant at the Geological Survey (or both). He told me that Bailey worked at Anglesey before the War, and he was very familiar with Greenly's idea. (Shackleton knew about mélanges too, but he had difficulty envisioning the shearing mechanism in rock deformation.) It was thus not surprising that Bailey recognized after the War the Ankara rocks as mélange at the first sight (B and McC 1950 [*Bailey and McCallien, 1950a*]). Kündig may have used the term mélanges, but he was a Swiss oil man and he probably did not know of Greenly's work. I did not pay much attention to Kündig's 1959 or 1960 comparison of the Franciscan with the Ankara Mélange, when I had not yet had a chance to study the Franciscan. After I started working on the Franciscan in 1963, I remembered Kündig, but forgot the Ankara Mélange. I "re-invented the wheel" independently, but I was taught the word mélange by Badoux in 1966, and did not appreciate the meaning of the word until I read Greenly in 1967. In a way, Greenly's memoir was a retrodiction of my postulate of a deformation mechanism of rocks that are called mélanges. Going back to the problem of rediscoveries. I think that McCallien may have worked independently without foreknowledge and that he may or may not have recognized the Ankara rocks are mélanges [*Hsü did not know about the story related herein when he wrote this, so he could not be sure that McCallien and Erol indeed independently discovered mélanges*]. Bailey recognized the mélanges, but he certainly did not make a rediscovery, because he had been familiar with Greenly's work at Angelsey. I did make an independent rediscovery, but my use of the term was inspired by Badoux, who was familiar with the argille scagliose which had been compared with the Ankara Mélange. The postulate of Franciscan as subduction mélanges was made, as far as I knew, perhaps by Ben Page to explain the odd relation between the Franciscan and the Great Valley rocks. I would like to give credit, however, to Warren Hamilton as being the first to come out explicitly in print, with his GSA article, which preceded my AGU article (written in 1970 and published in 1971), and which was cited by me in that article.

This turns out to be a rather interesting little story. I believe that I have given everyone their due credit. Don't hesitate to write me again if you need further clarifications.

With best regards, sincerely,

Ken

A few days after he sent me this letter, on 15th June 2002, Ken sent an appendix to his earlier missive in an e-mail, which was no longer possible to put into Şengör (2003). I reproduce it below, slightly edited to sieve out some personal notes and a few errors that had crept into his e-mail:

'Greenly's idea of mélange was much ahead of his time. Mélanges are very common, and they are found in every mountain belt. They have been encountered by many geologists before and after Greenly. They have been called beds of exotic blocks, Wildflysch, "boulder shale beds," argille scaglise, even after Greenly coined the term mélange. Some Alpine geologists did attribute an origin of penetrative shearing to Wildflysch, but none, not even Steinmann, have recognized ophiolite rocks as a unit of tectonic mélange[5] . They prefer the term "eugeosycline." Kündig used that term for Ankara Mélange. My 1971 AGU paper is entitled Franciscan mélanges as model for eugeosynclinal sedimentation and underthrusting tectonics, even when I pointed out that the so-called "eugeosyncline" was not an elongate trough of submarine volcanism, but half of the Pacific Ocean with the ocean crust formed by seafloor spreading.

I wonder why so few structural geologists ever read Greenly's monograph. I was at Angelsey, Greenly must have been helped by the fact that the outcrops on seaside cliffs do not allow any conclusion but penetrative shearing of rocks of different ductility. That his idea was forgotten for more than half a century could be attributed to several factors:

1) The unsound philosophy of geological research: A problem is supposedly solvable by observation only, and is considered solved when a name is attached to it. This approach provides *ad hoc* explanations of natural phenomena, especially unusual phenomena and does not provide a paradigm or comes up with a theory of everything.
2) With their ill-placed arrogance that science only uncovers fact, many establishment geologists fail to recognize that science has no value until it becomes a paradigm, i.e., a speculative theory of everything.

Mélanges are not uncommon, but they were given different names in different places, and the significance of the natural phenomenon from which they resulted was not sufficiently recognized.

Mélange, Wildflysch, argille scagliose were just names. The concept of mélange had little significance to earth scientists except for those who work in mélange terranes, until the idea was developed into a theory to link the penetrative shearing of the mélanges to the consumption of seafloor by subduction in a plate-tectonic model. I admired Janet Watson and her colleagues at the Geological Society of London because they recognized the significance of the theory. I was awarded the Wollaston Medal for my contribution to "the stratigraphy of mountains." This citation annoyed classical stratigraphers of mountains to no end.

Similarly Mendel made some observations as *ad hoc* explanations of some phenomena of inheritances. The Mendelian observations acquired significance only after the theory of combinations of alleles was developed by geneticists.'

It is clear from the above (and from his earlier accounts in 1985 and 1990) that Hsü had agonized over the Franciscan rocks much like Erol and McCallien had done over the Ankara rocks. He finally came up, independently of either Greenly or Bailey and McCallien, with the idea that he was looking at a tectonic jumble, essentially at a mega-fault breccia, but he thought it had been initially mixed in a submarine slide environment (because at the time, Hsü, like Shackleton before him, could conceive of no other process that could create such a thoroughgoing mixture). Once he conceived the process that had created the rocks he was looking at, he was able to map them in an intelligible way and developed a methodology of mapping (Hsü, 1968, 1990), which was essentially the same as Greenly's.

Let us remember that Greenly was able to understand the Mona mélanges owing to his experience in the Northwest Highlands of Scotland, where fault rocks had been mapped in great detail after the recognition of large-scale thrusting and Matley's earlier work, also inspired by the Highlands discoveries. In other words, Greenly first recognized (read: hypothesized!) *a process* and then sorted out the structure. Bailey and McCallien simply used Greenly's hypothesis and experience to explain Erol's and McCallien's observations and deductions. Hsü, having been educated in the more parochial (and more devoutly Baconian) U.S. environment of the fifties, had to reinvent all that through much agonizing only to be told at the end by an Alpine geologist that he had reinvented the wheel (see Hsü, 1985, 1990 and the letter quoted above).

But the great importance of the Franciscan mélanges came when it was realized that they represented the sweep of an ocean floor for at least 1000 km (Hamilton, 1969; Dewey and Bird, 1970; Dickinson, 1970; Hsü, 1971). Such fault displacements had not been hitherto recognized. Nor structural rocks, of the volume of the Franciscan Complex, had been recognized before (Şengör and Sakınç, 2001). With the recognition of this new process, namely subduction, mélanges acquired a great novel significance as indicators of past subduction and they began to be recognized the world over. However, this new significance has altered nothing on their characteristics as first enunciated by Greenly and repeated, independently of him, by the rediscoverers of mélanges.

5. CONCLUSION

In the paragraphs above I briefly reviewed the discovery and repeated independent rediscovery of mélanges. The first discovery occurred shortly after crush breccias were recognised as being associated with great amounts of thrusting in the Scottish Northwest Highlands. As Hsü points out in his second letter to me above, Greenly was led to the conclusion of tectonic mixing because his outcrops were so clear and because he had experience elsewhere with tectonic shearing and breaking up of rocks.

Noble was much in the same situation, though he had less experience than Greenly in the tectonics of orogenic belts. Despite that his observations finally constrained him to the hypothesis of tectonic mixing along large-displacement faults. He assumed that the faults were thrusts, simply because in those days all nearly flat, large-displacement faults were thought to be thrusts. Although his tectonic framework later turned out to be wrong, the process of tectonic disruption and mixing that he recognised has stood the test of time across changes of several theories concerning the tectonic framework of the Amargosa Range.

The situation was different with McCallien and Erol. Although McCallien had been familiar with convergent tectonics from his Highland experience, he had not earlier seen or known about mélanges. However he did have experience with moraines, i.e. with bouldery rocks with a fine-grained, in part sheared matrix (see Şengör, 2003). Erol had no experience whatever. Mélange was the first rock type he ever mapped. McCallien assumed, on the basis of his knowledge and earlier field experience, that in Erol's area a regular stratigraphy could be established. He urged Erol to go out and establish it. Erol struggled long and hard and finally had to tell his teacher that it was not possible. McCallien at first could not understand this. So, the two of them went out together and were baffled together. It was McCallien's insistence on meticulous mapping combined with iterative hypothesis generation and testing as the mapping advanced that finally delivered the key to the structure of the area: Erol mapped numerous blocks floating in a sheared matrix. From the varied composition of the blocks and the matrix McCallien and Erol recognised that they were looking at a terrible mixture. Neither of them had a clue as to how the mixing had been achieved. They played with different hypotheses. But all the hypotheses they tried had a component allowing the mixing associated with shearing to occur and from the north dip of the foliation they thought that a kind of north to south tectonic transport had to have occurred. Finally, when the better informed Sir Edward Bailey arrived, he told them that thrusting had been most likely responsible for the mixing. He knew that from his earlier experience in the United Kingdom: he knew about the Anglesey mélange.

Hsü was working in a very parochial environment in which nobody could advise him as to what he was grappling with. Generations of Californian geologists had been stumped by the Franciscan, simply because it was not a regularly layered rock sequence. Hsü worked hard just as his predecessors did. His situation was similar to those of Noble and McCallien and Erol before him. He had had no experience before in such difficult rocks and did not know the relevant literature. He had forgotten what Künding had told him in 1959 or 60, because he had not been familiar with the problem. That he did not read Bailey and McCallien until almost a decade after Kündig's visit, despite the fact that he had started working on the Franciscan in 1963, shows how completely he had forgotten what Kündig had said.

Hsü had grown up intellectually in an environment that had been not entirely enamoured with large horizontal motions. The American geological world had become less and less comfortable with large thrusts or with large strike-slip faults in the post-war years. The dean of North American tectonicists, Walter H. Bucher (1956), had devoted his presidential address to the Geological Society of America to showing how thrusts could be generated in mountain belts without much shortening and the Cordilleran master Armand J. Eardley had become more of a verticalist between the two editions of his classic *Structural Geology of North America* (1951, 1962, see esp. the preface to the second edition, p. xv). So Hsü had a completely different tectonic framework in his head from that of Greenly or Bailey. He thus opted for gravity gliding for mixing the mélange, because he had been earlier working for the Shell laboratories, where the fluid-pressure-dominated gravity-gliding hypothesis for the motion of thrust sheets had been developed in a celebrated couple of papers by Hubbert and Rubey (1959) and Rubey and Hubbert (1959).

Thus five different people (except for Sir Edward Bailey who knew about Greenly's work), working in different places, at different times and with different tectonic models came up with precisely the same concept of mélanges as rock bodis extensively disrupted and mixed by a mechanism involving great amount of shearing.

Two of these were Britons, one an American, one a Turk and one a Chinaman. At the time of the discoveries they made, none was aware of the others' discovery. As soon as they found out (except Noble, who did not comment on the similarity of his chaos structure to mélange) about the discovery of the others they admitted that it was the same as their own.

There was nothing in their social context that would have led them to their discoveries independent of the evidence. In fact, had they been dependent on cultural and social factors independent of their observations, they would have come up with very different interpretations of what they saw. Their discoveries were made against all the odds of their preconditioning. They all were trained to think that sedimentary rocks occurred in layers, even if folded and thrust. Instead they occurred as blocks of diverse sizes in the mélange. The geologists whose discoveries were related in this paper all were taught that sedimentary, igneous and metamorphic facies had a certain logic to them: certain facies precluded the presence of some others nearby, whilst requiring the proximity of yet others. Instead, mélange violated all that and all facies occurred everywhere without much of a logic. Our geologists all were taught that principle of superposition applied in sedimentary rocks. It did not in mélanges. They all were taught that sedimentary layers continued up to the basin margins allowing correlations. In mélanges sedimentary strata stopped abruptly in mid outcrop with no immediately obvious reason. Yet all the familiar rock types did occur in mélanges. Not all parts of mélanges were disrupted in a similar degree. In places in a mélange region one is commonly misled into thinking that one is not in a mélange terrain. Despite all these extremely adverse conditions, five different people came up with exactly the same interpretation of mélanges in diverse intellectual and cultural environments in different places and times.

Moreover, the mélange interpretation survived all changes of theory: both local and global. It was first conceived broadly within the framework of the contraction theory globally and in a setting of large-scale thrusting locally. When the chaos structure was reinterpreted as being a result of extensional tectonics instead of shortening, the interpretation of chaos as a sheared and disrupted rock body along large-displacement faults remained intact. When the contraction theory was given up and plate tectonics replaced it as the reigning theory of global tectonics, mélanges became one of its key elements in applying plate tectonics to old rocks. Nothing was changed on how they were interpreted on the basis of their internal evidence, except that the faults along which they form now were recognised to have much larger displacements. The mélange concept of Greenly still lives on as a very close approximation to the mélange concept as it is now used. From the retrospect of plate tectonics we now see what a fruitful concept it has been and it remains so. If we could call Greenly, Noble and Sir Edward Bailey back today, they would readily recognise their brainchild in full use. McCallien lived long enough (he passed away in 1981) to see the full-fledged development of plate tectonics and the great rôle mélanges played in it. Erol is still alive and well and enjoys the prominence of mélanges in contemporary geology. Finally, Hsü, one of the greatest geologists alive, fully appreciates Greenly's great discovery and how often it was later rediscovered, as he wrote in his letters I reproduce above. It was he, following the lead of his old friend Warren Hamilton (another giant of contemporary geology), who was responsible in placing mélanges in a plate tectonic context.

Geology is full of such examples of rediscoveries in diverse settings and theoretical contexts. I could have easily chosen the extensional interpretation of mid-oceanic ridges from Molengraaf (1928) through Stille (1937) and Niemczyk (1943) to Heezen (1960) or the discovery of strike-slip faulting from old miners to Arnold Escher von der Linth, Köhler and Eduard Suess (see esp. Köhler, 1885, p. 84, where he wrote that he had been the first to recognise strike-slip faulting in an 1880 paper; but Suess, 1883, pp. 153-154 gives clear evidence that Escher had recognised it already in 1854. Also see Escher von der Linth, 1878, p. 71). On a more theoretical plane the discovery of convection currents in the earth's mantle is a similar story. I chose to talk about mélange because it is such a celebrated discovery and such a widely used concept. In 1990, Hsü's 1968 paper on the third rediscovery of mélanges became a citation classic (Hsü, 1990). Most likely only very few of the authors who cite him now are familiar with the long history of the discovery of mélanges. But the good thing about it is that neither do they need to be to get on with their science. The concept of mélange is independent of its discoverers. It has long become a property of the World III in Popper's sense (Popper, 1979 and 1982, addendum I). It will stand or fall on its own merits and its merits shall be determined by how much longer it will satisfy the ever multiplying observations made on mélanges world-wide.

The story reviewed here shows, I think quite unequivocally, that objective evidence cannot be avoided in generating scientific knowledge. If it could have been, if there were, in essence, no reality independent of our minds, if reality, facts, knowledge were mere social constructions, such repeated independent discoveries in diverse contexts would have been truly miraculous.

Science does not advance by revolutions. In creeps forward by piecemeal discoveries. Now and then these piecemeal discoveries accumulate to a point of making a reigning large-scale interpretation obsolete and a new large-scale interpretation is invented to replace it. The new large-scale interpretation very commonly inherits a good deal of the intellectual content of its overtaken predecessor. As I once wrote (Şengör, 1998, p. 123), this is similar to royal succession within a dynasty. A king dies and is replaced by his successor, not by popular acclaim, but by the rules of succession. This not only generally leaves the structure of the kingdom intact, but often the new sovereign continues to use the personnel of his predecessor. These are also replaced in time, but generally not all at once. After a certain time, the kingdom may acquire a completely new personnel and if one is not attentive to the history of the changes, one may think that all happened at the time of the change of the sovereign. Complete and rapid overhauls, i.e. revolutions, do happen in the lives of states and societies, but not in science, because the aims and the fundamental method of science have remained unchanged since when it was first invented in Miletus by Thales and Anaximander nearly 2700 years ago.

İ.T.Ü. Maden Fakültesi Jeoloji Bölümü ve Avrasya Yerbilimleri Enstitüsü, Ayazağa 34469 Istanbul, Turkey

6. NOTES

[1] I thank Gürol Irzık and Kostas Gavroglu for inviting this contribution and waiting for its submission with endless patience. Irzık also read its typescript and offered helpful comments.

[2] I am not familiar with a popular account of mélanges that I could recommend to my readers who are not geologists. McCall (1983) gives the best introduction and reproduces the key papers concerning the development of the mélange concept including the *Nature* version of the Bailey and McCallien paper (1950a) and Hsü's 1968 paper (but, unaccountably, not the relevant passages of Greenly's book!) Taira et al. (1992) is a superbly produced photographic atlas of the Shimanto Belt in Japan, where mélanges occur in abundance and are being produced today immediately to its southeast in the Nankai Trench.

[3] Laudan (1977, pp. 147 ff.) attacked this position with the claim that in abandoning one theory for supposedly a better one, some real scientific problems are eliminated from the agenda of the science. He tried to support his claim on an example from the history of geology: he asserted that geology from 1830 to about 1900, 'particularly with the emergence of stratigraphy,' produced no serious geological theory concerning how deposits get consolidated into rocks; how the earth originated from celestial matter and slowly acquired its present form; when and where the various animals and plants originated; how the earth retains its heat; the subterraneous origins of volcanoes and hot springs; the origin and constitution of rocks; how and when various mineral veins were formed. This is a statement of truly stunning inaccuracy! First of all, stratigraphy had emerged long before 1830, in the mid-17th century, and lithostratigraphy was fully and commonly in use in the 18th. Even if Laudan here meant biostratigraphy (as a result of a misreading of Gillispie, 1960, p. 299?), it too had already emerged so fully by the end of the teens of the 19th century, that Alexander von Humboldt could level at it a criticism in 1823 (pp. 41f.) that by all standards remains modern in our own age. After 1830, for each and every one of the problems listed by Laudan there were scores of theories proposed and debated, many more than ever proposed before! In fact, the theory of metamorphic rocks really came into its own after its initiation in the hands of Hutton in the 20's and the 30's of the 19th century (see Şengör and Sakınç, 2001) and continuously developed thereafter, receiving two very competent reviews by Daubrée in 1860 and Vogelsang in 1867. The consolidation of rocks was addressed immediately after Lyell by de la Beche in his *Researches in Theoretical Geology* (1837, see esp. chs IV and V), to cite only one example out of many. The origin of the earth continued occupying theoreticians, both those with a geological bent and those with a geophysical bent and on such considerations the entire theory of contraction and Lord Kelvin's arguments on the young age of the earth were based. Earth's heat was so central a concern that in 1832 a prize was announced in Holland for a thesis that would explain the origin of the augmentation of temperature with depth in the earth and its relation to hot springs and volcanoes. Bischof's famous book (1837) that won the prize was a response to this announcement. Bischof's later great and influential book on physical and chemical geology (*Lehrbuch der Chemischen und Physikalischen Geologie*, of which two editions were published, plus an English translation in 1854 and 1855) was a continuation of such studies. In fact, Bischof received the prestigious Wollaston medal of the Geological Society of London mainly for this great work (Bischof, 1863, p. VIII). Von Humboldt's *Kosmos* (1845-1859) is almost entirely devoted to such questions. As to the origins of plants and animals, I shall confine myself to a single reference: Darwin (1859)! For mineral veins, I mention Forster's (1883) well-known book that went through more than one edition. I have given these few examples out of literally hundreds so as not to give the impression that my sharp criticism was concocted out of thin air. But the student of the history of geology would at once realize how utterly wrong Laudan's statement is (for the historian or the philosopher of science not familiar with the history of geology, I would recommend the following surveys of the 19th century geology: Zittel, 1899; Greene, 1982; Oldroyd, 1996; for the history of ideas on internal heat, melting and volcanism in the 19th century, see esp. Sigurdsson, 1999). Most of Laudan's criticism of the idea of progress resulting from successive theories, each with greater explanatory power than its predecessor, is similarly ill-informed and void.

[4] Erol's critique of Chaput is here somewhat unjust, but it is not his fault. He had at his disposal only a poor Turkish translation by Prof. Hâmit Nâfiz [Pamir], of Chaput's article, which had been published bilingually, for Erol could not read French. What Chaput actually had written was that the ancient massif of Elma Dağı and its Mesozoic cover had been in places strongly broken and imbricated indicating a push towards the north-west (Chaput, 1931, p. 28). What Erol had found was the opposite vergence.

[5] This is not true. Émile Argand recognized the jumble of ophiolites from ocean floors with both shallow and deep water sedimentary rocks. In *La Tectonique de l'Asie*, he wrote: 'A geosyncline will generally result from a horizontal traction that stretches the raft of sial.... Until compensation, the sima rises under the thinned sial; this behavior accounts for the frequent association of green rocks with bathyal and abyssal sediments. The mixture ['*Le mélange*'] of abyssal with shallow water sediments takes place through submarine sliding on a slope.' (Argand, 1924, p. 299). In Şengör (1998, p. 87, footnote 125) I have pointed out that it was clearly just a happy coincidence that here Argand used the French word *mélange* (=mixture) for a rock association that we today also call by the same term in all the world's languages after Greenly's earlier usage. But *conceptually* what he here describes is clearly the same thing that we call ophiolitic mélange. Argand knew of ophiolitic mélanges from his Alpine experience.

7. REFERENCES

Bailey, E. B. (Sir), 1935, Tectonic Essays Mainly Alpine: Clarendon, Oxford, xii+200 pp.+1 foldout map.

Bailey, E. B. (Sir), 1936, Sedimentation in relation to tectonics: Bulletin of the Geological Society of America, v. 47, pp. 1713-1726.

Bailey, E. B. (Sir), and McCallien, W. J., 1950a, The Ankara Mélange and the Anatolian Thrust. Nature, 1660, p. 938.

Bailey, E. B. (Sir), and McCallien, W. J., 1950b. Ankara Melanji ve Anadolu Şaryaji (The Ankara Mélange and the Anatolian Thrust). Maden Tetkik ve Arama Enstitüsü Mecmuasi, no. 40: 12-16; English version pp. 17-21[5]

Bailey, E. B. (Sir), and McCallien, W. J., 1953. Serpentine lavas, the Ankara Mélange and the Anatolian Thrust. Transactions of the Royal Society of Edinburgh, v.62, part II (No. 11), pp. 403-442, with eight photographic plates and two maps.

Bailey, E. B. (Sir), and McCallien, W. J., 1961, Structure of the Northern Apennines: Nature, v. 191, pp. 1136-1137.

Bailey, E. B. (Sir), and McCallien, W. J., 1963, Liguria Nappe: Northern Apennines: Transactions of the Royal Society of Edinburgh, v. 65, pp. 315-333.

de la Beche, H. T., 1837, Researches on Theoretical Geology: F. J. Huntington & Co., New York, xv + 342 pp.

Bertrand, M., 1884, Rapports de structure des Alpes de Glaris et du basin houiller du Nord: Bulletin de la Société géologique de France, 3e, série, v. 12, pp. 318-330.

Bischof, G., 1837, Die Wärmelehre des Innern unsers Erdkörpers ein Inbegriff aller mit der Wärme in Beziehung stehender Erscheinungen in und auf der Erde. Nach physikalischen, chemischen und geologischen Untersuchungen: Joh. Ambrosius Barth, Leipzig, XXIV+312 pp.

Bischof, G., 1863, Lehrbuch der Chemischen und Physikalischen Geologie, erster Band, zweite, gänzlich umgearbeitete Auflage: Adolph Marcus, Bonn, XV+865 pp.

Bucher, W. H., 1956, The role of gravity in orogenesis: Bulletin of the Geological Society of America, v. 67, pp. 1295-1318.

Chalmers, A. F., 1990, The limitations of falsificationism: in Chalmers, A. F., What is This Thing Called Science? 2nd edition, University of Queensland Press, pp. 60-76.

Chaput, E., 1931, Esquisse de l'Evolution Tectonique de la Turquie, traduction en Turc par Hamit Nafiz — Türkiyenin Tektonik Trahçesine Umumî Bir Bakiş, Tercüme eden: Hamit Nafiz: İstanbul Darülfünunu Geologie Enstitüsü Neşriyatindan, no. 6, 107 pp.

Darwin, C., 1859, On the Origin of Species by Means of Natural Selection, or the Preservation of Favoured Races in the Struggle for Life: John Murray, London, ix+[I]+502 pp.

Daubrée, A., 1860, Études et Éxpériences Synthétiques sur le Métamophisme et Sur la Formation des Roches Crystallines: Imprimerie Impériale, Paris, VII+127 pp.

Dewey, J. F. and Bird, J. M., 1970, Mountain belts and the new global tectonics: Journal of Geophysical Research, v. 75, pp. 2625-2647.

Dickinson, W. R., 1970, Second Penrose Conference: The new global tectonics: Geotimes, v. 15(4), 18-22.

Eardley, A. J., 1951, Structural Geology of North America: Harper & Brothers, New York, xiv+624 pp.

Eardley, A. J., 1962, Structural Geology of North America, second edition: Harper & Row, New York, xv+743 pp.
von Eicken, H., 1887, Geschichte und System der Mittelalterlichen Weltanschauung: Verlag der Cotta'scher Buchhandlung, Stuttgart, XVI+822 pp.
Einstein, A., 1981, Mein Weltbild - Ullstein, Frankfurt a. M., 210 pp.
Erol, O., 1949. Ankara Güneydoğusundaki Elma Daği ve Çevresinin Jeoloji ve Jeomorfolojisi Üzerinde Bir Arastirma. Unpublished Ph.D. dissertation, Dil ve Tarih-Coğrafya Fakültesi, Fizikî Coğrafya ve Jeoloji Kürsüsü, Ankara, 84 + VI pp. + 3 maps and one plate of figure.
Escher von der Linth, A., 1878, Die Sentis-Gruppe: J. Dalp, Bern, XVIII+262 pp.+VI plates
Feyerabend, P., 1988, Against Method, revised edition: Verso, London, viii+296 p.
Feyerabend, P., 1991, Three Dialogues Concerning Knowledge: Basil Blackwell, Oxford, 167 p.
Forster, W., 1883, A treatise on a section of the strata from Newcastle-upon-Tyne to the mountain of Cross Fell in Cumberland, with remarks on mineral veins, 3rd ed.: Newcastle lvi+208 pp. +11 plates.
Gavroglu, K., 1994, Types of discourse and the reading of the history of the physical sciences: in Gavroglu, K. and others, editors, Trends in the Historiography of Science, Kluwer Academic Publishers, Dordrecht, pp. 65-86.
Gilbert, G. K., 1875, Report upon the geology of portions of Nevada, Utah, California, and Arizona, examined in the years 1871 and 1872: in Report upon Geographical and Geological Explorations and Surveys West of the One Hundredth Meridian, in Charge of First Lieut. Geo. M. Wheeler..., v. III.—Geology, Government Printing Office, Washington, pp. 17-187.
Gilbert, G. K., 1896, The origin of hypotheses. Illustrated by the discussion of a topographic problem: in W. Cross and C. W. Hayes, editors, The Geological Society of Washington, Presidential Address...-with Constitution and Standing Rules, Abstracts of Minutes and List of Officers and Members, 1895, pp. 3-24.
Gillispie, C. C., 1960, The Edge of Objectivity—An Essay in the History of Scientific Ideas: Princeton University Press, Princeton, ix+562 pp.
Graham, L. R., 1993, Science in Russia and the Soviet Union—A Short History: Cambridge University Press, Cambridge, X+321 pp.+17 (unnumbered) photographic plates.
Greene, M. T., 1982, Geology in the Nineteenth Century. Changing Views of a Changing World: Cornell University Press, 324 pp.
Greenly, E., 1919a, The Geology of Anglesey, v. I. Memoirs of the Geological Survey, Her Majesty's Stationary Office, London, xl+388 pp.+XXVI plates.
Greenly, E., 1919b, The Geology of Anglesey, v. II. Memoirs of the Geological Survey, Her Majesty's Stationary Office, London, pp. 390-980+plates XXVII–LX and 17 folding plates.
Greenly, E., 1922. A short summary of the geological history of Anglesey: Transactions of the Anglesey Antiquarian Society and Field Club, 1922, 20 pp.+2 plates.
Greenly, E., 1928-1932, Benjamin Neeve Peach: a study: Transactions of the Edinburgh Geological Society, v. 12, pp. 1-12.
Grice, I. G. R., 1970, Hume's law: The Aristotelian Society, Supplementary Volume XLIV, The Symposia Read at the Joint Session of the Aristotelian Society and the Mind Association at the University of Aberdeen, 10-12 July, 1970, pp. 89-103.
Hamilton, W., 1969, Mesozoic California and the underflow of Pacific mantle: Geological Society of America Bulletin, v. 80, pp. 2409-2430.
Heezen, B. C., 1960, The rift in the ocean floor: Scientific American, v. 203, pp. 98-110.
Hsü, K. J., 1966, Mélange concept and its application to the interpretation of the California Coast Range geology: Geological Society of America Abstracts for 1965, p. 82.
Hsü, K. J., 1967. Mesozoic geology of the California Coast Ranges—A new working hypothesis: in Etages Tectoniques—Colloque de Neuchatel 18-21 Avril 1966, Institut de Géologie de l'Universite de Neuchâtel, A la Baconnière, Neuchâtel, pp. 279-296.
Hsü, K. J., 1968. Principles of mélanges and their bearing on the Franciscan-Knoxville paradox: Geological Society America Bulletin, v. 79, pp. 1063-1074.

Hsü, K. J., 1971, Franciscan mélanges as a model for eugeosynclinal sedimentation and understhursting tectonics: Journal of Geophysical Research, v. 76 pp. 1162-1170.
Hsü, K. J., 1974, Mélanges and their distinction from olistostromes: in Dott, R. H., Jr. and Shaver, R. H., eds., Modern and Ancient Geosynclinal Sedimentation, SEPM Spec. Pub., 19 pp. 321-333.
Hsü, K., 1985. A basement of mélanges: A personal account of the circumstances leading to the breakthrough in Franciscan research. Geological Society of America, Centennial Special Volume 1, pp. 47-64.
Hsü, K. J., 1990. Mélanges and non-Smithian stratigraphy. Current Contents, no. 26 (June 25), p. 24.
Hsü, K. J. and Ohrbohm, R., 1969, Mélanges of San Francisco Peninsula: geologic reinterpretation of type Franciscan: American Association of Petroleum Geologists Bulletin, v. 53, pp. 1348-1367.
Hubbert, M. K. and Rubey, W. W., 1959, Role of fluid pressure in mechanics of overthrust faulting I. Mechanics of fluid-filled porous solids and its application to overthrust faulting: Bulletin of the Geological Society of America, v. 70, pp. 115-166.
de Humboldt, A., 1823, Essai Géognostique sur le Gisement des Roches dans les Deux Hémisphères: F. G. Levrault, Paris, viij + 379 pp.
Jenings, C. W., 1959, Geologic Map of California, San Luis Obispo sheet: California Division of Mines.
Koertge, N., ed., 1998, A House Built on Sand—Exposing Postmodernist Myths About Science: Oxford University Press, New York, xi+322 pp.
Köhler, G., 1880, Ueber die Störungen im Westfälischen Steinkohlengebirge und deren Entstehung: Zeitschrift für das Berg-, Hütten-und Salinen-Wesen im Preussischen Staate, v. 28, pp. 195-210 and plates XVI and XVII.
Köhler, G., 1885, Verschiebungen von Lagerstätten und Gesteinsschichten: Zeitschrift für das Berg-, Hütten-und Salinen-Wesen im Preussischen Staate, v. 33, pp. 84-92.
Kuendig, E., 1959, Eugeosynclines as potential oil habitats: Proceedings of the 5th World Petroleum Congress, Section I, 461-479, New York.
Kuhn, T. S., 1970, The Structure of Scientific Revolutions, second edition, enlarged: International Encyclopedia of Unified Science, Foundations of the Unity of Science, v. II, nr. 2, The University of Chicago Press, Chicago, XII+210 p.
Kuhn, T. S., 1977, Objectivity, value judgement, and theory choices: in Kuhn, T. S., The Essential Tension—Selected Studies in Scientific Tradition and Change, The University of Chicago Press, Chicago and London, xxiii+366 pp.
Kuhn, T. S., 2000, The Road Since Structure—Philosophical Essays, 1970-1993, with an Autobiographical Interview, edited by Conant, J. and Haugeland, J.: The University of Chicago Press, Chicago, viii+335 pp.
Lakatos, I., 1970, Falsification and the methodology of scientific research programmes: in Lakatos, I. and Musgrave, A., editors, Criticism and the Growth of Knowledge, Cambridge University Press, Cambridge, pp. 91-196.
Lapworth, C., 1883, The secret of the Highlands: Geological Magazine, new series, decade II, v. 10, pp. 120-128.
Lapworth, C., 1885, The Highland controversy in British geology: its causes, courses and consequences. Nature, v. 32, pp. 558-559.
Laudan, L., 1977, Progress and Its Problems—Towards a Theory of Scientific Growth: University of California Press, Berkeley, x+257 pp.
Laudan, L., Donovan, A., Laudan, R., Barker, P., Brown, H., Leplin, J., Thagard, P. and Wykstra, S., 1986, Scientific change: Philosophical models and historical research: Synthesis, v. 69, pp. 141-223.
Martin, B., Baker, C. M. A., Manwell, C. And Plugh, C., eds., 1986, Intellectual Suppression: Angus & Robertson Publishers, North Ryde, NSW, ix+[i]+304 pp.
Matley, C. A., 1913. The geology of the Bardsey Island. Quarterly Journal of the Geological Society of London, v. 69 pp. 514-533.
Maxwell, J. C., 1959, Turbidites, tectonics and gravity transport, northern Apennine Mountins, Italy: Bulletion of the American Association of Petroleum Geologists, v. 43, pp. 2701-2719
McCall, G. J. H., editor, 1983, Ophiolitic and Related Mélanges: Benchmark Papers in Geology/66, Hutchinson Ross Publishing Company, Stroudsburg, Pennsylvania, xiii+446 pp.

Molengraaf, G. A. F., 1928, Wegener's continental drift: in Theory of Continental Drift. A Symposium on the Origin and Movement of Land Masses both Inter-Continental and Intra-Continental, as Proposed by Alfred Wegener, The American Association of Petroleum geologists, Tulsa, pp. 90-92.

Niemczyk, O., 1943, Spalten auf Island—Geologische, geodätische und geophysikalische Forschungsarbeiten der Deutschen Island-Expedition des Jahres 1938: Konrad Wittwer, Stuttgart, [IV]+180 pp.+4 foldout plates.

Noble, L. F., 1941. Structural features of the Virgin Spring area, Death Valley, California. Bulletin of the Geological Society of America, v. 52, pp. 941-1000.

O'Hear, A., editor, 1995, Karl Popper: Philosophy and Problems: Supplement to 'Philosophy' Royal Institute of Philosophy Supplement: 39, Cambridge University Press, Cambridge, 297 pp.

Oldroyd, D. R., 1990, The Highlands Controversy—Constructing Geological Knowledge Through Fieldwork in Nineteenth-Century Britain: The University of Chicago Press, Chicago, ix+438 pp.

Oldroyd, D. R., 1996, Thinking About the Earth: A History of Ideas in Geology: Harvard University Press, Cambridge, xxx+410 pp.

Page, B. M., 1963, Gravity tectonics near Passo della Cisa, northern Apennines, Italy, geological Society of America Bulletin, v. 74, pp. 655-672.

Popper, K. R., 1935, Logik der Forschung: Springer Verlag, Wien, VI + 248 pp.

Popper, K. R., 1957, The Poverty of Historicism: Routledge & Kegan Paul, London, xiv+166 pp.

Popper, K. R., 1966, The Open Society and Its Enemies, v. 1, The Spell of Plato, fifth, revised edition: Princeton University Press, Princeton, xi+361 pp.

Popper, K. R., 1974, Replies to my critics: in Schilpp, P. A., editor, The Philosophy of Karl Popper, Library of Living Philosophers, v. 14, book II, pp. 961-1197.

Popper, K. R., 1979, Epistemology without a knowing subject: in Popper, K. R., Objective Knowledge, Clarendon, Oxford, pp. 106-152.

Popper, K. R., 1980, The Logic of Scientific Discovery, tenth impression (revised): Unwin Hyman, London, 480 pp.

Popper, K. R., 1982, The Open Universe—An Argument for Indeterminism: from the *Postscript to the Logic of Scientific Discovery*, edited by W. W. Bartley, III, Hutchinson, London, xxii+185 pp.

Popper, K. R., 1983, Realism and the Aim of Science: Rowman and Littlefield: Totowa, xxxix+420+[2] pp.

Popper, K. R., 1994, Logik der Forschung, zehnte, verbesserte und vermehrte Auflage: J. C. B. Mohr (Paul Siebeck), Tübingen, XXIX+481 pp.

Rubey, W. W. And Hubbert, M. K., 1959, Role of fluid pressure in mechanics of overthrust faulting II. Overthrust belt in geosynclinal area of western Wyoming in light of fluid-pressure hypothesis: Bulletin of the Geological Society of America, v. 70, pp. 167-206.

Şengör, A. M. C., 1998, Die Tethys: vor hundert Jahren und heute: Mitteilungen der Österreichischen Geologischen Gesellschaft, v. 89, pp. 5-176.

Şengör, A. M. C., 2001, Is the Present the Key to the Past or the Past the Key to the Present? James Hutton and Adam Smith versus Abraham Gottlob Werner and Karl Marx in Interpreting History: Geological Society of America Special Paper 355, x+51 pp.

Şengör, A. M. C., 2002, On Sir Charles Lyell's alleged distortion of Abraham Gottlob Werner in *Principles of Geology* and its implications for the nature of the scientific enterprise: Journal of Geology, v. 110, pp. 355-368.

Şengör, A. M. C., 2003, The repeated rediscovery of mélanges and its implications for the possibility and the role of objective evidence in the scientific enterprise: Geological Society of America Special Paper 373, pp. 385–445.

Şengör, A. M. C. and Sakinç, M., 2001, Structural Rocks: Stratigraphic Implications: in Briegel, U. and Xiao, W. J., eds., Paradoxes in Geology (Hsü Volume), Elsevier, Amsterdam, pp. 131-227.

Sezgin, F., 1987, The Contribution of the Arabic-Islamic Geographers to the Formation of the World Map: Veröffentlichungen des Institutes für Geschichte der Arabisch-Islamischen Wissenschaften, Rheide D Kartographie, Bd. 2, 50 pp. + 48 maps.

Sezgin, F., 1993, Ma'mûnische Weltkarte (Faksimile): in, Focus Behaim Globus Teil 1: Aufsätze, Teil 2: Katalog, Germanisches Nationalmuseum, Nürnberg 2. Dezember 1992 bis 28. Februar 1993, Verlag des Germanischen Nationalmuseums, pp. 650-651.

Sezgin, F., 2000, Geschichte des Arabischen Schrifttums, Band X, Mathematische Geographie und Kartographie im Islam und Ihr Fortleben im Abendland—Historische Darstellung, Teil I: Institut für Geschichte der Arabisch-Islamischen Wissenschaften an der Johann Wolfgang Goethe-Universität, Frankfurt am Main, XXX+634 pp.

Shackleton, R. M., 1954, The structure and succession of Anglesey and the Lleyn Peninsula: British Association for the Advancement of Science, v. 11(41), pp. 106-108.

Shackleton, R. M., 1969, The Pre-Cambrian of North Wales: in Wood, A., ed., The Pre-Cambrian and Lower Palaeozoic Rocks of Wales, University of Wales Press, Cardiff, pp. 1-22.

Sigurdsson, H. ,1999, Melting the Earth: The History of Ideas on Volcanic Eruptions: Oxford University Press, New York, x+260 pp.

Stille, H., 1937, Geotektonische Probleme im atlantischen Raume: in Bericht über die 250jährige Jubiläumsfeier der Kaiserlich-Leopoldinisch-Carolingisch-Deutsc Akademie der Naturforscher, Halle, pp. 129-139.

Strahan, A., 1919. Preface by the Director: in Greenly, E., The Geology of Anglesey, v. I. Memoirs of the Geoloical Survey, Her Majesty's Stationary Office, London, pp. v-vii.

Suess, E., 1883, Das Antlitz der Erde, v. Ia (Erste Abtheilung): F. Tempsky, Prag and G. Freytag, Leipzig, 310 pp.

Taira, A., Byrne, T. and Ashi, J., 1992, Photographic Atlas of an Accretionary Prism—Geologic Structures of the Shimanto Belt, Japan: University of Tokyo Press, [I]+124 pp.+ 23 pp. of inlaid Japanese text.

Trümpy, R., 1991, The Glarus nappes: A controversy of a century ago: in Modern Controversies in Geology (Proceedings of the Hsü Symposium edited by D. W. Müller, J. A. McKenzie, and H. Weissert: Academic Press, London, pp. 385-404.

Vogelsang, H., 1867, Philosophie der Geologie und Mikroskopische Gesteinsstudien: Max Cohen & Sohn, Bonn, [iii]+229+10 plates

Wernicke, B. and Burchfiel, B. C., 1982. Modes of extensional tectonics. Journal of Structural Geology, 4:105-115.

Wood, D. S., 1974, Ophiolites, mélanges, blueschists, and ignimbrites: early Caledonian subduction in Wales? in Dott, R. H., Jr. and Shaver, R. H., eds., Modern and Ancient Geosynclinal Sedimentation, SEPM Spec. Pub., 19 pp. 334-344.

Wright, L. A. and Troxel, B. W., 1969, Chaos structure and Basin and Range normal faults: evidence for a genetic relationship. Geological Society of America Special Paper 121, pp. 580-581.

Wright, L. A. and Troxel, B. W., 1984. Geology of the Northern Half of the Confidence Hills 15-Minute Quadrangle Death Valley Region, Eastern California: The Area of the Amorgose Chaos. State of California, The Resources Agency, Department of Conservation, California Division of Mines and Geology, Map Sheet 34, Plate I, explanatory text, vi+31 pp.

von Zittel, K. A., 1899, Geschichte der Geologie und Paläontologie bis Ende des 19. Jahrhunderts: R. Oldenbourg, München, XI+868 pp.

S. BAĞÇE

A STUDY ON THE HEURISTIC OF SACCHERI'S *EUCLIDES*

A Methodological-cum-Historical Approach[1]

1. INTRODUCTION

The discovery of non-Euclidean geometries was one of the most important developments in mathematical thought during the 19th century. This discovery has a long and very complicated history, which has given rise to many crucial questions. To answer those questions several accounts of the history of this discovery have been written from various perspectives.

Among these accounts, Bonola's (Bonola, 1955) seems to have set the pattern for later writers, such as Coolidge (Coolidge, 1947), Boyer (Boyer, 1968) and Kline (Kline, 1972) and subsequent discussions. Although there are some disagreements between the historical expositions of the aforementioned writers, they share a good deal common attitude to their subject.[2] What they all have in common is the tendency to see geometrical studies from Saccheri (Saccheri, 1733) to Beltrami (Beltrami, 1868) as a prolonged attempt to answer one question: is the parallel postulate necessarily true, given the rest of Euclid's postulates and axioms? As Gray rightly observes, "the standard account frequently ends with references to the logical independence of the postulate from the rest of geometry (Bonola, Chap. V, §94, Appendix V; Coolidge, pp.84-88; Kline, Chap. 38, §4 and Chap. 42)" (Gray, 1989, 169).

The problem is then considered as a problem of foundations. As a result of the discovery of non-Euclidean geometry, a negative solution was provided to the problem, thereby solving the problem. According to the "standard account", geometrical researches from Saccheri to Beltrami should therefore be best understood as foundational studies, and the history of geometry as a linear compilation of the results of these studies.

The standard account provides a good treatment of some certain aspects of these geometrical studies. For example, the division of the history of non-Euclidean geometry into qualitatively distinct periods, i.e., the *forerunners* and the *founders*, is made by Bonola (Bonola, 1955, xii). Although he does not provide any reason for this distinction, he seems to have intended to draw our attention to a historical fact: the forerunners, while trying to defend Euclid's geometry, discovered, of course without realising, some of the very important theorems of non-Euclidean geometry, whereas the founders did their work intentionally to construct a new geometry.

G. Irzık & G. Güzeldere (eds.), Turkish Studies in the History and Philosophy of Science, 137-150.

Thus, Bonola by this distinction aims at highlighting differences not between the methods, techniques and approaches of these geometers but between the results, i.e., theorems these geometers had put forward. That is a natural result of Bonola's original interest, i.e., he was concerned with only foundational and axiomatic problems.

However, this distinction was left unelaborated and thus, its heuristic value has not been understood properly. The forerunners had operated within the ideology of the old geometry. The founders, on the contrary, by employing new geometrical ideas, methods and approaches created a new geometry. However, the forerunners having brought about some novel applications of the old geometrical ideas and methods paved the way towards the creation of a new geometry. Thus, fundamental differences between these geometers' works should also be sought in the approaches of these geometers. It is those differences that played a crucial role in the discovery of non-Euclidean geometry. So through this distinction the standard account, to some extent, directs our attention to differences in mathematical methods: in the 18th century Saccheri and J. H. Lambert (1728-1777) used classical geometry in order to solve the problem of parallels; in the early 19th century J. Bolyai (1802-1860) and N. I. Lobachevsky (1793-1856) employed analysis and in the mid-19th century B. Riemann (1826-1866) and E. Beltrami (1835-1900) turned to the techniques of differential geometry. And it is also true that in the early 19th century several mathematicians came to think that what had been inconceivable might come true: a geometry different from Euclidean one might be logically and physically possible, i.e., the formal structure of the new geometry is capable of expressing empirical findings about our actual physical universe.

However, by emphasizing foundational studies and axiomatic problems alone, the standard account leaves untouched several important aspects of the history of geometry; for example, it does not offer an explanation of why particular geometrical methods were used at the time in such a way they were used, but not earlier, nor of how particular geometrical methods led to the use, and development of further geometrical methods. Furthermore, the standard view is not concerned with any changes in the nature of problems and with what brought such changes about, if any took place. This is because the proponents of the standard account do not see as significant the way geometrical results were achieved. This account thus fails to discriminate what I shall call the "heuristics" of geometrical development. By "heuristic", I mean the following:

(i) the methods employed by geometers in problem solving;
(ii) the characterisations of geometrical language, problems and theories;
(iii) the intentions of geometers themselves.

In particular, the exposition according to the standard account of Saccheri's work, *Euclides ab omni naevo vindicatus*, which was published in 1733,[3] is no exception: on that account, Saccheri's work was an ambitious defence of the parallel postulate, and thus, a study on a problem in foundations, which led to the discovery of non-Euclidean geometry through the consequences of the acute-angle hypothesis, which are today recognised as important theorems of non-Euclidean geometry (Bolyai-Lobachevsky's geometry)[4]:

> Still, though it failed in its aim, Saccheri's work is of great importance. In it the most determined effort had been made on behalf of the Fifth Postulate; and the fact that he did not succeed in discovering any contradictions among the consequences of the *Hypothesis of the Acute Angle*, could not help suggesting the question, whether a consistent logical geometrical system could not be built upon this hypothesis, and the Euclidean Postulate be impossible of demonstration (Bonola, 1955, 43-44).

Since the standard account does not view as significant the way geometrical results were achieved, the standard account leaves untouched several important aspects of Saccheri's work on the problem of parallels too. In this paper, in order to exhibit these important aspects, I shall read Saccheri's study in the light of the following questions:

(i) why was an indirect method of proving, i.e., *reductio ad absurdum*, used at the time in a particular way Saccheri used it, but not earlier?
(ii) was there any change in the nature of the problem of parallels, and if any, what brought it about?
(iii) how did Saccheri's work lead to the use of other approaches, methods in geometrical studies after him, if it did at all?

My aim here is, by answering these questions above, to interpret Saccheri's work as having brought about a new heuristic. Establishing this claim shall, in turn, enable us to see the history of geometry, not as a linear compilation of the results of geometrical theories, in which there is no natural continuity and progress, but on the contrary, as an exemplar of continuity and progress. Accordingly, I claim that the history of geometry should be best understood through the heuristics of geometrical theories.

Before answering the questions above in order to illustrate the value of the heuristic approach, it would be useful to outline Saccheri's geometrical work in connection with the parallel postulate, which is presented in his *Euclides*.

2. OUTLINE OF *EUCLIDES*

According to Saccheri, there are three "flecks" in Euclid's *Elements*, one of which is the parallel postulate (Saccheri, 1733, 5-7). In his *Euclides*, Saccheri sets himself the task to free Euclid's *Elements* from this "fleck"; for he observes of the parallel postulate that "no one doubts [its] truth" (Saccheri, 1733, 5). To do so, he first assumes the first 28 propositions of Euclid's *Elements*, whose proofs do not depend on the fifth postulate. He then tries to justify the fifth postulate by an indirect proof, i.e., *reductio ad absurdum*: by assuming that the proposition to be proved is false, and deriving a contradiction, one concludes that it is true. Saccheri wanted to establish the postulate as true, but not as a theorem derived from some question begging assumptions such as he detected in earlier commentators.

In other words, he wants to show that the denial of the existence and uniqueness of parallels as a postulate is incompatible with the rest of Euclid's postulates and axioms. In order to put in a convenient form the claim that the fifth postulate is false, he employs a certain plane figure, now known as a *Saccheri quadrilateral* that has right angles at A and B, and AD=BC:

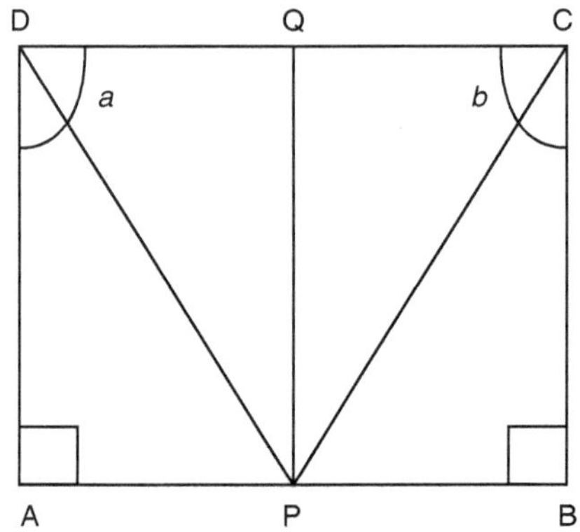

Figure 1. *A Saccheri quadrilateral*

On the basis of Euclid's first four postulates, Saccheri first wants to prove that the angles $\angle ADC$ and $\angle BCD$ are equivalent to each other. For if P and Q denote the midpoints of the segments AB and CD respectively, the two right triangles ADP and BCP are congruent (Heath 1956, 247-250). Thus, he concludes that

$$\angle ADP = \angle BCP, \text{ and } PD = PC. \tag{1}$$

Then the sides of triangle DPQ are equal, respectively, to the sides of triangle CPQ. And, consequently, these two triangles are congruent (Heath 1956, 261, 284).[5]
Thus,

$$\angle ADC = \angle ADP + \angle PDC = \angle BCP + \angle PCD = \angle BCD \tag{2}$$

Calling the equal angles at C and D, b and a, the following three possibilities are then exhaustive and exclusive:[6] It is either

$$a + b = \pi \tag{3}$$

or

$$a + b > \pi \tag{4}$$

or

$$a + b < \pi. \tag{5}$$

Saccheri calls these possibilities respectively as:

(i) the right angle hypothesis; for these are right angles,
(ii) the obtuse angle hypothesis; for these are obtuse angles, and

(iii) the acute angle hypothesis; for these are acute angles [abbreviated to (i) HRA, (ii) HOA and (iii) HAA].

Saccheri first assumes that each hypothesis holds in one given quadrilateral, but not necessarily in others; namely, we are at this stage still free to imagine an obtuse-angled quadrilateral in one region of the plane and an acute-angled quadrilateral in somewhere else. In the propositions V–VII he proves that if any of these three hypotheses is true for one of his quadrilateral, it is true for every such quadrilateral. These theorems show that space is homogeneous on each hypothesis, i.e., space is geometrically the same everywhere.

Saccheri further proves the following tripartite result with regard to triangles (proposition IX):

> In any right-angled triangle the two acute angles remaining are, taken together, equal to one right angle, in the hypothesis of right angle; greater than one right angle, in the hypothesis of obtuse angle; but less in the hypothesis of acute angle (Saccheri 1733, 41-43).

Saccheri then proves in proposition XI that his axioms together with the HRA are equivalent to the postulates of Euclidean geometry. In proposition XII, he proves that any two lines under the HOA always meet at a finite distance.[7] Thus, this proposition establishes that his axioms together with the HOA equal to the axioms of elliptic geometry. He then in proposition XXV and XXXII proves that, under the HAA, through a point, A, outside of a straight line, *l*, there are always two parallels, which are asymptotic straight lines to, *l*. Saccheri, thus, with these propositions establishes that his axioms together with the HAA are equivalent to the axioms of hyperbolic geometry.

Although one could define, as seen above, through Saccheri's quadrilateral three classical geometries, Euclidean, hyperbolic and elliptic, Saccheri went to establish his original claim that only Euclidean geometry was the true one. Saccheri, in order to achieve his aim, tries first to dispose of the HOA, and he concludes in proposition XIV that "the hypothesis of obtuse angle is absolutely false, because it destroys itself" (Saccheri, 1733, 59-61).

Saccheri uses his propositions XI–XIII in order to establish the conclusion stated in proposition XIV. However, Saccheri's disposition of the HOA might be regarded as problematic because he relies on proposition 16 of Euclid's *Elements*, which is not valid in the HOA; for the proof of this proposition depends on the infinite length of the straight line.[8] It is clear that the straight lines in the HOA cannot be assumed as having infinite length. This assumption nullifies the possibility of the existence of a pair of obtuse angles in a quadrilateral, and to that extent, proves the HRA.[9] However, this assumption does not block the possibility of such angles being acute.

On the other hand, there is a puzzling statement by Saccheri:

> I will never use from those prior propositions of Euclid's First Book, not merely the 27th or 28th, but not even the 16th or 17th, except where it is clearly a question of a triangle every way restricted" (Saccheri, 1733, *Preface*).

This statement of Saccheri's explains why he uses I.16 by making a distinction between these two pairs of propositions, I.16-17 and I.27-28. As Dou rightly observes, the propositions in the former are valid for restricted class of triangles, while the propositions in the latter, which establish the existence of parallel straight lines, are not (Dou, 1970, 389).

The propositions I.27-28 necessarily require the infinity of straight lines in length. However, the propositions I.16-17 are valid, provided only that the segments, interior to the triangle and going from one vertex to the mid-point of the opposite side, are smaller than the half of the length of the straight line (Dou, 1970, 388-389). So what we have here is that the propositions I.16-17 are *locally* valid whereas the propositions I.27-28 are necessarily false in the HOA.

What Saccheri had in mind in the case of the HOA is, as Dou points out, that two perpendiculars AD, BC on AB should intersect with each other, which seems to be contained in proposition XII.[10] Exactly for this reason, Saccheri says that he will never use propositions I.16-17, "except where it is clearly a question of a triangle every way restricted".

As said above, Saccheri paves the way with propositions XI-XIII for proposition XIV. Proposition XIII states that:

> If the straight line XA (of any designated length however great) intersecting two straight lines AD and XL, makes with them on the same side interior angles XAD and AXL less than two right angles: I say that these two (even if neither of those angles is a right angle) will meet each other at length in some point on the side of those angles, and indeed at a finite or terminated distance, if holds true the hypothesis of the right angle or of the obtuse angle.

This proposition only brings together the preceding propositions XI and XII, and establishes the HRA, from which it follows that the Saccheri quadrilateral must be a rectangle, i.e., the sum of the angles of the quadrilateral is equal to four right angles. Consequently, the usual theorems that are deduced from this postulate of Euclid's must also hold. Thus, the HOA is then false, that is, this proposition allows Saccheri to conclude in proposition XIV that:

> The hypothesis of obtuse angle is absolutely false, because it destroys itself.

Since this does half the job, he needs to show that the HAA is false as well. However hard he tries, he fails to find any logical contradiction in the HAA. He makes a decision by relying upon his belief in the truth of the HRA rather than upon logic itself. In this way, Saccheri appealed to a body of beliefs about lines: they could not do certain things.

Thus, Saccheri's attempt to disprove the HAA is not free from imperfections too; for he makes a decision, trusting intuition in the validity of the fifth postulate rather than logic. In proposition XXXIII Saccheri states that:

> the hypothesis of the acute angle is absolutely false, because it is repugnant to the nature of the straight line (Saccheri, 1733, 208).

Saccheri relies upon five lemmas and two corollaries in order to prove his proposition XXXIII (Saccheri, 1733, 173-207). His demonstration depends upon the extension to *infinity* of certain properties which are valid for figures at a finite distance, in particular the idea that two lines might have a common perpendicular at the point at infinity where they meet (Bonola, 1955, 43; Gray, 1989, 68). Thus, Saccheri's proof of the impossibility of the HAA is invalid. According to Bell, Saccheri was misled "by an improper use of infinitesimals" (Bell, 1972, 327).

However, being aware of the unsatisfactory character of his demonstration of the invalidity of the HAA, he again attempts in the rest of his book to derive the desired result by readopting the old idea of equidistance, which had been unsuccessfully employed by his predecessors.

3. *EUCLIDES* ON THE HEURISTIC APPROACH

Now, I would like to consider the aforementioned three questions in order to illustrate the value of "heuristic" approach.

The first question was why that method was used at the time in a particular way Saccheri used it, but not earlier. Saccheri wrote his second book, *Logica demonstrativa*, which was published in Turin in 1697. This book has a great importance: it is a book on logic which is modelled on Euclid's *Elements*. His interest in this book is clearly with the importance of definitions in mathematics, and arguments by *reductio ad absurdum*.[11]

Saccheri introduces a distinction between *nominal definitions* (*definitiones quid nominis* or *nominales*) and *real definitions* (*definitiones quid rei* or *reales*). The former are intended only to explain the meaning that is attached to a given term, whereas the latter are not only intended to declare the meaning of a word but also to affirm the existence of the thing to be defined, or in terms of geometry, the possibility of constructing it.[12]

Definitiones nominales are in themselves quite arbitrary, and they neither require nor are capable of proof. We have them simply because we want to turn them as quickly as possible into *reales*. A *definitio quid nominis* becomes a *definitio quid rei* "by means of a *postulate*, or when we come to the question whether the thing *exists* and it is answered affirmatively" (Saccheri, 1733, xviii).

According to Saccheri, the confusion between the *nominal* and the *real* definitions is one of the most fruitful sources of illusory demonstration. And the danger is greater in proportion to the "complexity" of the definition, i.e. the number of variety of the attributes belonging to the thing defined; for the greater then is the possibility that there may be among the attributes some that are *incompatible*, i.e., the simultaneous presence of which in a given figure can be proved to be impossible by means of *other* postulates, etc., forming part of the basis of the science.

However, additional elaboration by Saccheri broadens the matter, leading to the recognition of the more general question relative to the necessity of excluding the existence of incompatibility among the fundamental postulates. This approach is to make the basis of a demonstrative science. The existence of incompatibility among the fundamental postulates can be established not only by showing that these propositions are directly in contradiction with each other but also by showing that if the falsity of one of them could be proved by means of the others, a thing not directly recognisable.

This approach has a great importance with regard to the function of mathematics: viz., the logical compatibility between the fundamental propositions and the development, in a logically coherent way, of the consequences flowing from a given system of premises, without being susceptible of a direct interpretation or empirical verification. The question of whether such compatibility really exists is an essential one, since postulates are only subject to the condition of being mutually compatible. The problem of logical compatibility of postulates is the subject matter of the final chapter of *Logica*.

Given this work of Saccheri, it is not surprising to see Saccheri employing in his *Euclides* the results he achieved in his *Logica*.[13] That is, there is an exact correspondence between the use he makes in *Logica* of this demonstrative procedure, and the use he subsequently attempted to make of it in his *Euclides*:

> In this place however some one perchance may inquire, why I am so solicitous about proving exact the refutation of each false hypothesis. To the end, say I, that thence may be more completely established that not without cause was that famous axiom assumed by Euclid as known *per se*. For chiefly this seems to be as it were the character of every primal verity, that precisely by certain recondite argumentation based upon its very contradictory, assumed as true, it can be at length brought back to its own self. And I can avow that thus it has turned out happily for me right on from early youth in reference to the consideration of certain primal verities, as is known from my *Logica demonstrativa*" (Saccheri, 1733, 237).

The use of the results of *Logica* in *Euclides* enables Saccheri to achieve two objectives by drawing explicit attention to the question of compatibility between postulates when he considers the parallelism between two lines:

(i) in general, to show that geometry in Euclid's *Elements* is consistent, and
(ii) moreover, in particular, to show that the truth of the parallel postulate is derivable from the other first postulates and the first 28 propositions; namely, by appealing to this indirect type of reasoning Saccheri wanted to reformulate the problem of parallels in terms of the three hypotheses, the HRA, HAA and HOA. This realisation permits the use of *reductio ad absurdum* to yield exactly the desired result: if geometries based upon the second and the third hypothesis can each be shown to be self-contradictory, then the first (Euclidean) hypothesis must be true.

So, being equipped with all this logical machinery, it is natural to see Saccheri applying this particular type of reasoning, *reductio ad absurdum*, in his treatment of the problem of parallels; for he attempted to explain the principles of geometry in terms of logic. Although he was not the first to use this reasoning, which had been used by Euclid himself, he was the only one to employ this method in a systematic treatise on logic and geometry. His *Euclides* was, probably, the first attempt by employing *reductio ad absurdum* to redefine the problem of parallels in terms of the three hypotheses so that an effective solution to be offered.

As for the second question, which was whether there was any change in the nature of the problem of parallels, and if any, what brought it about, we need to know the history of the problem of parallels.

Long before Saccheri, the problem of parallels had puzzled Greek geometers a great deal: it is not intuitive, and asserts things about meeting lines indefinitely far in the distance. Several attempts were made in order to eliminate the doubts about the parallel postulate. Although the attempts from Euclid to about 1800 run concurrently, they may be categorised into the following three approaches:

(i) attempts directly to derive the parallel postulate from the rest of Euclid's postulates and axioms;
(ii) attempts to replace the parallel postulate by a more self-evident postulate;
(iii) attempts to explore what consequences would follow from denying the parallel postulate.

One of the very first major attempts historically known was made by Ptolemy (2nd Century A.D.), who tried to prove the fifth postulate from the rest of postulates and axioms and, the first 28 propositions of *Elements*. But he implicitly assumed that two straight lines did not enclose a space, which automatically gave the desired result (Bonola, 1955, 3-4; Heath, 1956, 204-206).

Another attempt came from Proclus (410-485), who based his proof upon the proposition that the distance between two points upon two intersecting straight lines could be made as great as we pleased by prolonging the two lines sufficiently (Bonola, 1955, 4-7; Heath, 1956, 206-208; Gray, 1989, 36-38). Proclus's proof was correct, but he substituted one questionable postulate for another. His proof was criticised by Clavius, and this criticism was supported by Saccheri (Heath, 1956, 208).

Nasr Eddin al-Tusi (1201-1274) likewise gave a proof of the postulate by assuming that the curve everywhere equidistant from a given straight line was in itself a straight. He deduced the postulate from this assumption (Bonola, 1955, 9-12; Gray, 1989, 49-53; Heath, 1956, 208-210). The possibility that lines which appeared to be converging for a while but then diverging was tacitly ruled out by al-Tusi; for it seemed to contradict any intuitive idea of straightness. However, the question is not the plausibility of any assumption but the logical necessity of the parallel postulate. All al-Tusi established is that the postulate is a theorem if one assumes a certain property of lines that appear to converge.

John Wallis (1616-1703) gave up the idea of equidistance, which had been employed by previous mathematicians without any success. He put forward a demonstration of the postulate by assuming that, given a figure, another figure was possible which was similar to the given one and of any size whatever (Bonola, 1955, 15-17; Heath, 1956, 210-211). Wallis, in fact, assumed this for triangles only (Bonola, 1955, 15-17; Gray, 1989, 57-58; Heath, 1956, 210-211). Although intuition may support the conception of form, independent of size, the idea is not more self-evident than the postulate itself. As Saccheri points out, the assumed postulate as to the existence of similar triangles is equivalent to unconditionally assuming the "hypothesis of the right angle" and consequently the parallel postulate (Saccheri, 1733, 105-109, where Saccheri points out that it is not necessary to assume so much, and that it is quite enough to postulate that there exist unequal triangles with equal angles).

Saccheri, instead of adopting either of these two strategies, employed *reductio ad absurdum* to establish the parallel postulate: the anticipation of deriving the parallel postulate as true from its negation enabled Saccheri to formulate the problem in terms of the three alternative hypotheses–the hypotheses of *the right, obtuse* and *acute angle*.

Saccheri's other interests in geometry, out of which his first book, *Quaesita geometria*, published in Milan in 1694,[14] arose, gave him the capacity to put the negation of the parallel postulate in a convenient geometrical form. Saccheri's interest in coordinate geometry was reflected in his approach to the problem. In other words, Saccheri formulated the negation of the parallel postulate in terms of a particular plane figure, by which he cast the problem of parallels in terms of quadrilaterals, triangles and angles, as different from the previous definitions, given in terms of lines and angles. He thereby made explicit the essential connection between the theory of parallels and the sum of the angles of a triangle. As we shall see below, this is a very important step for later developments on the subject.

To sum up: Saccheri is the first one to formulate the problem of parallels in terms of three mutually exclusive hypotheses. This is the first change in the nature of the problem. The second change is with the definition of the problem: it is defined in terms of quadrilaterals, triangles, and angles. This is to make an essential connection between the theory of parallels and the sum of the angles of a triangle.

These changes are brought about in the course of Saccheri's use of *reductio ad absurdum*, and of Saccheri's quadrilateral. So the changes in the nature of the problem run parallel with the changes of Saccheri's mathematical approach to the problem. Saccheri's choice of methods and the exhaustiveness of his approach to the subject were "modern" in the same way Lambert's were, in contrast to al-Tusi's and Wallis's.

Now, let us consider the third question, namely whether Saccheri's work led to the use of other mathematical methods, in particular, the introduction of analytic techniques in attitudes to the problem of parallels; if so, how it did.

As it is shown in Gray's valuable studies on the history of non-Euclidean geometry (1979, 1987, 1989), the crucial step in the solution of the problem of parallels, and the discovery of non-Euclidean geometry, was the introduction of analytic techniques, notably those of hyperbolic trigonometry, and later of differential methods. The introduction of analysis is shown to be effective because it allows a covert use of the concepts of differential geometry.

As is well known, Lambert extended Euler's treatment of sine and cosine, and made explicit the similarities between the hyperbolic and circular functions. And the latter led to the use, for the first time, of analysis in this area of geometry by F. K. Schweikart, F. A. Taurinus, and K. F. Gauss. When Lambert made the connection explicit between the hyperbolic and circular functions, by transcribing the spherical trigonometry formulae into formulae involving hyperbolic functions, he did not deduce that the new formulae apply to a geometry based on the HAA–a result which follows immediately on consideration of the formulae in the special case of an equilateral triangle (Gray, 1979, 240, 248).

Instead he used the formulae of hyperbolic trigonometry in his astronomical works where the sides of triangles could be taken to be imaginary. But he did not ask himself what kind of triangle obeyed the laws of hyperbolic spherical trigonometry. Gray writes that "with hindsight this turns out to have been a near miss, we have to wait 60 years before anyone else was to consider the connection between parallel postulate and spherical trigonometry. That man was F. A. Taurinus, a nephew of Schweikart, like him, a lawyer" (Gray, 1979, 248; Gray, 1987, 47).

Lambert did this work after 1766 as his interest in the problem of parallels disappeared. However, he did not see the significance of his discussions of trigonometric studies, hyperbolic functions.[15]

Lambert's interest in the problem of parallels came through G. S. Klügel's dissertation, *Conatuum praecipuorum theoriam parallelarum demonstrandi recensio* (1763). Lambert then wrote a book *Theorie der Parallellinien* in 1766 which contained his own investigations about the subject, but the book was published posthumously by J. Bernoulli and C. F. Hindenburg in 1786.

Saccheri's work influenced later geometers, especially Lambert (Segre, 1903, 535-547). Lambert was certainly familiar with Saccheri's work, because a review and criticism of Saccheri's work was given in Klügel's thesis, and Lambert quoted and praised this thesis in his *Theorie der Parallellinien* though never mentioned either Saccheri, or his book (Gray, 1979, 241; Dou, 1970, 396).

It should be stressed that Lambert employed not only exactly the same three hypotheses of Saccheri, but did not depart far from Saccheri's method in his treatment of the mentioned hypotheses. Thus, the essential connection between the theory of parallels and the sum of the angles of a triangle, which was originally put forward by Saccheri, was steadily kept in view. This can be taken as a first step towards the introduction of trigonometric methods in attitudes to the problem of parallels, and to the discovery of non-Euclidean geometry.

Moreover, Lambert in his *Theorie der Parallellinien* made two important observations:

(i) that, on the HAA, one line segment can be uniquely associated with one angle and, thereby, transferring the absolute measure of angles onto lengths. So, in such a new geometry, length becomes absolute. In Euclidean geometry, we do not have the absolute measurement of length. Since similar triangles exist in Euclidean geometry, we cannot transfer the absolute measure of angles onto lengths. Lambert seems to have wanted to dispose of this new geometry, but he did not.

(ii) The area of a plane triangle, under the second and the third hypothesis, is proportional to the difference between the sum of the three angles and two right angles. Thus, under the HAA it is

$$\Delta = k(\pi - A - B - C) \qquad (6)$$

and under the HOA it is

$$\Delta = k(A + B + C - \pi) \qquad (7)$$

where k is a positive constant (Heath, 1956, 212-213; Gray, 1989, 74).

As Bonola rightly points out, Saccheri, when discussing the HAA, had met the *defect* here referred to, and also noted implicitly that a quadrilateral, made up of several others, had for its *defect* the sum of the *defects* of its parts [proposition XXV]. However, he did not draw any conclusion from this as to the area being proportional to the *defect* (Bonola, 1955, 46).

4. CONCLUDING REMARKS

To conclude I would like to stress that Saccheri employed not only a particular Euclidean style of reasoning, but also the basic concepts of Euclid's geometry. Moreover, Saccheri believed in the truth of the parallel postulate and thus, in that of Euclid's geometry. He thought of Euclid's geometry as mathematically the only true and possible geometry as well as the necessary representation of our physical world. So, Saccheri was operating not only with the Euclidean style of reasoning and its basic concepts but also within the Euclidean ideology of geometry. This might not be inevitable, but rather a conscious decision, given the fact that a new geometrical method, i.e., Cartesian, was available. This might be because he was a Jesuit who might have enjoyed this old fashioned geometrical style. But as we have seen above there are some distinctive differences between Saccheri's mathematical methods of solving, and approach to, the problem of parallels and those of his predecessors'.

All the aforementioned differences and the novelty of Saccheri's study on the problem of parallels came through not the use of *reductio ad absurdum* but the novel application of this old Euclidean style of reasoning and Saccheri's quadrilateral both in redefining and solving the problem. By casting the problem in terms of quadrilaterals, triangles and angles, as different from previous definitions, Saccheri made explicit the essential connection between the theory of parallels and the sum of the angles of a triangle. That is the heuristic of Saccheri's approach to the problem; namely, his heuristic consists of the new employment of *reductio ad absurdum* and that essential connection. Saccheri, thus, brought a new heuristic concerning the nature of the problem of parallels, although he was still working within the Euclidean "research programme".

Saccheri's study could be seen as a first step towards the realisation of a new formulation of the problem of parallels in trigonometric language, which thus led to the solution of the problem and the discovery of non-Euclidean geometry. Saccheri's may, then, be regarded as constituting, in Lakatos's terminology, a "progressive problem shift".

The new heuristic, which had been brought about by Saccheri, was later developed by Lambert and Legendre by deriving some more new consequences from the HOA and the HAA, and by nearly discovering the connection between the parallel postulate and spherical trigonometry. But this connection was seen and explicitly stated by Taurinus through Gauss, who read Lambert's book. By making this connection explicit, the new heuristic originating in Saccheri's work, and later developed by Lambert and Legendre, took another form, i.e., being a different heuristic, the peak point of which were the studies of Bolyai's and Lobachevsky's.

These interesting aspects of Saccheri's study–of the history of geometry in general–are brought to the surface by paying particular attention to the changes made by Saccheri in the geometrical approach to, in the nature of, the problem of parallels, i.e., by the "heuristic" approach to the history–and theories–of geometry. Historical-cum-methodological studies on the heuristic aspects of geometrical theories can bring out the continuity and progress existing in the history of geometry. That is why I claim that the history of geometry is not a linear compilation of those geometrical results, and that the history of geometrical theories should be best understood through their heuristics.

Department of Philosophy, Middle East Technical University, 06531 Ankara, TURKEY, sbagce@metu.edu.tr

5. NOTES

[1] I would like to thank Timothy Childers, Marco Del Seta, Albert Dou, Donald Gillies, Jeremy Gray, Andrew Powell, Stathis Psillos and Elie Zahar, for their penetrating and stimulating comments and discussions on an earlier version of this paper.

[2] Gray classifies these expositions as the "standard account" since they employ a very similar approach to the subject (Gray, 1979, 237).

[3] This book is referred as *Euclides* in this article. Let me make some brief notes about Saccheri's life and his works.

Girolamo Saccheri, who was born in 1667 at San Remo, then the Republic of Genoa, was a Jesuit geometer. He studied at the Collegio di Brera in Milan where his teacher was another Jesuit geometer, Thomas Ceva, through whom Saccheri made acquaintance with Giovanni Ceva. Saccheri taught philosophy and apologetics in Turin, and then moved to Pavia, at the university of which he taught mathematics. Saccheri died in Pavia in 1733.

Giovanni Ceva had got Saccheri interested into coordinate geometry and encouraged Saccheri to write his first book on coordinate geometry: *Quaesita Geometria* (Milan, 1694), in which Saccheri solved a number of problems in elementary and coordinate geometry one of which is a problem in analysis known as the "window of Viviani".

Saccheri's second book was on logic: *Logica Demonstrativa* (Milan, 1697). He transformed the scholastic logic through a critical elaboration into the form of a series of demonstrations interconnected in a way analogous to the methods of geometers.

During his years in Pavia, Saccheri wrote *Nea-Statica* in 1708. He was inspired by–and partly a polemic against–Thomas Cava's *De Natura Gravium* (Milan, 1669).

Saccheri's last book is *Euclides ab Omni Naevo Vindicatus* (Milan, 1733), in which he attempts to vindicate Euclid's geometry.

[4] Saccheri's quadrilateral defines only one real elementary plane geometry either Euclidean or hyperbolic or elliptic according to the sum of the angles *a* and *b* being right, acute or obtuse (see *Figure 1*). It can easily be seen that Saccheri's axioms may lead to not only hyperbolic but elliptic too (Dou, 1970, 387).

[5] Moreover, these propositions are proved without invoking the parallel postulate.

[6] In Euclidean geometry, every triangle, no matter what its *size* and *position in space* are, has angles adding up to 180°. The contrary statement is that somewhere in space, a triangle of some size has angles adding up either to more or to less than 180°. Saccheri took it for granted that *all* triangles are obtuse, which is a stronger alternative.

[7] Saccheri in this proposition seems to be assuming that two lines meet the transversal with angles, the sum of which is less than 180°. However, as Dou indicates (Dou, 1970, 407), if under the HOA two lines meet a transversal making the angles equivalent to π, the case is reduced to the previous one and thus, the two lines meet in both directions.

[8] The assumption that the straight lines are infinite in length is also used in the proofs of propositions I.17, I.21, I.26-28, in the last of which Euclid for the first time uses the fifth postulate. The proposition I.16 states that "in any triangle, if one of the sides be produced, the exterior angle is greater than either of the interior and opposite angles" (Heath, 1956, 279-281). And the other five propositions, I.17, I.21, I.26-28, depend on I.16. In addition, Saccheri employs for the first time proposition I.16 in the proof of his proposition III.

[9] Since Saccheri uses the assumption of the infinite extendibility of straight lines in the proofs of propositions from III to XIII, the validity of Saccheri's proofs of those propositions is questioned. For example, Stäckel is reported as remarking in his book, *Die Theorie der Parallellinen von Euklid bis auf Gauss* (Stäckel and Friedrich, 1895, 52 and 62), that the proof in the HOA is insufficient (Dou, 1970, 388; Segre, 1903, 535). However, if Euclid's 2nd postulate is read as guaranteeing the infinite extendibility of straight lines,

then Saccheri is right in disposing of the HOA (Dou, 1970, 387-388; Sklar, 1977, 17-18). Moreover, Dou argues that "Saccheri will work without propositions 27th and 28th, and even without I.16 and I.17, unless that they are applied to a triangle that is bounded by all sides" (Dou, 1992, 534).

[10] For this see note 7 of this paper.

[11] Saccheri claims that his most important contribution is to be the first one to have given a systematic treatment of this type of reasoning in this book (Saccheri, 1733, *Introduction*; Dou, 1970, 395].

[12] Saccheri gives an example of the construction of a square in Euclid's *Elements*, Book I. proposition 46: it might be argued that Euclid has no right to define a square, as he does, when we do not know whether such a figure exists. Saccheri replies that the objection could have force, if Euclid, before proving and making the construction, assumed such a figure as given. He goes on to say that Euclid never presupposes the existence of the figure as defined until after Book I. proposition 46 (Saccheri, 1733, *Introduction*).

[13] It is pointed out that one of the reasons of Saccheri's attachment to the proof of the parallel postulate was to test his figure of reasoning through the law of Clavius (Saccheri, 1733, *Introduction*; Dou, 1970, 405).

[14] For this see note 3 of this paper.

[15] The reason might be, as Gray points out, that he was working within the Euclidean metaphysical ideology of geometry (Gray, 1979, 240).

6. REFERENCES

Bell, Eric T. *The Development of Mathematics*. New York: Dover, 1972.

Beltrami, Eugenio. "Saggio di interpretazione della geometria non-Euclidea." *Giornale di Matematiche* **6** (1868): 248-312.

Bonola, Roberto. *Non-Euclidean Geometry: A Critical and Historical Study of Its Development*. New York: Dover, 1955.

Boyer, Carl B. *A History of Mathematics*. New York: Wiley, 1968.

Coolidge, Julian L. *A History of Geometrical Method*. Oxford: Clarendon Press, 1947.

Dou, Alberto M. "Logical and Historical Remarks on Saccheri's Geometry." *Notre Dame Journal Formal Logic* **11** (1970): 385-415.

Dou, Alberto M. "The "Corollarium II" to the proposition XXIII of Saccheri's Euclides." *Publicacions Matemàtiques* **36** (1992): 533-540

Gray, Jeremy. "Non-Euclidean Geometry: A Re-Interpretation." *Historia Mathematica* **6** (1979): 236-258.

Gray, Jeremy. "The Discovery of Non-Euclidean Geometry." In *Studies in the History of Mathematics*, E. R. Philips, ed. MAA Studies in Mathematics, **26** (1987): 37-60

Gray, Jeremy. *Ideas of Space: Euclidean, Non-Euclidean, and Relativistic*. 2nd ed. Oxford: Clarendon Press, 1989.

Heath, Thomas L. *The Thirteen Books of Euclid's Elements*. New York: Dover, 1956.

Kline, Morris J. *Mathematical Thought From Ancient to Modern Times*. London: Oxford University Press, 1972.

Lambert, Johann H. *Theorie der Parallellinien*. In *Die Theorie der Parallellinien von Euklid bis auf Gauss*, F. Engel and P. Stäckel. B. G. Teubner: Leipzig, 1895.

Saccheri, Girolamo. *Euclides ab omni naevo vindicatus: sive conatus geometricus quo stabiliuntur Prima ipsa universae Geomertiae Principia*. Milan, 1733, translated into English by George B. Halsted, *Girolamo Saccheri's Euclides Vindicatus*. Chicago and London: Open Court, 1920.

Segre, Corrado. "Congetture intorno all' influenza di Girolamo Saccheri Sulla formazione della geometria non-euclidea." *Atti Acc. Scienza di Torino* **38** (1903): 535-547.

Sklar, Lawrence. *Space, Time, and Space-Time*. Berkeley, LA, and London: California University Press, 1977.

Stäckel, Paul an Engel, Friedrich. *Die Theorie der Parallellinien von Euklid bis auf Gauss*. B. G. Teubner: Leipzig, 1895.

PART III

PHILOSOPHY OF LANGUAGE AND MIND

ILHAN INAN

DISCOVERY AND INOSTENSIBLE *DE RE* KNOWLEDGE

1. DISCOVERY

Perhaps the most striking aspects of successful scientific theories are the predictions they make that lead to new discoveries. This has led some philosophers to consider this feature of a theory to be the unique criterion of its scientificity; the better predictions a theory makes, the more scientific it becomes.[1] Even if this view is considered to be a bit exaggerated, successful predictions seem to be among the central indicators of scientific progress.

There are of course different types of discoveries, and each type gives rise to its own philosophical problems; there are discoveries that are not based on any predictions, discoveries based on some accidental predictions, and discoveries that result from careful predictions derived from observations and theory. The last category involves certain significant questions especially with respect to the discovery of important unobserved objects such as new planets or chemical elements.

It seems to me that there is a certain epistemic bias presupposed by many authors about such discoveries. This is the position that before an object is discovered there can be no genuine *de re* knowledge of that object.[2] On this view, before Neptune was discovered all we had were predictions about this planet which some may have believed to be true, though this did not amount to *knowledge* until the discovery. After all, this is what seems to make discoveries so important; they lead to significant progress in knowledge. The day that Neptune was discovered it does seem that scientists learned something new about our solar system that they did not know earlier. This I take to be the received view. In this short essay I wish to argue that there is a strong case to be made for the opposite view, i.e. the view that it is possible to have genuine *de re* knowledge of undiscovered objects, and hence at least some predictive discoveries do not extend our knowledge, at least not in the way that is traditionally understood.

2. ACCIDENTAL VS. PREDICTIVE DISCOVERY

Both the scientist and the layman run into things they have never seen or even imagined. Let us call such a phenomenon an "accidental discovery", in case the

G. Irzık & G. Güzeldere (eds.), Turkish Studies in the History and Philosophy of Science, 153-162.

discovery is not justifiably and correctly foreseen in advance. Not every accidental discovery however is purely accidental. One may by accident run into an old historic monument totally unexpectedly; but one may also run into such a monument with some expectation, by knowing that there are such structures around in that area. The same goes for scientific discoveries. Let us then set aside discoveries that are totally unexpected since they are for our present purposes the least important ones. What remains allows for an interesting contrast: discoveries that are expected but based on accidental predictions, and discoveries that are expected based on non-accidental predictions. Of course the skeptic would challenge the view that there are genuine examples of the latter kind, or that we can know that there are such examples in the history of science. I do not claim to have an argument against the skeptic here. This would presumably require a solution to Hume's induction problem. Rather I will assume what common sense says: not all scientific predictions are accidental. If anyone is inclined towards the skeptic's position, then perhaps he should take all that I will say as an argument for a conditional such as "If there are non-accidental scientific discoveries, then. . . . .".

I wish to take two examples from the history of science in order to contrast the two types of discoveries. I will assume that the discovery of Neptune is an example of a non-accidental discovery, and the discovery of Pluto is an example of an accidental one.[3]

The discoveries of Neptune and Pluto were both based on predictions. The discrepancy between the observed and the theoretically calculated orbit of Uranus gave rise to the prediction of the existence of Neptune, and the discrepancies between the observed and the theoretically calculated orbits of both Neptune (after it had been discovered) and Uranus gave rise to the prediction of the existence of Pluto. Given that it was later established that Pluto is too small a planet to perturb Neptune and Uranus the discovery of Pluto is considered to be accidental. By contrast, Neptune in fact (observably) perturbs Uranus. An analysis of the difference between the two cases, I believe, will reveal that the existence of Neptune and certain facts about it, were knowable *de re* before this planet was discovered.

3. INOSTENSIBLE REFERENCE

In order to make a prediction about an undiscovered object it seems that there has to be a special kind of reference made to that object. In the primary sense of the word 'reference' is a two-place relation that holds between an expression in a language and some object (given a certain context), for example between the name 'Pluto' and the planet Pluto, the general kind term 'carbon' and the carbon atom, or the definite description 'the youngest person in the room here now' and whoever happens to be the youngest person here now. There is also a sense of the term 'reference' that is derived from the first (though some authors take this to be the primary and even the only sense) according to which reference is a three-place relation between a person, an expression and a object (given a certain context). For example I can use the name 'Athens' to refer to this city we are in now. So in this sense of the term, it is not the expression but rather a speaker that refers. And this is also a legitimate way of using the term 'reference'. Nevertheless the relation between the name of Athens and the city of Athens is the primary sense of the notion; it is this two-place relation between the name and the city that allows me to refer to this city by using that name.[4]

The semantic relation of reference between an expression and an object is established in part by human convention. Under normal circumstances we have an object that we experience, and we decide to give it a name, or use some description to talk about that object. I call the reference made by a speaker to an experienced object 'ostensible reference', and the expression the speaker uses in making that reference an 'ostensible term' for that speaker. This I take to be the most frequent way of referring.[5]

There is also a second type of reference that signifies a very important aspect of human language. That is the reference made to unexperienced objects, or more generally, reference to an object without knowing to which object the term one is using refers. I call such reference 'inostensible reference', and I shall say that a term is inostensible for a speaker if the speaker has not experienced the referent of the term or does not know to what or to whom it refers.[6] The ostensible/inostensible distinction has its application in everyday life experiences. The expression 'the man with the red tie' when there is such a man in front of me is an ostensible term for me, but the description 'the youngest person here now' is inostensible until I find out who that person is. Obviously the same name may be inostensible at one time and ostensible at another, and the transition from one to the other may be an indicator of growth of knowledge. Since we are not particularly interested in personal experiences now, let us relativize the distinction to the whole scientific community. For example the name 'Venus' is an ostensible name for the scientific community, given that this planet has been experienced (and we know which heavenly body it is), but the name 'the smallest red dwarf in the universe' is inostensible assuming that it is not known to which star this expression refers. It is this latter application of the distinction to the scientific community in which we are primarily interested.

Now it seems that the prediction of the existence of an unobserved object requires inostensible reference to that object. If there were no such reference, then the predictions would not be specifically about the predicted object. In the case of the discovery of Neptune, careful calculations were made by Le Verrier and Adams about the mass, position and orbit of an unobserved eighth planet in order to explain the perturbations in the orbit of Uranus. Before the discovery took place both scientists together with other co-workers were already using certain phrases that may be taken to refer at the time to Neptune. So the common-sense view is that just as we refer to tomorrow before having experienced it, Le Verrier referred to Neptune before it was discovered. For example we can imagine Adams having used the phrase 'the planet perturbing Uranus' to talk about Neptune (and Le Verrier may have used some French equivalent). If it is indeed a fact that Neptune is that planet, then Adams by using this inostensible description was referring to Neptune before the discovery. Though it may not be historically accurate we can imagine Adams and Le Verrier (or some other scientist) deciding to name the planet before the discovery. After all, there are numerous examples of such attempts in the history of science. The name 'Planet X' is suggested as the name of a tenth planet that is responsible for the discrepancies in the orbits of certain planets and Halley's comet. Similarly, based on the hypothesis that our solar system is a part of a binary star system, the name 'Nemesis' has been suggested as the name of the so-called companion star to our sun.

It may be argued that before the discovery takes place, we do have sentences about the object to be discovered, but there is no reference to the object in such sentences. For instance, on this view, before Pluto is discovered we have sentences *about* Pluto (in some sense of 'aboutness' that I will explain shortly), but there is no reference to Pluto in those sentences. This may be the case if those sentences are all general sentences, i.e. sentences with the existential quantifier in the front. For example, we may take the following sentence to be a prediction about the existence of Pluto:

(1) There is an unobserved ninth planet.

If there are exactly nine planets, there is a good sense in which (1) is about Pluto (before Pluto is discovered), since it would be Pluto and no other object that satisfies the property of being an unobserved ninth planet. However there is no reference to Pluto in (1) since there is no singular term, such as a proper name or a definite description, that makes reference to Pluto.

But obviously the prediction need not be stated in this form. In fact the scientists who predicted the existence of Pluto did believe that there was a particular planet that is responsible for the alleged perturbations in the orbits of Neptune and Uranus. They were looking for such a planet, and thus the following would capture their prediction better:

(2) There is a unique unobserved planet causing the perturbations in the orbits of Neptune and Uranus.

Still, (2) is a general proposition. Nevertheless, given (2), now a definite description can be formed such as 'the planet causing the perturbations in the orbits of Neptune and Uranus'. It is not unreasonable to suppose that some description similar to this was in fact employed to express other predictions. One of them may have been this:

(3) The planet causing the perturbations in the orbit of Neptune and Uranus is more distant from the sun than Neptune is.

However given that it was later established that Pluto is too small to cause the so-called perturbations, and that the discovery was in fact, at least partially, an accident, (3) does not have a singular term that refers to Pluto.

In the case of predictive discoveries that are not accidental we have singular terms that are in fact satisfied by the to-be-discovered object. Perhaps the discovery of Neptune is such a case. Both Le Verrier and Adams tried to calculate the orbit and the mass of a planet that would explain the discrepancies between the observed and theoretical data regarding the orbit of Uranus. They may have predicted not only that

(4) there is a unique planet causing the perturbations in the orbit of Uranus, but also that

(5) the planet that is mainly responsible for the perturbations in the orbit of Uranus is a giant planet.

Given that Neptune was discovered on the basis of these calculations and that it is the major reason for the so-called perturbations, then (5) does have a singular term that refers to Neptune.

I take it to be obvious that (5) is about Neptune, given that its presupposition is true, i.e. that Neptune is in fact the planet that is dominantly responsible for the perturbations in the orbit of Uranus. And if that is so, perhaps, anyone who would have assertively uttered (5) (or its synonym in some other language) would have referred to Neptune even before the discovery.[7]

Similarly there may have been reference to Pluto before its discovery, despite the fact that it was discovered by accident. The singular term in (3) does not refer to Pluto, however a description such as 'the ninth planet in order of distance from the sun' does. And there is no reason not to believe that such descriptions that in fact do refer to Pluto were employed by scientists before the discovery. Hence the difference between the two cases is not one of reference.

4. INOSTENSIBLE *DE RE* KNOWLEDGE

The fact that we can refer to something just by using a term that designates that thing is, in and of itself, not significant for the philosophy of science. What is significant is whether such reference does allow for knowledge of the object in question, and if so what type of knowledge this is. The fact that it is possible to use an inostensible definite description to refer to an entity does not necessarily guarantee that such reference enables us to have certain type of beliefs about the object. I can refer to the next winning lottery number, if reference to future contingencies is at all possible, and perhaps express certain propositions that I may believe to be true of this number. And this may, at times, amount to knowledge. For instance, it does not take much to know that the next winning number in the lottery will make some person rich. Such a piece of knowledge is what is usually called *de dicto* knowledge, and does not allow for *de re* exportation: I do not know *of* a certain number that it will make someone rich. And it is quite obvious that after the drawing, when I find out what number won, I cannot truthfully say, by pointing to the number, that I knew of *this* number that it would make someone rich.[8] The latter is *de re* knowledge whereas the former is *de dicto*, and not every *de dicto* piece of knowledge allows for *de re* exportation.[9]

What is interesting, and I believe quite controversial, is the question of whether we can have *de re* knowledge of an undiscovered object. The fact that Le Verrier or Adams had some knowledge about Neptune before the discovery is by itself not very striking. If they were justified in their beliefs that there is some unique unobserved planet causing perturbations in the movement of Uranus, then they did have at least some trivial knowledge about Neptune *de dicto*. But did Le Verrier or anyone else know before the discovery anything about Neptune that could amount to *de re* knowledge? What would show that this in fact was the case?

A semantic indicator of *de re* reference is that the term making such a reference is open to substitution even in oblique contexts. Sentences containing such terms express propositions that are generally called 'singular propositions'. Such a sentence contains a singular term that directly refers to a specific entity. Genuine proper names and pronouns that are not merely abbreviations of descriptions are prime candidates for such terms. I know that the youngest person here now is younger than everyone else, and that piece of knowledge seems to be trivial when taken *de dicto*.

Suppose I decide to name that person 'Young', without knowing who it is. I introduce the name as a genuine name of that person and not as an abbreviation of the description 'the youngest person here now'. Using Kripke's terminology, I use the description (which expresses a contingent property) merely to "fix the reference" of the name.[10] The sentence 'Young is younger than everyone else' should then express a singular proposition, and if I do claim to know this, it would be a non-trivial piece of *de re* knowledge. If I know something *de re* of someone, then it should not be wrong for me to go up to that person and tell him or her this by using the personal pronoun.[11] So when I discover who the youngest person here is, if it is not wrong for me to go up to that person and say "I knew that *you* were the youngest", that would be an indicator that I have *de re* knowledge of this person before I find out that he was the youngest. Most of us have the intuition, I guess, that it would be wrong for me to go up to that person and say such a thing, and therefore that there is no *de re* knowledge in such a case. Such intuitions have been used in the literature to generalize, I believe wrongly, that there can never be *de re* knowledge expressed by an inostensible term.[12]

Now we have a name for a possible companion star, 'Nemesis'. If there is such a star, then perhaps we can take the name to be a genuine name. So if there is any piece of knowledge expressed by a sentence with the name in it, it would be *de re* knowledge. Of course the fact that we have a genuine name of an object does not imply that we can use that name to refer to that object directly and express *de re* knowledge. The mere fact that we have a name of a star is not sufficient to claim we have *de re* knowledge of it.

Do we have *de re* knowledge about an object before its discovery?

In the case of accidental discoveries, I believe this definitely is not the case. It is implausible to hold that scientists had *de re* knowledge of Pluto before it was discovered.[13] Nevertheless they may have been able to refer to it and even know things about it *de dicto*. The reason being no observed impact of Pluto was detected *as the impact of Pluto*. What was thought to be an impact of Pluto, namely that it perturbs the orbits of Neptune and Uranus, turned out to be not the case. Thus what could have been known about Pluto could not have been anything coming from observations that causally relate to this planet. So there was no noticed *impact* of this planet on us, and what could only have been known were some properties of Pluto that we already have about all planets, or what can be deduced from descriptions that we have that refer to Pluto. This is similar to the type of knowledge I can express now about the youngest person here; I know that he or she is young and younger than anyone else is. I may deduce from my background knowledge that this young person may not have published anything yet etc. But none of this could amount to *de re* knowledge.

Similarly it is indeed true that we do not have *de re* knowledge of the shortest spy (or the first baby of the next century, etc.) Any knowledge we may have of such an individual will be deduced from our knowledge of short people or spies in general. I know that the shortest spy is a spy and that the shortest spy is a liar (given that I know that all spies are liars), but I do not have any specific knowledge about this person that causally relates to that person. The motivation for us to refer to such an individual does not have anything to do with that individual. We have not observed a *causal* impact of the person. The situation is quite different when there is such a noticed impact. Consider the case of Unabomber before he was caught (assuming that he has been caught).[14] The name 'Unabomber' then was being used to refer to an "unknown" man, and thus was an inostensible name. Such inostensible reference is quite different from the reference made to the shortest spy, for what motivated people to name and refer to the Unabomber were the terrorist acts that they had witnessed which he had been responsible for. The best way to contrast the two types of cases is to observe how we come to know the existence of the entity in question. I know that the shortest spy exists because I know that there are spies and no two people are the same height. This is quite different from the way in which we came to know that Unabomber exists. No general knowledge about bombers (or people in general) was sufficient to come to know such a thing. It was his bombings that gave rise to the knowledge of his existence. In general, the fact that we know that something exists is not sufficient for its having been discovered. We know that there exists a cause of the extinction of dinosaurs, but that cause is not yet discovered. We know that the 10 billionth digit after the decimal point of π exists, but what that number is remains undiscovered. In certain cases the knowledge of the existence of the entity is a result of a causal impact of that entity on us. The discovery of the element helium is also such a case. I believe that it is correct to say that the existence of helium was discovered before *it* was discovered. A certain yellow line in a light spectrum that was observed first by Pierre Jansen (who initially thought it to be a sodium line) was later correctly interpreted by Edward Frankland and Joseph Lockyer as a line caused by some undiscovered element. They knew then that the element causing this yellow line existed, which only later was discovered to be helium. I believe, at this point, that both scientists knew *de re* of helium that it existed (regardless of whether they used the name "helium" or not), for they had observed a causal impact of this element. It is exactly this feature of existential predictive discovery that enables us to have genuine *de re* knowledge of the undiscovered object.

5. CONCLUSION

There is an important epistemic difference between accidental and non-accidental discoveries. On the assumption that the discovery of Pluto and the discovery of Neptune are examples of the former and latter respectively, we should be able to conclude that the discovery of Neptune was based upon justified true evidence whereas the discovery of Pluto was based upon false (but perhaps justified) evidence. Scientists observed some phenomenon that was indirectly but causally related to Neptune. They experienced a causal impact of this planet, which led them to predict its existence and allowed them to form certain beliefs about it before its discovery. They believed that it was located at such-and-such a location in the sky, they believed that it had such-and-such an orbit, and that it was of a certain size, had a certain mass, etc. All this turned out to be (roughly) correct. They had sufficient justification for their beliefs, for they were based on observations that were causally related to this planet's physical impact. For reasons like these I conclude that there was genuine *de re* knowledge of Neptune before its discovery. If this is true, then it would be wrong to say that the discovery of this planet extended our knowledge, at least not in the way it is typically assumed. Though of course the discovery later led to new knowledge about certain properties of Neptune, certain fundamental things were known by at least a few scientists before the discovery took place. None of this of course is true in the case of Pluto.

In general my main thesis is this: simply having the means to refer to an entity is not sufficient to have *de re* knowledge of that entity; however in case insotensible reference is motivated by an observed impact of that entity, then genuine *de re* knowledge is possible; thus there can be genuine (inostensible) *de re* knowledge of undiscovered objects.

6. NOTES

1 This view is most explicitly stated in Imre Lakatos, Introduction to *Scientific Research Programmes: Philosophical Papers, Volume I.*

2 I am not sure whether within the philosophy of science this issue has ever been explicitly debated. Within the philosophy of language the debate over Kripke's argument for the existence of contingent *a priori* propositions, and also some of the literature on the *de re-de dicto* distinction is definitely closely related. There are references to this literature in the following footnotes.

3 Not everyone may agree with this. Some may wish to consider the discovery of Neptune as being accidental as well, given that Newton's theory was used to predict its existence, and that this theory is false. Such an objection however does not need to be answered here. I would ask the reader then to find another case of a non-accidental discovery to be substituted in place of my example.

4 Kripke makes a distinction between "semantic reference" and "speaker's reference" (which he extracted from Grice's more fundamental distinction between "utterer's meaning" and "sentence meaning") and argues that the two may at times diverge from one another. See his "Speaker's Reference and Semantic Reference." The two-place relation of reference that I have talked about in the main text is the relation of semantic reference.

5 This is an empirical claim of course. We may imagine a language and some users of that language who do not normally refer in this manner. Perhaps there is such a natural language that is in use now, though I strongly doubt that this is the case.

6 By "experience" I do not mean just sensory experience. I also wish to include experience of abstract entities such as numbers, properties, mental states etc.

7 This would be the case even if Leverrier and Adams were not justified in their beliefs about Neptune before the discovery. Reference to an object does not require justification in believing that the object exists. If, for instance, our sun does have a companion star, then we do refer to it when we use the description 'the companion star', or some such term, regardless of whether we are justified in believing this to be the case. At times, those who deny the existence of an object that in fact exists, actually refer to the object in question (even if they believe that they haven't). If God does exist, not only the theists but also the atheists do refer to some being. So if we really have a companion star, then both parties to the debate over the existence of that star do in fact refer to the same entity. My view is that speaker reference does not require epistemic justification, and I do presuppose this thesis in the main text. However the main argument can be given without it.

8 As Donnellan once put it, the fact that I know that the 98th prime number is not divisible by 3, is by itself not sufficient to conclude that I know *of* the 98th prime that *it* is not divisible by 3. See his "The Contingent A Priori and Rigid Designators" for this example and a very interesting discussion of it.

9 Keith Donnellan [1979] made use of this consideration to argue against Kripke's argument for the possibility of contingent *a priori* knowledge. He correctly argues, by using David Kaplan's example, that it is too soon to have any *de re* knowledge of the first baby to be born at the turn of the century. Kaplan had claimed earlier that he could dub such a child "Newman 1" and use this name to refer to this unborn baby. Donnellan appeals to our intuition that it would be wrong to go up to the first baby of the next century

when we find out who he or she is (and he or she has grown up) and say to him or her "I knew that you would be the first baby of this a century long before you were born." I agree with Donnellan that it would be wrong to say such a thing. However I do not believe that this by itself shows conclusively that we do not have such a piece of knowledge now. At times saying something that is indeed true may be wrong from a pragmatic consideration. We have Moore's paradox as a very good example of this. But even if we grant Donnellan this claim, there still is an overgeneralization here. From one type of case Donnellan has jumped to all such cases. His implicit thesis is simply this: Inostensible reference never gives rise to *de re* knowledge.

[10] See Kripke's *Naming and Necessity.*

[11] Donnellan's argument very briefly sketched above in footnote 10 suggests this as a test to show whether there is *de re* knowledge in such cases.

[12] Quine, in his classic "Quantifiers and Propositional Attitudes" was one of the pioneers to argue for the philosophical significance of the *de re/de dicto* distinction. However after having changed his mind on the matter several times he decided that the *de re/de dicto* distinction is "empty". (See his "Intensions Revisited.") I believe that Quine's argument for this position is not convincing. For a discussion of this see my Ph.D. Dissertation, Chapter 3.

[13] There is a special case however: one may have knowledge of the entity in question without knowing it is that object. This would be the case for instance if Pluto was in fact observed before its discovery but was thought to be a star or some other heavenly body.

[14] 'Unabomber' was the name given, I believe by the American media, to the person who was responsible for many bombings in the United States before he was caught.

7. REFERENCES

Donnellan, K. "The Contingent A Priori and Rigid Designators." In French, Uehling, and Wettstein, (eds.), *Contemporary Perspectives in the Philosophy of Language*, Minneapolis: University of Minnesota Press, 1979, 45-60.

Inan, I. *Inostensible Terms: Epistemological and Semantical Issues in the Theory of Reference*, University of California, Santa Barbara, 1997.

Kripke, S. *Naming and Necessity*, Cambridge: Harvard University Press, 1972.

Kripke, S. "Speaker's Reference and Semantic Reference." In French, Uehling, and Wettstein, (eds.), *Contemporary Perspectives in the Philosophy of Language*, Minneapolis: University of Minnesota Press, 1979.

Lakatos, I. *Scientific Research Programmes: Philosophical Papers, Volume I*, J. Worrall & G. Currie (eds.), New York: Cambridge University Press, 1978.

Quine, W. "Intensions Revisited", In French, Uehling, and Wettstein, (eds.), *Contemporary Perspectives in the Philosophy of Language*, Minneapolis: University of Minnesota Press, 1979, 268-274.

Quine, W. "Quantifiers and Propositional Attitudes." *Journal of Philosophy* 53 (1956): 177-187.

A. KARANFIL SOYHUN

IMPLICATIONS OF THE DISTINCTION BETWEEN SEMANTICS AND PRAGMATICS FOR PHILOSOPHY OF SCIENCE

One of the main principles of the Pragmatic Defense[1,2] (PD) is the theory of Direct Reference[3] (DR), which can be stated as follows:

> (DR) is the view that the only semantic-linguistic function of a proper name is to pick out a unique individual as its reference.

DR has direct implications for philosophy of science. If DR is correct, the meaning of proper names cannot be theory-laden, even in scientific contexts.[4] For proper names have stable, simple meanings, which are not theory dependent. But, if meanings of proper names are not theory-laden, then the support for the claims of meaning incommensurability becomes negligible with regard to proper names within scientific contexts. Hence a semantic issue has interesting implications for a thesis in philosophy of science.

However, DR faces a serious problem in the context of belief reports, for it seems unable to account for the apparent context sensitivity of proper names in this context. Given that belief reports were taken as a good indicator of semantic synonymy, this inability played an important role in the rejection of DR. But this difficulty also led to at least two further claims: (i) that meanings of proper names are, at least in part, context dependent, and (ii) that no real meanings can be found in belief reports, even though one cannot do without them.[5] These results, especially the rejection of DR and (i), pave the way for the claims of theory-ladenness and meaning incommensurability. In the first section, I discuss the problem DR faces, and explain how the responses to this problem might lead to the claims of theory-ladenness and meaning incommensurability.

In the second section, I introduce PD in detail, and show, by separating semantic and pragmatic components and giving semantic and pragmatic accounts of belief reports, that PD solves the problem of belief reports without introducing context dependency or skepticism.

In the third section, I discuss various implications of PD, which might be of interest to philosophy of science. If the account of PD is correct, one of the things we need to give up is using propositional attitude contexts as a test for semantic synonymy. For, according to PD, propositional attitude contexts are not semantic-

G. Irzık & G. Güzeldere (eds.), Turkish Studies in the History and Philosophy of Science, 163-176.

ally innocent, that is to say there is much pragmatic overlay in the ways we use the propositional attitude contexts, and thus they cannot serve as good indicators of semantic synonymy.

Moreover, the main interest of PD is in its separation of pragmatic and semantic components of belief reports. Any theory that tries to put pragmatics in semantics ends up with difficulties in accounting for semantics, communication, and various claims of context dependency, which in turn are responsible for many of the claims of theory-ladenness and incommensurability.

1.

In this section, I introduce the problem of belief reports which DR faces, and other similar difficulties in natural language which are traditionally solved with claims of context dependency. I will then show the kind of problems these kinds of responses face, and the way in which they open the way to the claims of theory-ladenness and incommensurability.

1.1. The Problem of Belief Reports

DR has unintuitive implications for the semantics of belief attributions. It has been argued that DR implies that the following two belief reports express the same proposition, and thus that they have the same truth value in every context, given that the names 'Clemens' and 'Twain' are coreferential.

(1) Lois believes that Twain is an author

(2) Lois believes that Clemens is an author

It is difficult to see how DR, by itself, a theory about proper names, implies anything about the semantics of belief attributions. After all, DR says nothing about belief attributions, or even how we get to propositions from sentences that contain proper names.

In fact we need one more principle in order to see the consequences of DR in belief contexts:

> (PC) Principle of Compositionality: what a sentence expresses (i.e., a proposition) is a function of the semantic contributions of its parts

It seems that one of the above principles must be wrong, because the two principles together imply that (1) and (2) must express the same proposition[6], and thus they should have the same truth-value in all contexts. For, according to PC, propositions are a function of the semantic contributions of (1) and (2)'s parts. But the only sentential difference between (1) and (2) is that 'Clemens' occurs in the latter where 'Twain' occurs in the former. However, according to DR, coreferring names have the same semantic-linguistic function, since they both pick out the same individual. Given that 'Twain' and 'Clemens' pick out the same individual, these two names must make the same semantic contribution to sentences in which they occur. Thus both (1) and (2) should express the same proposition:

(3) < Believes, Lois, < Being an author, Twain>>

So, according to the two principles[7], (1) and (2) are semantically equivalent. Thus if (1) is true, so is (2) (and vice versa).

However, contrary to DR's claims, it seems that in some contexts (1) and (2) may have different truth values. For example, if Lois does not know that the names 'Twain' and 'Clemens' corefer, (1) and (2) seem to have different truth-values. If (1) and (2) have different truth-values, then they must express different propositions. Given PC, the differences between (1) and (2) can only be due to the differences of the semantic contributions of their parts. Since the only sentential difference between (1) and (2) is that 'Clemens' occurs in the latter where 'Twain' occurs in the former, it follows that, contrary to DR's claims, the proper names 'Twain' and 'Clemens' differ in their semantic contributions.[8,9]

I discuss PD's response to this problem in detail in the second section. There are interesting similarities between the responses to some of the difficulties in other areas of semantics, and the various responses that philosophers have made to the difficulties in belief attributions. It seems that the kind of responses that do not keep the boundaries of semantics and pragmatics separate face similar problems.

1.2. Other Problems in the Semantics of Natural Language

Providing a semantic theory of belief reports has been difficult. However, providing semantics for other components of natural language has also been difficult. I argue that the reasons are similar in both cases. The difficulties concerning the semantics of the word 'and' are well known. It seems that 'and' is sometimes used to mean 'and then,' the temporal operator. But other times, it seems to work as a truth functional connective. Levinson shows that when faced with such apparent changes in the meanings of words, theorists have resorted to two types of solutions:[10]

(i) some claim that the words in natural language are *ambiguous*,

(ii) some claim that the meanings of words are vague, and versatile, and are often influenced by 'collocational' environment.

These two solutions lead to extremely complex semantics and further difficulties. The main problem with the ambiguity solution, (i), is the fact that there will be a need to grant an apparently endless increase of senses to the simplest words, such as 'white'. For sometimes 'white' is used to mean 'wholly white', as in the following example:[11]

(4) The flag is white.

while other times to mean 'partly white':

(5) The flag is white and red.

Furthermore, it becomes increasingly difficult to find a systematic way to explain and predict which words are ambiguous. It seems that there is no set criterion according to which we can claim a word is ambiguous.

One of the main problems with the latter approach, (ii), is just how the hearers know which variation in the value of 'white' is involved in (4) or (5). Furthermore, when somebody, after uttering (4), and thus adopting the 'wholly white' interpretation, may add as an afterthought that 'and in fact it is also red and blue', then the outcome (6) should be an outright contradiction, yet (6) is a perfectly plausible statement.

(6) The flag is white, and in fact it is also red.

Notice that both types of solutions will be context dependent. For disambiguation usually requires contextual information. Both of these accounts require a complex, context sensitive semantics for almost every term. But giving context so much power either to disambiguate or to determine the meaning of the term in question may lead to the claims of theory-ladenness and meaning incommensurability. For the context of a scientific term will in large part be the theory in which it is used. I am not arguing that one has to go in either direction. However, I think that both claims become more plausible when one is faced with such difficulties. I argue in the second section that one can avoid all these difficulties by separating the semantic and pragmatic components of utterances. But, before going into this solution, I will show in the next subsection that claims of theory-ladenness and meaning incommensurability become even more plausible for proper names when one rejects DR in favor of the competing theory of meaning.

1.3. Rejecting DR

The theorists who rejected DR usually adopted some version of the descriptivist account for proper names[12]. According to descriptivism, a definite description, or a set of descriptions, gives the meaning of a proper name. This description or set of descriptions has to be sensitive to the context of the utterances; otherwise it cannot account for the apparent context sensitivity of proper names in the context of belief reports. So the meaning of a proper name which is used in a scientific theory will be determined by the theory, for the theory is going to be an important part of its context. Hence the meaning of the name will be theory-laden. More importantly, when the theory changes, the meaning of the name may also change, for the new theory may assign a different description or, more likely, a different set of descriptions to the name in question, thus lending support to the claims of meaning incommensurability. Good support for my claims comes from Kuhn's writings.[13] In the following passage, Kuhn claims that pre-Copernican astronomers and Copernican astronomers both use the name 'Earth', but they assign incommensurable meanings to it.

> Part of what they meant by 'earth' was fixed position. Their earth, at least, could not be moved. Correspondingly, Copernicus' innovation was not simply to move earth. Rather, it was a whole new way of regarding the problems of physics and astronomy, one that necessarily changed the meaning of both 'earth' and 'motion'. Without those changes the concept of moving earth was mad.[14,15]

In other words, let A be a proper name used in the scientific theories T_1 and T_2. Let the-F and the-G be sets of descriptions. If in T_1, the-F gives the meaning of A, and in T_2 the-G gives the meaning of A, then despite the fact that these two theories use the same name, the meaning of the name changes from one theory to another. It is clear from the example above that the F and the G can have very different and in fact contradictory properties. More importantly, the words in the sets may also have very different meanings in those theories as well. Thus, especially if the theories are very different from each other, there will be the rampant meaning incommensurability which so excited Kuhn and others.

To sum up, rejecting DR and accepting descriptivism as well as the claim that names have context sensitivity, leads to the claims of theory-ladenness and meaning incommensurability. I think a good way to avoid these results is to take a good look at the semantic-pragmatic distinction. Separating these two components of language solves both the puzzles created by simple terms like 'white' and 'and' without leading to complex, context sensitive semantics and the problem of belief reports. The latter solution also avoids a complex context sensitive semantics, as well as the claims of theory-ladenness and meaning incommensurability for proper names. This solution will be introduced in detail in the next section.

2.

In this section, I introduce the Gricean tool of conversational implicatures in detail, and show how we can solve both the problem of belief reports and the problem of the simpler terms. In the last subsection, I will briefly discuss the implications of PD for philosophy of science.

Grice distinguishes semantic and pragmatic components of utterances. The Gricean solution is based on his analysis of "implicatures". Grice argues that utterances of sentences, with the help of the context, can convey information which is not expressed semantically by the sentence. Hence, there can be situations in which what the sentence semantically expresses is true, but the conveyed information, which is called a conversational implicature, is misleading. According to this account the semantic contribution of 'and' and 'white' is the same in most cases. However, (4) conversationally implicates that the flag is purely white, and this implicature is cancelled in (6). The semantic meaning of the term 'and' is simply its normal truth function. A sentence like "He fell asleep, and took a sleeping pill" does not have a semantic anomaly. It has a pragmatic anomaly, for it is violating some of the pragmatic rules that I will introduce shortly. As Levinson shows, within the Gricean theory, the difficult problems that trouble the other two solutions are actually foreseen and resolved; furthermore, the theory allows us to have a simple semantic core with expressions mostly having simple and stable semantic contents. It seems that this pragmatic tool, added to a simple semantic base, can successfully account for the complex, context sensitive, and unstable nature of language use, without providing much ground for claims of incommensurability.

PD will argue that the same Gricean tool can solve the problem of belief reports. According to PD, belief reports of the sort:

(7) A believes that S

where A is the believer, and S is an English sentence, are always used to report the propositional content of the sentence S. However, the sentence S conveys further information, and thus there is a closely associated conversational implicature to most belief reports. Thus PD claims that although (1) and (2) in our earlier examples express the same proposition, they have different conversational implicatures. (2) seems false because its implicature is misleading in some contexts. By keeping the semantics of belief attributions as simple as possible, and by showing that the difficulties arise because of conversational implicatures, which fall within the realm of pragmatics, PD avoids the various difficulties that arise as a result of not separating the realm of pragmatics from the realm of semantics for belief attributions.

2.1. Conversational Implicatures

In this section, I will explain in more detail Grice's Theory of Implicatures. This will help me construct the implicatures arising in belief reports. Grice argues that there are basic principles that direct the conduct of conversation. These principles are called "Maxims of conversation" by Grice, and according to him, they are based on rational considerations. Taken together these Maxims express a general "Co-operative Principle" which is specified by Grice as follows: **"Cooperative Principle:** Make your conversational contribution such as is required, at the stage at which it occurs, by the accepted purpose or direction of the talk exchange in which you are engaged."[16]

For our purposes the relevant maxims of conversation are expressed as follows:

> **Maxim of Quality:** try to make your contribution one that is true, specifically:
> (ii) do not say that for which you lack adequate evidence
> **Maxim of Relevance**: make your contributions relevant
> **Maxim of Manner:** be perspicuous, and specifically:
> (i) avoid obscurity (ibid.)

Maxims provide the guidelines for efficient, rational conversations. More importantly, they produce inferences which are well beyond the semantic content of the uttered sentences. Grice calls these pragmatic inferences "conversational implicatures" in contrast with inferences such as entailments, consequences, etc.

The implicatures may be generated either by observing the maxims, hence giving rise to standard implicatures[17], or by flouting the maxims. For an example of the former consider the following exchange:

(8) A: I am out of petrol.
(9) B: There is a garage around the corner.[18]

B's remark implicates that A may find petrol in the garage. B would have been violating the maxim of relevance if he knew that the garage was closed or had no petrol to sell. Thus the implicature, that A will find the garage open, and it will have petrol.[19]

Grice defines conversational implicatures in terms of the Cooperative Principle and the maxims:

> The speaker S, by saying that p, conversationally implicates q
> iff
> 1) S's audience presumes that S is observing the maxims (at least the cooperative principle)
> 2) In order to make this assumption consistent with S's saying that p, the audience must presuppose that S thinks that q.
> 3) S thinks that the audience can determine that (2) is true, and has done nothing to stop the audience from thinking that q.[20]

When we apply this analysis to our example we can explain why B conversationally implicates that the garage has petrol for A: if we presume that B is observing the maxims, and assume that B thinks that the garage will be open, and that A can figure out that he thinks that the garage will be open etc.

There is another way of classifying implicatures, which will be important in the next section: generalized vs. particularized implicatures. The former is a kind of implicature that arises without a particular context, and thus does not require particular contextual information in order to be inferred. The particularized implicatures require specific context in order to arise, and to be inferred. The implicature of (4) is an example of the generalized form. The implicatures in (8) and (9) are particularized. Most of the floutings and exploitations of the maxims are particularized, for they require specific contextual settings.[21]

2.2. Semantic and Pragmatic Components of Belief Reports

In this section I will introduce both the semantic and pragmatic account of PD. In order to do this, I will first introduce Salmon's[22] account of the semantics of belief reports. This will help to clarify the distinction between what is semantically entailed, and what is pragmatically conveyed. According to Salmon, the analysis of belief attributions can be given in the following manner[23]. Belief reports express existential generalizations over the ternary relation that holds between a proposition and an agent and a way the agent is familiar with the proposition. So, taking Q to be the proposition expressed by the sentence S, the semantic content of (7) can be analyzed as:

(10) p_{literal}: $(\exists x)$ [B grasps Q by means of x & BEL(B,Q,x)]

Notice that in this account, even though it uses a three-place relation, since the way the agent is familiar with the proposition is not specified, but only quantified over, both (1) and (2) are semantically equivalent. In fact, they both express the following proposition:

(11) $(\exists x)$ [Lois grasps < Being an author, Twain> by means of x & BEL(Lois, < Being an author, Twain>, x)

Now, we can try to specify the conversational implicatures of belief reports. I will call these *b-implicatures*. There are different kinds of conversational implicatures of belief reports: some are generalized and some are particularized. The standard, and generalized, implicature of the following belief report uttered by the speaker, A:

(12) B believes that S

where B is an agent, and S is an English sentence, is roughly the proposition expressed by:

(13) if B were to be presented with the sentence S, B would agree (assent) to the sentence S.

(13) is not the only implicature of (12), for there can be other implicatures of (12) arising from the particular contextual background. However, here I deal only with the generalized implicatures of belief reports.[24]

We can apply Grice's analysis to the above example in the following way. To begin with, we would presume that A is obeying the Maxim of Quality, and the sub-maxim of Manner that says to avoid obscurity. First, if the speaker is obeying the Maxim of Quality, and not saying something for which she lacks evidence, then she must have evidence that supports her utterance. Furthermore, a common sort of evidence for one's belief in a proposition is described by Kripke's "Disquotational Principle," DP:

> (DP): "If a normal English speaker, on reflection, sincerely assents to an appropriate standard English sentence, 'P', then he believes that P."[25]

Since A's evidence for her belief report is commonly B's assent, or assertion, or some other behavior that shows a favorable disposition towards the sentence, A will likely use that sentence or a similar one, as the embedded sentence.[26] For by doing so, she keeps her report as close to truth as she can.

Also, as important as the Maxim of Quality, the submaxim of Manner has a serious contribution here as well. According to the submaxim that says to avoid obscurity, the speaker should use a sentence with which the agent is familiar, as the embedded sentence. To show the importance of this requirement, I will provide an example where the requirement is not met. Let's say S expresses the proposition P, and so does S'. It happens to be the case that the agent, B, can recognize P via S, for S involves familiar guises, and S is the sentence the agent would use. But B cannot recognize P via S', for the guises involved in S' are unfamiliar to B. We can easily see that in such circumstances, if A uses S' instead of S, without signaling anything to her audience, then A is setting the stage for an obscure situation. For, in such circumstances, because of the Maxim of Quality, and DP, the audience will expect the agent to assent to S'. This expectation will fail, for B cannot recognize P, which she believes, via S'. Then the audience will be left wondering what went wrong. The speaker may have intentionally misled them. The speaker may have been misinformed. Or the agent may not be very rational. Hence, the obscurity. So making a belief report, using terms other than the ones the agent would use, can easily lead to difficult situations, and violate both the Maxim of Quality and the submaxim of Manner that says "avoid obscurity".

We can put this discussion in a more Gricean manner in the following way. We can assume that A obeys both the Maxim of Quality and the submaxim of Manner that says "avoid obscurity." Furthermore, because of DP, A knows that the audience will presuppose that A has good evidence that B is favorably disposed towards S (I'll call this presupposition "q"), and A has done nothing to stop the audience from thinking this way. Hence A, by saying that p, conversationally implicates q, for all three of Grice's conditions are satisfied:

1) A's audience presumes that A is observing the maxims.
2) In order to make this assumption consistent with A's saying that p, the audience presupposes that A thinks that q.
3) A thinks that the audience can determine that (2) is true, and has done nothing to stop the audience from thinking that q.

If this rough picture is correct, then we have a *standard implicature* here. Furthermore, as I have already argued, it is a *generalized implicature*, i.e., it arises without a need for a particular context. For, as our discussion above indicates, it is a well-established practice to use belief attributions to convey, not only the proposition to which the believer has an agreeable disposition, but also the way the believer is familiar with the proposition, which is often determined by the sentence. Thus the implicature arises without a need for a particular context.

The implicatures that are both standard and generalized are difficult to distinguish from the semantic content of linguistic expressions, for they are routinely associated with the relevant expressions in ordinary contexts. This phenomenon explains why there has been so much confusion in deciding on the semantics of belief reports, and determining the truth value of belief reports. However, it also necessitates that we show b-implicatures are indeed implicatures. This I will show in the next section, by showing that b-implicatures do indeed possess the characteristics of conversational implicatures.

Now we have some understanding of the ordinary b-implicatures, as well as the semantic content of belief reports. The semantic content of the belief report specifies only the agent and the proposition to which the agent stands in the believing relation. However, most often, the report implicates that B will assent to the sentence S.

2.3. The Testing of B-implicatures

So far I have claimed that there are certain implicatures that are ordinarily attached to belief attributions, and I have tried to show what they are, as well as how they work. I have already pointed out that generalized, standard implicatures are easily confused with the semantic content of linguistic expressions, for they are routinely associated with the relevant expressions in ordinary contexts. This helps to explain the confusion, but of course it also puts on me the burden of proof that b-implicatures are indeed implicatures. To show this, I will introduce Grice's tests for identifying implicatures, and show that b-implicatures pass these tests.

(1) Cancellability (defeasibility): Implicatures are deniable without a sense of contradiction. There are two ways of denying an implicature: (i) cancellation: the speaker is committed to the falsity of the implicature; (ii) suspension: the speaker is not committed to the truth or the falsity of the implicature. This characteristic is a direct consequence of implicatures being pragmatic tools in contrast to logical entailments, or part of the semantic content of the sentence. In our earlier example B said the following:

(9) There is a garage around the corner.

This has the implication that A will find the garage open, and it will be selling petrol. However, B could easily add the following remark.

(14) There is a garage around the corner, but I am not sure if it is still open.

(14) does not have the same implicatures as (9), and it does not create a sense of contradiction like the following:

(15) There is a garage around the corner, but I am not sure if there is.

Since there being a garage is part of the semantic content of (14), and is thus entailed by (14), it cannot be canceled without causing contradiction.

B-implicatures are cancelable. For example, the implicatures of (16):

(16) Martie believes that Lori is bright.

can be canceled in the following way:

(17) Martie believes that Lori is bright, though he would not have put it this way.

There is no contradiction in (17). However, the implicature that Martie would assent to the embedded sentence in (16) is canceled in (17).

Furthermore, implicatures can drop out in certain contexts, and so can belief implicatures. Assume that both the reporter, David, and the hearer of the belief attribution, Lori, are well aware that Martie listened to Lori's talk without knowing her name, and that he agreed with most of Lori's points. In this context, David's utterance of (16) will not commit him to the implicature ordinarily attached to (16), roughly:

(18) Martie will assent to "Lori is bright".

For it is clear from the context that Martie lacks an important piece of information that will enable him to respond to "Lori is bright" in a favorable manner (namely, her name).

(2) Non-detachability: Implicatures are typically attached to the semantic content of the utterance and not to the linguistic form. Hence, in most cases, if we replace the words of the utterance with their synonyms, there won't be any change with regard to the implicature of the utterance. This is due to the fact that implicatures arise from the proposition expressed and from the truth conditions. Hence most expressions with the same semantic content have the same implicatures.[27]

However, this does not hold for belief reports. DR is committed to coreferential proper names being synonyms, but when we replace one proper name with another coreferential one, the implicature attached to the utterance changes. To see this, we only need to change our previous example a little. Assume that Martie knows Lori under another name, 'Chris', and that Mike's audience, as well as Mike himself, have no idea about this lack of information. Then Mike, upon hearing Martie says

(19) She is bright.

while pointing at Lori, will utter (16), instead of the following:

(20) Martie believes that Chris is bright.

Even though (16) and (20) have the same semantic content, it seems that they have different implicatures. At least some of the implicatures of (16) are definitely misleading, though (20)'s implicature is not misleading. For Martie would assent to the embedded sentence in (20), but not to the embedded sentence in (16).

Thus, it might be argued that what I call b-implicatures are not conversational implicatures, for they are detachable. But notice that not all implicatures are detachable. Implicatures arising under the two submaxims of Manner, 'avoid obscurity' and 'avoid ambiguity', are *detachable*. For the two submaxims of Manner make essential reference to the surface form of the utterances, and hence they are important *exceptions* to the claim that implicatures are determined by semantic content, and not by surface structure.[28]

I have already argued that b-implicatures arise because of the Maxim of Quality and the sub-maxim of Manner that says to avoid obscurity. Thus, in order to avoid obscurity, they make reference to the surface structure. The fact that the surface structure sometimes plays an important role in specifying the b-implicatures, seems to account for their detachability.

(3) Calculability: Implicatures are calculable. That is to say, we can construct an argument showing how the literal meaning of the utterance, the co-operative principle, and the maxims, taken together, will lead the addressee to the implicature in order to preserve the principle of co-operation. The argument will be of the following sort:

(21) "**(i)** S has said that p
(ii) there is no reason to think that S is not observing the maxims, or at least the co-operative principle
(iii) in order for S to say that p and to be indeed observing the maxims or the co-operation principle, S must think that q
(iv) S must know that it is mutual knowledge that q must be supposed if S is to be taken to be co-operating
(v) S has done nothing to stop the addressee from thinking that q
(vi) therefore S intends me to think that q, and in saying that p has implicated q" [29]

Let's take (26) and give an argument similar to (21):

(22) **(i)** Mike has said that Martie believes that Chris is bright
(ii) there is no reason to think Mike is not observing the maxims (especially the maxim of manner)
(iii) in order for Mike to say that Martie believes that Chris is bright and be indeed observing the maxims, Mike must think that Martie will assent to the embedded sentence
(iv) it must be supposed that Mike must know that it is mutual knowledge that Martie will assent to the embedded sentence if Mike is to be taken to be co-operating
(v) Mike has done nothing to stop the addressee from thinking that Martie will assent to the embedded sentence
(vi) therefore Mike intends me to think that Martie will assent to the embedded sentence, and in saying that Martie believes that Chris is bright he has implicated that Martie will assent to the embedded sentence

Hence we may conclude that b-implicatures are calculable.

(4) Non-conventionality: Implicatures are not part of the conventional meaning of words. This is shown by the fact that they are calculable, and that an utterance can be true while its implicature is false.

I have already argued that b-implicatures are calculable, and that a belief report may express a true proposition, despite having a misleading implicature. Hence we can say that b-implicatures are non-conventional.

2.4. Solution to the Problem of Belief Reports

I have so far argued that one can hold DR and PC, accept that (1) and (2) are semantically equivalent, and still solve the apparent difficulties concerning truth value by appeal to conversational implicatures. In other words, both reports express the same proposition, and they are both true. However, they have different conversational implicatures. (1) implicates that Lois will assent to the following sentence:

'Twain is an author'

while (2) implicates that Lois will assent to the following sentence:

'Clemens is an author'.

(2)'s implicature is misleading and can be canceled by adding

'But she would not put it that way'.

This solution can be extended in different ways to account for the various versions of the problem of belief reports. However, for the purposes of this paper, this example seems sufficient. I have showed that b-implicatures have the characteristic properties of conversational implicatures, thus providing support for PD. In the next section, I will conclude by briefly discussing the implications of PD which may be interesting for the philosophy of science.

2.5. Conclusion

The pragmatic defense provides substantial simplifications in both the structure and the content of the semantics of belief reports. Furthermore, the pragmatic defense bridges the gap between what is literally said and what is conveyed, and thus it provides an account of how we communicate using belief reports, which explains away the difficulties that arise in belief reports. These would be good reasons for a philosopher of language to adopt it. However, I think that DR, which is a major component of PD, as I have already discussed, makes the claim of meaning incommensurability of proper names indefensible. For, if PD is correct, the name 'Earth' means Earth, and the descriptions associated with it by scientific theories play no semantic role. Thus there is no theory-ladenness, or meaning incommensurability when it comes to proper names. Kuhn was simply wrong about this. Granted, by using the name 'earth' different scientists may convey different information. But, this information is not part of the semantic meaning of the name, and it can be discussed independently. Moreover, nothing I have said so far has any implications about whether the term 'motion' means the same thing in different scientific theories. After all, DR is a theory about proper names. Nevertheless, the realm of proper names seems free of the claims of incommensurability, and this may help in anchoring conversation between different scientific communities adhering to different paradigms.

Also, despite Quine's skepticism about there being any real meaning in the context of belief reports, belief reports are used as a test for semantic synonymy by some philosophers of science. If PD's account is correct, it is obvious that one should not use this context to test for semantic synonymy, for it has also a rich pragmatic component.

3. NOTES

[1] This view is first introduced by Soames (1987a, 1987b) and Salmon (1986,1988). I simply expand their suggestions, and give a detailed explanation of the conversational implicatures of belief attributions.

[2] This name is first used by Crimmins (1992). I have been warned that as a name 'the Pragmatic Defence' is misleading. It would be a better name for a response to an objection, rather then a name for a theory that has both pragmatic and semantic components. In fact, in my dissertation, I use another name for PD (Soyhun 1999). However, I think the most ingenious, if not essential, characteristic of PD is its pragmatic component, and the defence this component provides. Thus, I think, the name fits. I hope to avoid any confusion by defining the theory in detail.

[3] DR is also called 'Millian View', 'the naive view', or 'Russellianism'. More detailed discussion can be found in Braun (1996), Crimmins (1992), Richard (1990), Salmon (1986, 1988), Soames (1987a, 1987b).

[4] DR is a theory about proper names, and thus nothing discussed in this paper has any implication with regards to kind terms like 'water'. However, theory-ladenness claims and claims of incommensurability are general claims, and they concern proper names. Whether the terms 'Sun' and 'Venus' can have the same meaning in both the Ptolemaic and Copernican paradigms is a good example of the generality of the claims of incommensurability and theory-ladenness.

[5] Quine (1960, 1966).

[6] I take propositions to be singular (Russellian) propositions. The notation I use is a matter of convenience. I don't think anything important depends on it.

[7] Ted Sider's comment on an earlier version of this paper was influential in formulating the problem in this way.

[8] Of course, one could give up PC in order to avoid this conclusion. This move is made by Crimmins (1992), and I argue against it elsewhere.

[9] The earliest versions of the problem of belief reports can be found under various names in both Frege and Russell.

[10] Levinson (1983), p. 98.

[11] This discussion in much more detail can be found in Levinson (1983), p. 98-100.

[12] Frege, Russell, Searle, and Strawson are among the many who adopted a version of descriptivism.

[13] I am thankful to Robert Nola for using this passage in his talk at Bogazici University.

[14] Kuhn (1970) p. 149.

[15] In this paper, I am interested in meaning incommensurability. But if one has a Fregean type of view, where the descriptions associated with the name determines the reference, one can see that reference incommensurability would be the next logical step. See Nola and Kroon (2001).

[16] Grice (1975) pp. 152-153.

[17] Levinson (1983) p. 104.

[18] Grice (1975) p. 155.

[19] If there is a garage around the corner, B's statement will be true. However, if the garage looks like it has been closed for years, A will have every right to be angry with B. For B, despite the truth of his statement, seems to have misled A knowingly.

[20] Levinson (1983) p. 113.

[21] For more detailed discussion see Levinson (1983) p. 126.

[22] Salmon (1986).

[23] This is not exactly what a belief report expresses; however it is logically equivalent to it.

[24] For instance, when we report the beliefs of people who speak different languages, the implicatures in question are surely not like (13). More detail on this can be found in the third chapter of my dissertation.

[25] Kripke (1979) p. 241.

[26] Evidence can come in many ways, but we can safely assume that A either heard such a statement, or has good evidence of some sort, such as hearing from a reliable third party and so on.

[27] Levinson (1983) p. 122.

[28] Levinson (1983) p. 125.

[29] Levinson (1983) p. 114.

BIBLIOGRAPHY

Crimmins, M., and J. Perry. 1989. "The Prince and the Phone Booth: Reporting Puzzling Beliefs." *Journal of Philosophy* 86: pp. 685-711.

Crimmins, M. 1992. *Talk About Beliefs*. London: MIT Press.
Frege, G. 1892. "On Sense and Reference". In A. P. Martinich (ed.), *The Philosophy of Language*. New York: Oxford University Press.
Frege, G. 1918. "Thoughts." In P. Geach (ed.), *Logical Investigations*. New Haven: Yale University Press.
Grice, H. P. 1957. "Meaning." In A. P. Martinich (ed.), *The Philosophy of Language*. New York: Oxford University Press.
Grice, H. P. 1975. "Logic and Conversation." In A. P. Martinich (ed.), *The Philosophy of Language*. New York: Oxford University Press.
Grice, H. P. 1981. "Presupposition and Conversational Implicature." In A. P. Martinich (ed.), *The Philosophy of Language*. New York: Oxford University Press.
Kripke, S. 1979. "A Puzzle About Belief." In A. Margalit (ed.), *Meaning and Use*. p. 239-283. Dordrecht: Reidel.
Kripke, S. 1980. *Naming and Necessity*. Cambridge, MA: Harvard University Press.
Levinson, S. 1983. *Pragmatics*. Cambridge: Cambridge University Press.
Kuhn, T. S. 1970. *The Structure of Scientific Revolutions*. Oxford: Oxford University Press.
Mill, J. S. "System of Logic" In J. M. Robson (ed.), *The Collected Works of John Stuart Mill*. Toronto: University of Toronto Press. Volume VII.
Nola, R. and Kroon, F. 2001. "Ramsification, Reference Fixing and Incommensurability" In Hoyningen-Huene and Sankey (eds.), *Incommensurability and Related Matters*. Dordrecht. Kluwer Academic Publishers. p. 91-122.
Quine, W. V. 1960. *Word and Object*. Cambridge, MA: MIT Press.
Quine, W. V. 1966. "Quantifiers and Propositional Attitudes." In A. P. Martinich (ed.), *The Philosophy of Language*. New York: Oxford University Press.
Richard, M. 1990. *Propositional Attitudes: An Essay on Thoughts and How We Ascribe Them*. Cambridge: Cambridge University Press.
Russell, B. 1956. "On Denoting." In R. Marsh (ed.), *Logic and Knowledge*. London: Unwin Hyman.
Sadock, J. M. 1978. "On Testing for Conversational Implicature." In P. Cole (ed.), *Syntax and Semantics 9: Pragmatics*. New York: Academic Press. p. 281-98.
Salmon, N. 1986. *Frege's Puzzle*. Cambridge, MA: MIT Press.
Salmon, N. and Soames, S. 1988. *Propositions and Attitudes*. Oxford: Oxford University Press.
Salmon, N. 1995. "Being of Two Minds: Belief With Doubt." *Nous* 29: pp. 1-20.
Saul, J. 1993. "Still an Attitude Problem." *Linguistics and Philosophy* 16: pp. 423-35.
Saul, J. 1996. *The Problem with Attitudes*. Ph.D. dissertation, Princeton University.
Soames, S. 1987a. "Substitutivity." In J. J. Thomson (ed.), *On Being and Saying: Essays for Richard Cartwright*. pp. 99-132. Cambridge, MA: MIT Press.
Soames, S. 1987b. "Direct Reference, Propositional Attitudes and Semantic Content." *Philosophical Topics* 15: pp. 47-86.
Soyhun, K. 1999. *The Theory of Direct Reference, Belief Attributions, and Conversational Implicatures*. Ph.D. Dissertation. University of Rochester.
Stich, S. 1983. *From Folk Psychology to Cognitive Science: The Case Against Belief*. Cambridge, MA: MIT Press.

MURAT AYDEDE

COMPUTATION AND FUNCTIONALISM: SYNTACTIC THEORY OF MIND REVISITED

1. INTRODUCTION

There is a thesis often aired by some philosophers of psychology that syntax is all we need and there is no need to advert to intentional/semantic properties of symbols for purposes of psychological explanation. Indeed, the worry has been present since the first explicit articulation of so-called Computational Theory of Mind (CTM). Even Fodor, who has been the most ardent defender of the Language of Thought Hypothesis (LOTH) (which requires the CTM), has raised worries about its apparent consequences. The worry can be put in the form of a question, which Fodor called the "Eponymous Question" alluding to the title of a chapter in his (1994) book:

(EQ) If cognition is computational, how can psychological laws be intentional?

This question has been haunting people working in the field since the publication of a paper by Stich in 1978 in which he gave his celebrated "autonomy argument". Then, as everybody knows, came Fodor's notorious "Methodological Solipsism" in 1980, in which he argued for the *formality condition*: namely, thought processes are causal sequences of symbol tokenings in one's language of thought (LOT), and the causal processes are sensitive only to the syntactic/formal properties of its symbols. Hence, he argued against what he called a "naturalistic psychology," i.e. a psychology whose laws essentially advert to broad semantic properties of mental states they cover. The alternative, rationalist psychology, according to Fodor, was to advert only to formal characteristics of symbols, of which Fodor conceived as narrow computational roles of LOT symbols.

Stich's 1983 book, *From Folk Psychology to Cognitive Science*, was the culmination of the worries. He turned these into a sustained argument against the possibility of a scientific *intentional* psychology (along with the common sense belief-desire psychology), and at the same time, for a syntactic way of doing psychology, i.e., for his Syntactic Theory of Mind (STM). He defended an eliminativist stance: STM involves the elimination of all intentional idioms proposed to be used in a scientific enterprise, hence it envisions a scientific psychology free of semantics. STM has been around for almost two decades now. It has generated a lot of discussion because it has usually been taken to articulate the paradox alleged to underlie the LOTH,

G. Irzik & G. Güzeldere (eds.), Turkish Studies in the History and Philosophy of Science, 177-204.

which was to vindicate intentional folk psychology through computationalism. For this reason, I will concentrate on Stich's book in what follows, and argue that the worries are altogether baseless, that a computational theory needs a semantic individuative scheme to get off the ground, and that the envisioned alternative, i.e. a pure CTM (or, STM) is a non-starter, and cannot do the required job. Although nowadays there are probably few adherents of STM, so far no one, to my mind, has left a lasting impression of having refuted the theory. Indeed, as recently as 1994, Fodor raised the worry and tried to answer it by showing the feasibility of its alternative, and not by directly attacking the syntacticalist claim. In what follows, I intend to leave a lasting impression: I hope to refute the STM and show that the kind of semantic-free psychology it envisions is impossible, thus answering Fodor's Eponymous Question.

In his book, Stich has argued basically for two claims. First, the application conditions of such intentional common sense predicates as 'believes that *P*' and 'desires that *Q*' are essentially vague, context-sensitive, observer-relative, and thus are not suitable to be used as stable projectable predicates in the vocabulary of a scientific psychology. In particular, according to Stich, since observer relativity partly stems from the fact that content ascriptions are essentially based on similarity judgments along different dimensions (see below) between the ascriber and the ascribee, a consequence of ordinary content ascriptions is that a certain form of parochialism will profusely infect our psychological theories if we insist on having a content-based psychology which, according to Stich, essentially relies on ascribing such contents to agents covered by its generalizations. This means that content-based psychologies are bound to miss many important generalizations about the psychology of children, exotics, perceptually or cognitively handicapped people, higher animals, etc., since any content ascribed to these will necessarily reflect how cognitively similar the ascribee is to the ascriber. In short, Stich thinks that content-based psychologies won't make respectable science.

His second general claim is that we don't *need* to advert to the content of mental states in doing psychology, "syntax" will be enough. Furthermore, we had better advert only to the syntactic properties of mental states if we don't want to miss any psychological generalizations: STM offers a paradigm that has all the virtues of content-based psychologies and none of its vices.

These two claims are relatively independent of each other: in particular, the truth of the latter does not depend on the truth of the former. If Stich is right in his second claim, then the falsity of his former claim, i.e. his characterization of contentful mental states as scientifically problematic posits, would imply that content vocabulary is at best otiose in doing scientific psychology. It is therefore important to see whether Stich is right in his second claim. In what follows I will argue for three basic claims.

The first is that when we see more clearly the nature of STM, as presented by Stich, the claimed superiority of STM over content-based psychologies totally disappears. Put differently, I will be arguing for a conditional claim: *if* Stich is right in his claim that content-based psychologies have the disadvantages he enumerates, *then* STM-style theories have exactly the parallel problems; so it is false that the STM framework is scientifically superior to content-based psychologies as conceived by Stich. Therefore, Stich will loose his primary motivation to promote STM.

Secondly, I will argue that STM can't do the required job: it lacks the necessary resources to type individuate particular psychological states *qua* mapped onto particular "syntactic objects" as Stich puts it.

Thirdly, I will show that the STM-theorist is, at any rate, committed to intentional vocabulary at some stage of theory construction. In other words, if the STM strategy is taken to claim, as Stich seems to intend, that it is possible and advisable to develop psychological theories without using any intentional scheme whatsoever no matter what the stage of theory construction is, then STM is false: psychological theory construction cannot get off the ground if the strictures of STM are firmly complied.

I will end by moralizing on Stich's failure, and point out that if STM ($\approx$ Narrow Causal Account of typing symbols over which computation is defined, as I will show) and type-type identity theory are false, then a content-based psychology (= intentional psychology) is practically mandatory. Hence, if cognition is computational, psychological laws have got to be intentional!

Since all my arguments crucially depend on what exactly STM is, I will present it in a way that its purely functionalist structure becomes explicit. However, before embarking on my criticism, I need to say a little about how Stich views the problem space within which he criticizes content-based psychologies and thus motivates his own alternative, STM. In particular, it will turn out that the exact way in which Stich motivates his STM is very important, since my arguments against STM partly rely on his own strategy.

2. THE PROBLEM SPACE ACCORDING TO STICH

Stich takes what he calls the Mental Sentence Theories as his starting point, and assumes their basic framework throughout his discussion. After a lengthy presentation of a Fodor-style LOTH and computationalism in general he raises the following problem:

> for a Fodor-style account of belief sentences to hang together, we must have some workable notion of what it is for two distinct people, speaking different languages, to have in their heads distinct tokens of the same sentence type. (Stich, 1983: 43-4)

On behalf of Fodor, he offers three possible solutions: One is the *Narrow Causal Account* (NCA), according to which two sentence tokens count as type identical if and only if (iff) they have the same narrow causal/functional role. Since this is going to be of some importance, let me elaborate on it a bit. According to Stich, "[t]o adopt this view of... psychology is to exclude any reference to noncausal relations... There can be no mention of a subject's social setting, natural environment, or personal history, nor of the psychological characteristics of other people" (1983: 22). This is what makes this kind of individuation *narrow causal*. It is narrow because the causal role in question is defined in terms of generalizations that detail nomological connections among *proximal* stimuli, behavior (like motor commands), and other central cognitive states. Moreover, the causal relations are given by a set of *counterfactual supporting* generalizations. Thus, for a mental state of an individual to count, say, as the belief that *P*, it is not necessary that the state *actually* play a causal role in the individual's mental economy; all that is required to be true of the state is that it *would play* a certain causal role if some other conditions specified in the generalizations *were to obtain*. So the notion of functional/causal role of a mental state should be so understood as to include the *potential* causal interactions that the state would enter. Accordingly, two mental states of two distinct organisms count as of the same type if their potential causal interactions are the same, namely, if they are covered by more or less the same lawlike generalizations, despite the fact they may differ quite radically in their actual etiologies. Finally, the generalizations in question are hedged by *ceteris paribus* clauses.

The second account is what Stich calls the *Semantic Account* (SA) according to which two sentence tokens count as type identical iff they have the same semantic content. The third one is what might be called the *Physical Account* (PA) according to which the sentence tokens are of the same sentence type iff their physical properties, their shape, so to speak, are the same. After quickly dismissing the PA as hopeless, Stich makes the following remark:

> [The] interesting question is how causal accounts and content accounts compare with each other. Do they categorize mental tokens differently, or do they inevitably come out with the same categorization? On this issue, opinions divide. According to Fodor the two sorts of classification schemes coincide, "plus or minus a bit." Indeed Fodor sees this as "*the* basic idea of modern cognitive science." Any thoroughgoing [i.e., content] functionalist in the philosophy of mind will also end up on this side of the divide. On the other side, denying that causal and content accounts converge, are Field, Lycan, Perry, McDowell, and the truth. (1983: 48-9)

Here Stich conceives of the SA as fixing the type identity of mental sentences according to their broad content. Twin-earth cases show that functionally identical twins' mental states may differ in their broad content. However, this is not going to be very important for what follows. Since, for many people in the field, narrow content is a construct out of broad content, Stich has the same line of argument against narrow content.[1]

Here is Stich's argumentative strategy. Stich thinks that if folk psychology is to be scientifically vindicated through some version of a mental sentence theory, the SA of typing mental sentence tokens is indispensable. He then proceeds to show that the NCA and SA come up with radically different taxonomies. The way he does this is idiosyncratic. He constructs a series of thought experiments that are supposed to show that folk judgments about how to classify certain mental states do radically differ from the way the NCA would type them. Then relying on what these thought experiments seem to show, he proceeds to give an account, or rather a "descriptive analysis" of folk conception of belief as a paradigm case of a contentful mental state, i.e. as a paradigm case of mental state typed according to the SA.

According to Stich's analysis, the "content identity" of beliefs that is thought to be assumed by folk psychology is a myth. On the basis of the evidence he claims to have collected through his thought experiments, he claims that the notion of content according to folk psychology is such that it is only a similarity measure along three different dimensions that the folk implicitly assume. One dimension of similarity between contents is the *functional or causal-pattern* of contentful mental states: "A pair of belief states count as similar along this dimension if they have similar patterns of potential causal interaction with (actual or possible) stimuli, with other (actual or possible) mental states, and with (actual or possible) behavior" (Stich, 1983: 88-9). The second dimension draws on the *ideological (doxastic) background* of the agents. Since these can greatly vary from person to person, the relation between two beliefs in two different people can only be a matter of similarity: "The ideological similarity of a pair of beliefs is a measure of the extent to which the beliefs are embedded in similar networks of belief" (89). The third dimension of similarity measure concerns the *reference or truth-conditions* of beliefs. According to Stich, since they are dependent on the speakers' linguistic community, social embeddings, the causal history of the use of terms, the speakers/believers' physical as well as cultural environments, etc., the reference will vary as these vary without necessarily affecting the functional role of a mental state. To the extent that these factors are similar, to that extent the contents of beliefs will be similar. Stich thinks that this is essentially what the SA of typing mental sentences comes down to.

It is now relatively easy to see how the two taxonomic schemes diverge. The NCA can capture only the *causal pattern similarity* dimension assumed in the SA. It can't be sensitive to the other dimensions. Stich concludes that "the mental sentence theory of belief, if fleshed out with a narrow causal account of belief, just does not comport with our workaday folk psychological notion of belief—it is not an account of belief, as the term is ordinarily used" (1983: 49).

If the two taxonomic schemes differ, what scheme should a scientific psychology adopt? Stich argues that adopting the SA is ill advised, because mental states typed according to the SA will make bad science since a semantic taxonomy would only provide the psychologist with a theoretical vocabulary whose application is vague, unstable, context-sensitive, and observer relative. Who would want such a science, Stich argues, especially if there is a clear alternative that is free of such defects? According to Stich, the alternative is a psychology whose taxonomic scheme is based on the NCA. This is the STM paradigm. Hence Stich's main conclusion: if a mature cognitive science is and ought to be committed to the NCA ($\approx$ STM), then folk psychological notions like beliefs and desires are likely to be eliminated.

This is how Stich motivates and argues for his STM. It is therefore very important to see whether Stich is right in his claim that the STM paradigm is really superior in any of the respects in which he criticizes content-based psychologies. As I advertised, I will argue that Stich is wrong.

3. WHAT IS STM?

According to Stich, the core idea of STM can be captured in the following way:

> the cognitive states whose interaction is (in part) responsible for behavior can be systematically mapped to abstract syntactic objects in such a way that causal interactions among cognitive states, as well as causal links with [proximal] stimuli and behavioral events, can be described in terms of the syntactic properties and relations of the abstract objects to which the cognitive states are mapped. More briefly, the idea is that the causal relations among cognitive states mirror formal relations among syntactic objects. (1983: 149)

Stich here considers two networks, one of which is the network consisting of the causal relations among brain state types, proximal stimuli and behavioral events. This network is supposed to be mirrored by another network expressed by a syntactic psychological theory **T**. This theory consists of at least three kinds of generalizations: (1) the ones that nomologically connect proximal stimuli to B-states (belief-like states) with *particular* syntactic objects mapped to them, (2) the ones that describe causal relations among B-states and D-states (desire-like states), and (3) the ones that nomologically connects B-and D-states to motor-gestures. Following Michael Devitt (1990), I will call these kinds of generalizations: I-T, T-T, and T-O generalizations respectively.[2]

If we want to put **T** into some canonical form, we may write out **T** as a single conjunctive sentence, replacing all the occurrences of the theoretical predicates such as "x has a B-state mapped to δ_1" and "x has a D-state mapped to δ_1" with expressions of the form:

$$\text{x is in (some member of) } B(\delta_1),$$

$$\text{x is in (some member of) } D(\delta_1).$$

B and D are functions (in the set theoretic sense) that map a particular syntactic object, which the theorist had already specified for the job at hand, onto the set of x's first order physical state types that have the functional role that **T** associates with that syntactic object. We may now express **T** in the following way:

$$\mathbf{T}[s_1,\ s_2,\ \ldots,\ B(\delta_1),\ B(\delta_2),\ \ldots,\ D(\delta_1),\ D(\delta_2),\ \ldots,\ b_1,\ b_2,\ \ldots] \qquad \text{(i)}$$

where s_i's are proximal stimulus types and b_i's behavioral event types (motor gestures), and δ_i's are specific syntactic objects.

Roughly, this is the form an STM theory would take. Let us now see how STM is committed to the NCA of typing brain states hypothesized by the theorist, i.e., how we can get their explicit functional definitions.

From (i) it is easy to get the Ramsey sentence of **T** by quantifying over the functions B and D:

$$(\exists f_1)(\exists f_2)\,\mathbf{T}[s_1,\ s_2,\ \ldots,\ f_1(\delta_1),\ f_1(\delta_2),\ \ldots,\ f_2(\delta_1),\ f_2(\delta_2),\ \ldots\ ,\ b_1,\ b_2,\ \ldots] \qquad \text{(ii)}$$

We can now get the explicit functional definition of B:

B $=_{df}$. The function f_1, such that there is a function f_2, such that the two uniquely satisfy

$$\text{`}\mathbf{T}[s_1,\ s_2,\ \ldots\ ,\ x_1(\delta_1),\ x_1(\delta_2),\ \ldots,\ x_2(\delta_1),\ \ldots,\ b_1,\ b_2,\ \ldots]\text{'}$$

Similarly for the definition of D.

Although this is the formal procedure to get the explicit functional definitions of B and D, what we really want is explicit functional definitions of '$B(\delta_i)$' and '$D(\delta_i)$' for each *i*. The intuitive idea is this. Notice that in this formalism the existential quantification in getting the Ramsey sentence is over certain functions that map distinct syntactic objects to distinct sets of an organism's first-order physical states. Here, in a certain sense, syntactic objects are exploited as external indices that pick out certain states of an organism that have distinct functional roles as specified by theory **T**. Each specific syntactic object in virtue of its distinctive place in **T**'s generalizations specifies a unique functional role that the two functions B and D then map onto the underlying physical states of the organism. Intuitively, we may extract the functional definition of $B(\delta_i)$ for each *i* in the following way: since δ_i in the domain of B, in virtue of its place in **T**, is supposed to pick out a *unique* functional role that may be indexed by F_i,

$$B(\delta_i) =_{df.} \text{the set of first-order states that have } F_i \text{ as determined by } \mathbf{T}.$$

Similarly for D and for each particular δ_i.

Now Stich does not present his STM in this way. Here I have used a procedure very similar to the one developed by Brian Loar (1982a) in his presentation of his own content functionalism. This is not accidental of course. In fact, this is the point. For, as should be obvious, Stich's STM, *structurally* at least, is nothing but a de-intentionalized version of Loar's content functionalism, except that Loar takes the causal role of "observational" beliefs to be fixed on the basis of *distal* stimuli. Where Stich uses abstract syntactic objects, Loar uses ("fine-grained") propositions, intentional objects *par excellence*. The type identity of specific abstract syntactic objects is given by their place in the theory. This is the way they are purely functionally defined according to their narrow causal profile.

In fact, the similarities between STM and Loar's content functionalism are, in one respect, stronger than that. Loar uses propositions in the initial stage of getting the functional theory first (and, for good reasons—see below). He then proposes a procedure by which all the propositions are replaced by purely formal expressions. The theory in this ultimate form structurally is almost an STM! Loar, of course, is no eliminativist. His aim is to naturalize intentionality by offering a sophisticated functionalist theory. So he thinks at some stage he should get rid of the intentional objects like propositions he had initially used. Once the theory is completed, it is supposed to provide sufficient (and, necessary?) conditions for a mental state to have a semantic content, which can ultimately be specified without using any intentional terminology. This is his strategy, and as far as it goes it is perfectly kosher. But if I am right in what I am going to say, it does not go very far, at least in its narrow version.

4. SOME CURIOUS ASPECTS OF STM

My presentation of Stich's STM may be taken to be tendentious. I presented it as a purely functionalist theory and said that the abstract syntactic objects, which the brain states are mapped onto, may be viewed as indices that are external to the underlying first-order brain states. But, STM is supposed to be a formal/syntactic theory very similar to the Computational Theory of Mind (CTM) Fodor has developed and defended. STM is supposed to be a de-intentionalized version of what Stich calls Mental Sentence Theories. Indeed STM has been taken in this way in the literature by its friends and foes. But if my presentation is right, STM is not in fact theoretically committed to there being syntactically complex "sentences" literally realized in the brain. If so, how could it be very similar to Fodor's CTM? Stich writes:

> It is not, strictly speaking, required for an STM theorist to view hypothesized neurological state tokens as mental sentence tokens, though talking of them in this way is often an all but unavoidable shorthand. (1983: 152)[3]

This is curious but actually understandable. Remember Stich's question about how the tokens in different heads can be individuated as of the same sentence type. His solution is the NCA. But the NCA requires a theory first in which syntactic expressions figure as theoretical terms in the generalizations. However, once we have such a theory, it is easy to define the syntactic expressions functionally à la Loar. But once we do that, the question of whether the referents of such expressions do really have syntactic structure somehow realized in the brain becomes secondary and at best an open empirical question. For, if the functional theory is true, it seems that we can do everything we want that the Mental Sentence version of the theory can do.[4]

So STM as a purely functionalist theory is not committed to a semantic-free LOT. On the other hand, of course, whatever CTM is, it cannot be neutral with respect to the question of whether there are syntactically complex sentences realized in the brain. CTM should be so formulated that it essentially entails a positive answer to this question. The problem in fact stems from the widely shared conviction that the type identity of brain sentences can and should be given in terms of the NCA (for some, as well as in terms of other ways like the SA). Fodor, at least in his early writings, is explicit about how to formally type the LOT symbols: functionalism à la NCA is the answer.[5] Below, I will argue that this can't be done. So there is at least this dissimilarity between STM and CTM: whereas STM is non-committal about there being brain sentences, CTM, whatever it is, is essentially committed to it.

Having made the point, however, I want to talk of STM as if it were concerned with the functional individuation of syntactically complex brain sentences. Not only because, as Stich says, this is an all but unavoidable shorthand, but also because I want to see whether Mentalese expressions can be individuated on the basis of the NCA if the LOTH is true. Hence, my argument can equally be seen as an argument against Narrow Content Functionalism (NCF) in so far as it is pursued as part of a naturalistic semantic program run on a LOT story. So, in what follows, I will assume the framework of Mental Sentence Theories, and often treat STM in this form.

So far we have been talking about the functionalist nature of STM, and thus its commitment to the NCA of typing brain states. But what does this have to do with syntax? More particularly, how does Stich conceive of syntax when he talks about the syntactic type identity of brain sentences? Indeed, what makes his theory a "syntactic" theory? To this last question he answers in the following way:[6]

> We would have no reason to view brain states as syntactically structured unless that structure can be exploited in capturing generalizations about the workings of mind/brain's mechanisms. Attributing syntactic structure to brain state tokens—assigning them to syntactic types—is justified only if some

> interesting set of causal interactions among those tokens is isomorphic to formal relations among abstract syntactic objects. (1991: 244)

Notice that if Stich is right about this, Fodor can't have any reason for postulating a separate computational level in which intentional laws of psychology are implemented. In particular, what is puzzling about Stich's answer is that he doesn't mention at all the Turing legacy which is the main driving force behind Fodor's insistence that the computational story, according to which thought processes are defined over the formal/syntactic properties of representations, is our only plausible story about how semantically coherent processes can be physically/mechanically possible.[7] Stich's interest seems not to be in computationalism classically understood. This is understandable to a certain extent. For Stich doesn't think that there are any semantically coherent thought processes that need the attention of science because he doesn't think that there are any states with semantic content. Put this aside. He has a different line of answer.

When he talks about the syntactic type identity of brain sentences, he has a "rich" notion of syntax, according to which mere difference in lexical items (e.g. "Tully was bald" versus "Cicero was bald", or "*Fa*" versus "*Fb*") is enough to make the sentence tokens belong to different syntactic types.[8] In particular, for Stich, the criterion according to which two sentence tokens in two different heads count as of the same type is a syntactic one. But since this criterion is captured by the NCA, the syntactic type identity of brain sentences is a matter of functional identity:

> when mental states are viewed as tokens of syntactic types, the functional profile exhibited by a mental state can be equated what we have been calling its formal or syntactic properties. (Stich 1983: 190)

So it seems that, according to Stich, the very postulation of complex semantic-free sentences realized in the brain whose "syntactic" type identities are given purely functionally is what makes Stich's theory a syntactic theory. As I argued elsewhere, I don't think this notion of syntax is the one that is needed for a Fodorian Computational Theory of Mind: what is required for the LOTH is a combinatorial syntax that fixes the logical form of expressions.[9] Nevertheless, "syntactic" typing of LOT symbols has always been understood on the basis of the NCA. But put this aside. The important question I will address below is whether the type identity of brain sentences can be given in terms of their narrow causal profile, *whatever it is called.* Now let us see whether the STM paradigm is any superior over content-based psychologies.

5. THE ALLEGED SUPERIORITY OF STM

From Stich's analysis of folk conception of belief individuation, it follows that predicates like 'believes that P' (1) are vague and unstable, (2) depend on a (observer-relative) similarity matrix along three different dimensions for their applicability, and (3) their application involves many unnecessary "fine-grained distinctions which contribute nothing in the explanation [and prediction] of behavior." From (2), it also follows that there are likely to be many important cognitive generalizations that will not be stateable in terms of such predicates. So a content-based psychology will inherit all of these limitations. In contrast, the STM style theories, Stich claims, will have none of these defects.

In this section, I will argue that *if* Stich's criticism of content-based psychologies is right *then* exactly parallel problems equally plague STM. But for this, we first need to see, exactly, how Stich argues for the superiority of STM. In other words, since my claim is conditional, we need to see in some detail what makes its antecedent true according to Stich, and why he thinks that STM is free of similar problems.

5.1. How STM Is Supposed to Be Superior

In discussing how STM theories will succeed where the content theories fail, Stich again uses the thought experiments he had considered in showing how the content taxonomies radically differ from the ones based on the NCA. Much of the difference stems from the fact that whereas the individuation of content essentially depends on three different dimensions, the NCA is only committed to individuating mental states according to their narrow causal pattern. The other two dimensions, ideological and referential (or, truth-conditional) similarity are to be amputated. First, these last two are unnecessary and therefore contribute nothing to the explanation and prediction of behavior. Second, by getting rid of them, context-sensitivity is eliminated. That is because, as in every multi-dimensional similarity judgment, it is the context that decides which dimension is to be emphasized in deciding whether a given state in a particular situation counts as the belief that *P*. Sometimes referential similarity will count more, sometimes ideological similarity, or simply causal pattern similarity depending on the demands of the particular situation in which the question arises.

Stich puts the greatest emphasis on the problems created by the *ideological similarity dimension*. This involves what he calls the *holism problem* in the folk conception of belief. In order to bring out the problem vividly, let's focus on his most celebrated thought experiment: the case of Mrs. T. Mrs. T is an elderly woman who suffers from a progressive loss of memory. At the end, she does not "know" what an assassination is, what dying is, who McKinley was, etc. Nonetheless, she appears to remember/believe that McKinley was assassinated, because that is what she persistently says when asked "What happened to McKinley?" According to Stich, the folk psychology's clear verdict is that she does not believe that McKinley was assassinated. Stich's diagnosis is that when she ceased to have a certain set of relevant beliefs, she ceased to believe that McKinley was assassinated, despite the fact that she appears to respond correctly to the question. This, Stich says, shows that folk conception of belief attribution attends to the doxastic background of an agent. From this he infers that the type-identity of someone's belief is partly constituted by what other actual beliefs the individual happens to have. This is the notorious problem of content holism, according to Stich.

On the other hand, an individuating scheme based on the NCA, he claims, is and ought to be nonsensitive to the actual doxastic surrounding of a mental state it individuates. That is because the NCA taxonomizes her state underlying her utterance on the basis of its *potential* (narrow) causal interactions. Thus STM is able to account for her ability to infer, for instance, "McKinley was buried in Ohio" from her "acknowledgment" of "McKinley was assassinated" and "if McKinley was assassinated then he is buried in Ohio." So whereas content psychologies miss such important generalizations as those that cover Mrs. T, STM theories will be able to take such agents under their scope.

Stich also argues that with respect to the *reference* and *causal pattern similarity* dimensions STM is superior. But I do want to leave aside the discussion of Stich's arguments about these latter dimensions here since I think they are pretty weak and don't occupy any centrality in his discussion. Stich puts the greatest emphasis on the holism problem of content-based psychologies, and clearly thinks that STM theories are free of this serious problem. If it can be shown that the NCA is equally problematic in this respect then STM is undermined completely, and this is exactly what I intend to do next.

5.2. The Parallel Disadvantages of STM-Style Theories

5.2.1. Doxastic Similarity Dimension and the Holism Problem

Although most of Stich's arguments for his case against belief turns on the "holistic" nature of commonsense individuation of beliefs, it is unfortunate that he does not elaborate on what, exactly, holistic nature of content comes down to. Much of what he says on the matter is provided through a handful of examples like the case of Mrs. T.

In what follows, I will first try to explain as clearly as possible the sense in which Stich thinks that commonsense individuation of beliefs is holistic. My discussion will show that he is obscure and vague about what he thinks the "holism problem" is. I will then indicate exactly why Stich thinks that the holistic nature of belief is a problem for content based-psychologies, and why he thinks that an STM-style theory is totally free of it. We need to be as precise as possible about this, because my argument against Stich depends on his own premises.

According to Stich, the identity of a particular belief, say, the belief that McKinley was assassinated, depends on what other actual beliefs a person happens to have. The doxastic surround of the belief that McKinley was assassinated is constitutive of its content. But Stich does not say what this doxastic surround is, how it is determined, nor how it is supposed to be constitutive of a given content. For instance, at any moment, a person who believes that McKinley was assassinated has also a very large stock of other beliefs. Do all of them contribute to the content of the belief in question, or only some of it? From the way Stich writes and uses the expression 'doxastic surround of a belief' and its cognates, it seems that only some portion of a person's entire belief system is relevant to the determination of the content. Unfortunately, he gives almost no clue about how big the portion is supposed to be.

The case of Mrs. T is typical. She ceases to have many beliefs among which, for instance, is the belief that if someone is assassinated then she is dead. In fact, she no longer "knows" what dying is, what a presidency is, what presidents do, etc. In all of Stich's similar examples, the beliefs that make up the doxastic surround of a particular belief have rather certain "direct relations" to the belief. They are not only semantically close, but also, in some loose sense (to be discussed later), "conceptually" tied to the belief. In all such examples, the fact we are invited to observe is that when someone ceases to have *those* kinds of beliefs then someone ceases to have a particular belief.

Stich is crucially vague and not particularly careful here. If his claim is that the content of a particular belief is (partly) determined by the set of *all* beliefs one has, which I take it what holism at its extreme comes down to, then he has not provided any single reason, let alone a relatively elaborate argument, for his claim. On the other hand, if his claim is that only *some* beliefs determine content, as he seems to intend, then the identity conditions for belief are not holistic.

I think that part of the reason why Stich is so vague and careless is that he does not care about this distinction, some or all. According to Stich, it seems, the very fact that the content of a belief depends for its identity on at least some other actual beliefs the agent has is enough to make serious trouble for any psychology that hopes to essentially advert to content in its generalizations. For one thing, given Quine's influence on him, Stich clearly thinks that the distinction between those beliefs that determine content and those that don't can be anything but sharp and principled. If so, content can at best be a matter of degree. And this is enough to make trouble: a content psychology is possible, at best, for those who are doxastically similar. But even for such a psychology, vagueness will still continue to be a serious problem, since it is almost certain that doxastic similarity never actually achieves doxastic identity among people. Whatever the case is with Stich's analysis of belief, however, he certainly thinks that his alternative paradigm of doing psychology, STM, does not have any such problem.

Why does Stich think that the framework provided by STM has no such "holistic" problem. Here is a typical remark by Stich:[10]

> In chapter 7, section 3 our focus was on ideological similarity, and the persistent problem was that as subjects became increasingly ideologically distant from ourselves, we lost our folk psychological grip on how to characterize their beliefs. For a syntactic theory, however, ideological similarity poses no problem, since the characterization of a B-state does not depend on the other B-states that the subject happens to have. A B-state will count as a token of a wff [well-formed formula] if its potential causal links fit the pattern detailed in the theorist's generalizations, regardless of the further B-states the subject may have or lack. (1983: 158)

Stich, then, goes on to clarify how this can be so by working on the example provided by Mrs. T:[11]

> If we assume that before the onset of her disease the B-state which commonly caused her to say "McKinley was assassinated" obeyed generalizations like (4)-(6), then if the illness simply destroys B-states... without affecting the causal potential of the tokens which remain, the very same generalizations will be true of her after the illness has become quite severe. In chapter 7 we imagined a little experiment in which, shortly before her death, we tell Mrs. T, "If McKinley was assassinated, then he is buried in Ohio," and she replies, "Well, then, he is buried in Ohio." This is readily explainable by (5) [the syntactic version of psychologized Modus Ponens]... So if the generalization is there, it can be captured by a syntactic theory. But as we saw, there is no comfortable way to capture this generalization in the language of folk psychology... Thus a cognitive science that adopts the STM paradigm can aspire to broadly applicable developmental, clinical, and comparative theories, all of which are problematic for a content-based theory because of the constraints of ideological similarity. (1983: 158-9)

Is it true that ideological similarity poses no parallel problems for STM-style theories? I think not. It is time to see why.

5.2.2. Holism and STM

Here is the structure of the argument for my claim that STM, contrary to Stich's advertisement, has exactly the parallel problem.

(1) The STM framework is committed to the NCA of type individuating B-states *qua* mapped to particular syntactic objects like, say, '*Fa*', through the generalizations that cover them.

(2) The NCA is capable of individuating such states *only if* it has enough generalizations of a certain sort, which I will call, S-generalizations.

(3) If STM has S-generalizations among its stock of generalizations then it has all the parallel problems that Stich complains about content-based psychologies regarding the dimension of ideological similarity.

In the remainder of this section, I will make this argument stick. I take it that (1) is common ground (see above). Let me first argue for premise (2).

All the T-T generalizations Stich ever considers, by way of giving examples or otherwise, have rather a certain character: they all quantify over *particular* syntactic objects, i.e., they all use meta-variables to refer to classes of actually specified sentences that have a certain common "logical form". Even in the quote above, it is apparent that when he talks about the causal interactions of the token that underlies Mrs. T's utterance of 'McKinley was assassinated', the generalizations Stich has in mind are of this kind. Let me call the generalizations that quantify over *particular* brain sentences "L-generalizations", since these apply to any sentences that have a certain "logical" form. L-generalizations are all blind to the primitive non-logical vocabulary that the STM-theorist specifies.

It should be obvious that if all the T-T generalizations that go into the specification of the causal role F in the individuation of B(δ_i) for any *i* (see above) are of this type, i.e., if they are all L-generalizations, then there cannot be a *unique* causal role for each particular B(δ_i), which means that there can be no type individuation of B-states with particular syntactic objects mapped onto them. Here is why: with only L-generalizations in force, *any* sentence token has *potential* inferential (causal) connections to *any* other one. Put differently, since, on Stich's own admission, the generalizations in the theory detail not only the actual but also the potential causal interactions of any particular B-state, and since any sentence token can potentially be "inferred" from any other (i.e., causally connected through L-generalizations to any other), L-generalizations all by themselves cannot type individuate particular B-states.[12] All they can specify is at most the "logical" form or syntactic type of sentence tokens. As we will see in the next section, this situation does not change even when we add the I-T and T-O generalizations to L-generalizations: together they are still incapable of providing unique causal roles for particular B-states.[13] For one thing, as I will point out later, there can be no such (narrow) I-T/T-O generalizations. But, for our purposes here, more importantly, even if there are such generalizations they can help at best to identify a very small subset of particular B-states whose character is rather "observational". However, Stich himself is pessimistic about there being any such subset (see below). My point is that S-generalizations are necessary (not sufficient) for type individuating at least some B-states, and this will do for premise (2).

What is needed, of course, is a different kind of T-T generalizations in addition to L-generalizations, T-T generalizations that are not blind, so to speak, to the primitive non-logical vocabulary of the STM-theorist, generalizations that detail (part of) the causal role that is unique to, say, the B-state mapped to '*Fa*'. It is obvious that such "low-level" generalizations will typically be the syntactic parallels of such "content generalizations" (*C-generalizations*) as[14]

(i) For all subjects S and for all x, if S comes to believe that x is a cow, then S will typically come to believe that x is an animal,

(ii) For all subjects S and for all x, if S comes to believe that x is a bachelor, then S will typically come to believe that x is unmarried,

and so on.[15] Let me call the *syntactic parallel* of this kind of C-generalizations "*S-generalizations*". Stich is committed to such generalizations, otherwise there is no individuation of particular B-states. Hence, premise (2).

Let's now take up premise (3): If Stich is committed to S-generalizations, then his STM framework has exactly the same "holism" problem which he claims plague content-based psychologies. There are different ways of showing this, but at the end they all come to the same thing. Let me begin with the obvious version.

S-generalizations are *low-level* generalizations. What makes them low level is the following fact. Subjects that are covered by such generalizations are also covered by L-generalizations if the subjects have certain *actual* B-states. For instance,[16]

IF S has the belief* that #for all x, if x is a cow, then x is an animal#,
  and
  S comes to believe* that #Samantha is a cow#,
THEN
  S will typically come to have the belief* that #Samantha is an animal#.

What might license this inference* is, of course, the existence of high-level L-generalizations that Stich (*mutatis mutandis*) specifies among his examples:

(5) For all subjects S, and all wffs A and B, if S has a B-state mapped to A→B and if S comes to have a B-state mapped to A, then S will come to have a B-state mapped to B. (1983: 155)

As we may recall, according to Stich the "holism" problem that plagues the content-based theories consists in the fact that the type identity of a particular belief (partly) depends on what other actual beliefs the subject has. Stich thinks that this fact is the source of the problem. In contrast, he claims, the NCA of typing particular B-states has no such commitment to there being any actual B-states surrounding a particular B-state in terms of which its type-identity is determined.

But, *if* every subject who is covered by S-generalizations is also covered by the relevant L-generalizations, in the way I've just indicated, then the STM-theorist *is* committed to their being actual B-states for determining the type-identity of particular B-states, and thus committed to construct syntactic theories only for those who more or less share their doxastic* background. In other words, in the STM paradigm, the "syntactic" type identity of sentence tokens, contrary to Stich's advertisement, *is* acutely sensitive to the *actual* particular B-states that surround them. This is a problem that is exactly parallel to what Stich calls the "holism" problem of belief individuation. And so, STM must incur all the parallel problems which Stich claims seriously bother content psychologies: Sharing a particular B-state can only be a matter of degree, therefore, those that are doxastically* dissimilar to us cannot be covered by STM-theories. What are we to do with the children, exotics, cognitively handicapped, higher animals, etc.? Furthermore, unless Stich can come up with a principled distinction between those B-states that contribute to the syntactic type identity of a sentence token and those that don't, the vagueness that already exists in the conditions that type identify sentence tokens will be greatly aggravated. Again, we have exactly the parallel problem here. If Stich is right in his criticism of content-based theories regarding "holism" problem, it is false that STM theories are any superior in just that respect.

However, one might object: It is not necessary that for any subject who is covered by S-generalizations is also covered by the relevant L-generalizations in the way I have just indicated, and therefore, it is not necessary for an STM theory to be committed to there being *actual* B-states, which a subject must have, for the type individuation of sentence tokens. It may be that the S-generalization (i) above may be true of a subject even though she may not have any *actual* belief* that #all cows are animals#. In such a case, the syntactic type identity of a sentence token may be given in terms of such *dispositions* as the likes of (i) and (ii) specify without any recourse to high-level L-generalizations. How does this evade the problem? Well, let me show that it doesn't.

Although I don't have to make my point in the way I will do, I think it is important to cast the issue in terms of that perspective. But nothing important will hang on it. STM has usually been brought up as a de-intentionalized version of a language of thought story, or CTM. We have seen that STM is not in fact committed to there being (semantic-free) sentences literally realized in the brain. But it may be taken in this way, and this is the assumption we are now operating under.

Anyone who is sympathetic to the computational paradigm must keep in mind that CTM is a "rules and representations" framework: any relatively higher level mental processes consist in transformation of syntactically structured representations according to rules that are causally sensitive only to the formal properties of representations over which they are defined. In other words, the typical computational treatment of such inferences as expressed by (i) or (ii) will take the form of applying some relatively high level rule like Modus Ponens to *actually tokened* complex sentences.

Of course, this is only one possible implementation story that can be given for such generalizations like (i) or (ii) at the computational level. Another possibility is that the rules that govern the inference from #x is a cow# to #x is an animal# is rather more specific and low-level, rather like the syntactic analogues of Carnap's "meaning postulates" implemented as rules.[17] But, either way, according to CTM, you need rules to manage inferential processes defined over data structures.[18]

My point is simply this. On a computational paradigm S-generalizations can be cashed out *either* by postulating high-level laws and actual beliefs* upon which they operate *or* by postulating the syntactic analogues of meaning postulates in the form of low-level rules. So, once it is obvious that an STM-theorist is committed to such S-generalizations, the STM-theorist is no longer in the position that Stich claims is free of the problems confronting the content psychologist. Let me illustrate.

As we may remember, Stich claims that unlike content psychologies, the STM framework is capable of covering Mrs. T's mental states in its generalizations. That is because, he says, according to STM, the type identity of the state underlying her utterance "McKinley was assassinated" does not depend on further actual doxastic* states she has. In fact, as the example is constructed, she has almost none. The type individuation of Mrs. T's state proceeds according to its inferential**potential*, not according to what it *actually* inferentially* interacts with. So far so good. For instance, all the L-generalizations that cover her state detail just this potential. But, of course, with only L-generalizations, the STM-theorist cannot individuate the state. It is obvious that in addition to L-generalizations, the STM-theorist needs some such S-generalizations as

(iii) For all subjects S, if S comes to believe* that #someone is assassinated#, then S will typically come to believe* that #someone is dead#

In order to type individuate the state underlying Mrs. T's utterance of "McKinley was assassinated". But, of course, we see that it is precisely this kind of generalizations that become inapplicable to Mrs. T when we come to see that she ceases to have many relevant beliefs. This can easily be explained on the version of the computational story that derives the S-generalizations from actual beliefs* and high-level L-generalizations. But Stich would probably insist that this is the wrong version. Well, then, let us look at the other version where S-generalizations are implemented as specific "dedicated" rules rather like the syntactic analogues of "meaning postulates".

The question now is whether there are any such rules intact in Mrs. T's case. As we may remember, it becomes apparent *under questioning* that she does not "know" whether an assassinated person is dead, what dying is, who McKinley was, etc. When she is asked whether McKinley was dead, she answers "I don't know". What better evidence can there be for the fact that the above S-generalization is broken? In the case of S-generalizations, appeal to potential causal profile doesn't even begin to help since it is precisely this potential that is lost in her case. But then, if such generalizations do no longer cover the mental states of people like Mrs. T, of course, we can't tell the computational story along the lines we have been assuming given that the other version is out. But either way, the important point is that the S-generalizations do simply not hold in Mrs. T's case. If so, however, the STM theorist is in exactly the same boat as the content psychologist: there is simply no saying what "syntactic" state Mrs. T is in, since the STM-theorist is no longer able to type individuate her state.

The same is true, similarly, for people who are doxastically* dissimilar to us like children, exotics, cognitively handicapped, higher animals, etc. In so far as the S-generalizations that are true of them are not available or non-existent, there is no type individuation of their syntactic mental states, hence they are beyond the reach of STM-theories.

So here is the score. Contrary to Stich's claim and advertisement, because of their commitment to the NCA, STM theories are committed to the type individuation of particular B-states (depending on the computational story preferred) either according to what other actual B-states the subjects have, or according to what

S-generalizations are true of them. The first option makes STM equally sensitive to the *actual* doxastic background of subjects. The second option restricts the scope of STM-theories to those for whom S-generalizations exist, or are specifiable, thus, again, to those who are doxastically/disposiotionally similar. But the consequences of both options are just the same for the prospects of STM if the prospects of content psychologies are as Stich claims them to be.

6. THE NCA AND THE TYPE INDIVIDUATION OF BRAIN SENTENCES

The introduction of S-generalizations generates a fatal dilemma for the STM-theorist: Either (1) the type-individuation of syntactic objects is possible but only for individual systems (or clusters of systems) given separately for each, or (2) the individuation cannot be carried out interpersonally, i.e., for sufficiently large populations. Both horns are equally destructive for the prospects of STM. I argue for this conclusion elsewhere extensively (Aydede, 2000). But the gist of the argument can be conveyed intuitively.

Remember that the theory **T** consists of generalizations of three sorts (I-T, T-T, and T-O). The T-T generalizations are divided into two: L- and S-generalizations. We have also seen that the heaviest burden in the individuation of particular syntactic objects is carried by the S-generalizations. Now if STM is to provide solid foundations for a semantic-free scientific psychology, then, minimally, **T** must consist of only those generalizations that satisfy the following constraints simultaneously.

a) they together must secure a unique causal role for each distinct syntactic object (i.e. there must be sufficiently enough of them for securing the uniqueness),
b) they must be interpersonally applicable (i.e., they must be true of a sufficiently large intentional population, if not all intentional organisms),
c) they must be lawlike (minimally, they must go beyond being statistical summaries of what causes what).

Is there such a theory? If one reflects on the question for a moment, one can see that the answer is negative. Let's suppose that we want the theory to be true of what we might otherwise characterize as the common folk. (So **T** can be thought of as more or less the de-intentionalized and cleaned-up version of Folk Psychology.) This entails that there are S-generalizations in **T** that satisfy (a)–(c). But what could they be?

The most plausible candidates that come to mind are the syntactic parallels of content generalizations that detail "analytical connections" among concepts like:

(iii) For all *S* and *x*, if *S* comes to believe* that #*x* is assassinated#, then, *ceteris paribus, S* will come to believe* that #*x* is dead#.
(iv) For all *S* and *x*, if *S* comes to believe* that #*x* is a bachelor#, then, *ceteris paribus, S* will come to believe* that #*x* is unmarried#. Etc.

One immediate problem with this suggestion is that if the STM-theorist is to pick out these S-generalizations by appealing to analyticity, then he is being unfaithful to his own program and tenets: the semantic notions are being used in the construction of **T** in crucial ways. Secondly, if Quine is right about there being no principled distinction between analytic and synthetic statements (as Stich himself thinks), then this strategy is unavailable to the STM-theorist.

But perhaps Quine is wrong and the first difficulty can be circumvented by giving a syntactic (or, at any rate, non-semantic) criterion to pick out such S-generalizations. This is another way of asking: what are the criteria for choosing the S-generalizations that are to go into **T**? Perhaps, these are given all by (a)–(c), or at least (b)–(c). Indeed, it may seem plausible to think that (iii) and (iv) are lawlike and interpersonally applicable, whereas the following are not:

(v) For all *S* and *x*, if *S* comes to believe* that #*x* is a tiger#, then, *ceteris paribus, S* will come to believe* that #*x* is dangerous#.

(vi) For all *S* and *x*, if *S* comes to believe* that #*x* is a bachelor#, then, *ceteris paribus, S* will come to believe* that #*x* is a neurotic#.

Although I am not quite sure what exactly being lawlike comes to, nevertheless it seems intuitively plausible to regard (v) as violating (c), even if it satisfies (b), which I doubt. (vi) may be true of a few eccentrics, but clearly violates both (b) and (c). Perhaps this kind of approach can be made to work. That is a lot 'ifs' and 'perhaps', I admit, but let's be charitable and carry on.

But still, it is clear that on this strategy there won't be *enough* generalizations to secure *unique* causal roles for each possible syntactic object that an STM-theorist would posit. We can see this if for a moment we drop the ban on talking about analytic generalizations and see (iii) and (iv) as such, i.e. if we take (b)–(c) as reconstructing analyticity in non-semantic terms. So supposing that S-generalizations that are to be put in **T** on this proposal will intuitively detail the analytic conceptual connections, we may ask: are there enough of these? Let's first consider whether S- and L-generalizations all by themselves can secure uniqueness. It seems that they can't. For supposing that they can is tantamount to assuming that each concept can be defined completely. But given the failure of philosophy to define any concept of any significance in the last two millennia, as Fodor once pointed out in his characteristic way, this is simply not true. Second, it seems intuitively clear that even if there are indeed analytic connections, they are very scarce. This is admitted even by the proponents of content functionalism like Block.[19] The obvious remedy to this is to drop (c) and appeal to S-generalizations that detail "empirical" or "contingent" connections among concepts* while satisfying (b) nonetheless. So, with this move, generalizations like (v) are now in, but not (vi) and the like. This is already dangerous holistic terrain, but there is *still* no guarantee that **T** will secure a unique role for all syntactic objects. Indeed, what needs to be done is to poll all those beliefs of the form '$(x)(Fx \rightarrow Gx)$' that are more or less accepted by all those who share a "common psychology". Indeed with this move comes the admission that STM-theories will only be possible for those who are sufficiently doxastically similar. Putting aside the serious worries about the arbitrariness and vagueness this would create and the impossibility of developing STM-theories for those who are doxastically dissimilar, the main question still lingers: are there sufficiently enough of these generalizations to secure uniqueness? The answer is by no means obvious.

But perhaps with the help of I-T and T-O generalizations this problem can be overcome. But this is surely an illusion. Take, for instance, the causal generalizations that are supposed to connect lawfully a set of proximal stimuli to, say, #Clinton is not faring well#, or any similarly specific sentence. Whatever the laws of psychophysics may tell us with respect to a very restricted range of psychophysically available properties, they will certainly be silent for the vast majority of symbol types figuring in full-blown propositional attitudes*. The problem partly stems from stimuli being proximal. There are certainly no scientifically well-delineated sets of proximal stimuli nomically correlated with the objects of beliefs*. This is to say that no such set could constitute a natural kind which would lawfully correlate with the objects of beliefs*. The other part of the problem is the holism involved in belief* fixation. Which proximal stimuli will cause which symbol(s) to be tokened in the belief*-box is determined by what other symbols actually happen to be there and by the overall internal organization of the belief*-box (simplicity, conservatism, etc.).

The history of behaviorism also provides an overwhelming inductive evidence that there are no such laws to be stated. No one has ever succeeded actually stating a single such law! Similarly for the supposed generalizations that would lawfully connect basic motor-gestures to particular symbol types in the belief*- and desire*-boxes. To be sure, behaviorists were after lawful stimuli/behavior connections, which is different. But the moral must be the same, since their failure primarily stemmed from an inability to find projectable predicates to apply to all and only those proximal stimuli under physical descriptions that lawfully govern a given piece of behavior. They assumed that such stimuli directly and lawfully control the relevant piece of behavior: they wanted to bypass mediating internal states. They failed primarily because of the holism problem again. Nothing changes, however, if you assume that it is particular beliefs*, rather than behavior, that are directly under the lawful control of proximal stimuli: the routes from stimuli are equally holistic in each case.

Perhaps I am laboring this point needlessly. It should be clear that there are no laws to be stated with respect to proximal inputs/outputs for the full range of particular symbol types deployed in central cognitive processing as direct objects of propositional attitudes*.[20] And even if there may happen to be some, they will be so few and fragile that they will be of very little help in type individuating all the symbol types we may need in psychological explanations.

But if there is no unique role for each distinct syntactic object, there is no individuation of them suitable for the purposes of a common psychology. This is the second horn of the dilemma I stated above.

Given all this, an STM-theorist, in order to secure uniqueness and to meet the charge of arbitrariness involved in selecting the S-generalizations, may be forced to advert increasingly more to the kind of S-generalizations whose scopes are increasingly narrower, and at the limit, true of only single persons. **T** could then be stated for each individual with a certain distinctive psychology. But who would be so eager for the prospects of scientific STM-theories, if they won't be interpersonally applicable? This was the first horn.

Either way the NCA-cum-STM is in serious trouble. I am very doubtful whether it can ever be saved, and I am pretty sure it's not worth the try.

It is important to note that the requirements (b) and (c) pull in opposite directions. To the extent to which an STM-theorist can furnish interpersonally applicable generalizations and secure a unique role for each symbol type, to that extent he goes against the requirement that the generalizations be lawlike. And to the extent to which he can give lawlike generalizations and secure unique roles, to that extent he violates the condition that they be interpersonally applicable. I don't think that there is an optimum point in the continuum between these extremes such that you can meet both of these requirements and secure a unique functional role for each possible symbol type.[21]

Let me indicate two more problems for the kind of type individuation of mental sentences Stich envisages for STM. As remarked, Stich proposes a cluster view of identifying neural states as particular sentence tokens: "to count as a token of a sentence type, a neurological state must satisfy some substantial number of the cluster of generalizations included in a theory, without specifying any particular generalizations that must be satisfied, nor exactly how many must be satisfied" (1983: 152). He admits that this introduces vagueness into the identity conditions of mental sentences. However, the problem this may cause is more than just introducing vagueness. It risks downright misclassification. Consider again S-generalizations, for, in a certain sense, they are expected to do the heaviest work in the individuation of sentences on the STM framework. The problem is that there may be two sentence tokens satisfying almost the same generalizations but nevertheless differing in type because they satisfy a few different "essential" generalizations. Consider the token belief* that #...gay...# and the token belief* that #...lesbian...#, it is likely that they have very similar causal roles. What may be distinguishing them are just a few (counterfactual supporting) S-generalizations such as 'B*(#gay#) → B*(#male homosexual#)' and 'B*(#lesbian#) → B*(#female homosexual#)'. What reasons could Stich give us that such cases are not seriously troublesome or do not really arise? I can see none.

Another problem is one that Stich himself raised against content functionalism. On Stich's own admission, given two subjects with the same B- and D-states, the potential as well as actual causal patterns (concerning especially the ones captured by L-generalizations) that their B- and D-states will exhibit are very likely to differ. This is the problem parallel to the one that the content functionalism faces: the kinds and the degree of complexity of inferences that people can draw (i.e., their logical/analytical acumen) vary greatly from person to person. If any attempt to incorporate these different causal patterns into a functionalist theory in a principled way will be, as Stich says, "ad hoc and implausible", how could Stich think that an STM-theorist's parallel attempts will not similarly be *ad hoc* and implausible? Notice that insisting here that B-states are not beliefs cannot even begin to help: the explanation of a certain kind of mental activity on the basis of purely syntactic transformations of some complex abstract objects mapped onto B- and D-states is exactly what STM theories are supposed to be good at.

This completes my second argument against STM.

7. WHY A PURELY SYNTACTIC PSYCHOLOGY CANNOT GET OFF THE GROUND

Throughout Stich's 1983 book, there are various passages in which Stich seems to argue that an STM-theorist had better refrain altogether from using intentional notions even in the theory construction stage. Here is a typical one: "cognitive psychologists can and do develop the theory of mental processes without attending to the semanticity of formulae in the mental code" (1983: 193). In fact, Stich's discussion of what he calls the Weak Representational Theory of Mind (RTM) is an attempt to show that the assumption that the formulae have semantic content is frivolous at any stage of theory development.

Many people seem to think that functionalism in scientific psychology can be carried out without ever raising any semantic worries. In this section, I will argue that this in fact can't be done. In particular, I will show that the construction of an STM-style theory cannot be carried out without using intentional notions. This problem is one that seems to belong to "the context of discovery", but nevertheless it will be instructive to see why an STM-theorist is committed to using intentional notions in at least theory construction stage. I already detailed the reasons why STM is seriously problematic otherwise.

In fact, it is for a very simple reason that within a strictly STM paradigm theory construction cannot get off the ground without using any intentional idioms. STM is a purely functional theory. As such, all the theoretical predicates that denote functionally defined particular brain state types depend for their reference on the entire theory being *in situ*. In other words, within the STM paradigm, the only legitimate way to refer to the nodes of the causal network of brain states is by way of theoretical terms whose applicability entirely depends on the theorist's having almost the whole functional theory first. That is his point when Stich insists that the type identity of a sentence token (a brain state token) entirely depends which and how many generalizations cover it:

> It is only against the background of a systematic mapping of state types to sentence types that any given state token counts as a token of a particular sentence type... No one neurological state can count as a token of a sentence type unless many neurological states count as tokens of many different sentence types. But this holism... is quite distinct from the holism imposed on the folk psychological notion of belief by the embedded appeal to ideological similarity. For the status of a state as a token of a sentence does not depend on what other cognitive states a subject currently happens to be in. It depends only on the causal interactions that the state would exhibit with stimuli, with behavior, and with other states. (1983: 153)

But there is no way to start theory construction without having an initial and independent way of referring to the nodes of the causal network of the brain states about which nothing is known in the initial stages. In other words, when there is no theory yet, the prospective "theoretical terms" can't refer. This presents a dilemma. On the one hand, the STM-theorist wants to theorize about the functional organization of particular brain state types. For this she must have an independent way of referring to them, independent of a more or less completed theory. On the other hand, as far as she refrains from using an intentional scheme, she can't even guess what she is talking about when she uses terms like 'the B-state mapped to *Fa*'. That is because the theorist has no independent way of identifying the nodes of the network of brain state types. This network is completely unknown.

The problem stems from the STM paradigm itself. Notice that if there were an independent way of picking out the nodes (brain state types) in the causal network that does not presuppose a more or less complete specification of which nodes are connected to which others and how (i.e. their potential functional roles determined by the generalizations of the theory), then we would use this scheme in our way to saying what generalizations there are, i.e. in our theory construction. This is exactly what Brian Loar does (1982a) in presenting his semi-broad content functionalism: he uses propositions to pick out those brain states and state whatever generalizations there are that need to be stated. Once he gets the generalizations into place, he gets rid of propositions in favor of syntactic objects. Then, of course, he is in a position to specify, theoretically at least, all the functional roles there are without using any semantic terms. Once he does that the result is almost an STM theory very much like what Stich envisions.

So it should be obvious that the way out of this dilemma can only be semantic, not syntactic. The upshot is that pure "syntactic" (psycho)-functionalism in scientific theory construction à la Stich cannot be carried out without assuming the truth of content (semantic) functionalism (à la Loar). They stand or fall together, which is not to say that narrow content functionalism has got to be true (see above).

If what I have said so far is right, the lesson to be drawn is that syntactic functionalism is not an option in psychological *theory construction* somehow at the discretion of the psychologist. When we reflect upon the historical rise of functionalism in the philosophy of mind, that this is so should be obvious. Functionalism was developed as a response to the inherent difficulties in behaviorism and (type-type) identity theories. It was conceived as a metaphysical theory saying *what* mental states *are*. Functionalism identified mental states with functional states. But that was not enough. Functionalism had to be able to provide the identity conditions of mental state types. This required providing identity conditions for functional roles. Functionalists had to be able to say what functional roles uniquely define what types of mental states. But this required having a theory first. Some versions of functionalism took this theory to be folk psychology made explicit with all the intentional/mental terms employed as theoretical terms. Then, Ramsefying this theory was the major step in explicitly getting the identity conditions for functional roles. Similarly with psychofunctionalism: the theory to be Ramsefied was conceived to be a theory to be developed by scientific psychology. The underlying idea was the same. Once such a theory was at our disposal with all its intentional/mental terms employed as theoretical terms, we could explicitly get the identity conditions for the functional roles by Ramsefying it. In all this, the construction of the theory to be Ramsefied was conceived along with using all the intentional vocabulary available to the theorists. And that was OK, because functionalism was competing against dualism, eliminativism and reductionism (type identity theory). That is the reason why functionalism at its core is essentially an intentional realist theory. But Stich's STM tries to reverse the situation, it wants to develop functional theories without ever using intentional terms; in this, however, Stich is putting the cart before the horse. As we have seen, this turns out to be impossible, because the remaining vocabulary to be used in theory construction cannot do the required job. In a sense, in fact, Stich cuts the branch he is sitting on.

Admittedly, my point in this section is one that belongs to the context of discovery. It might be claimed that as such it is not that important: what matters is whether the ultimate STM-style theory, when completed, is committed to any intentional scheme. The STM theorist might use any tools (intentional or otherwise) that would help in getting the theory, i.e. in the context of discovery. But once the theory is completed and successful, it should not matter how it was gotten in the first place. For instance, as long as it belongs to a discovery stage, an STM-theorist might use a procedure like Loar's. It is the form of the ultimate completed theory that counts.

Well, I have two points to make against this. First, given Stich's criticism of content-based psychologies, it should be obvious that the brain states initially typed according to an intentional scheme will exhibit all the vagueness, context-sensitivity, and parochialism that Stich claims will pertain to a semantic taxonomy. So he can't avail himself to the SA of typing even in the context of discovery. Second, it is simply absurd to assume that a taxonomic scheme will be semantic-free if at the end it is essentially obtained by a SA and then gotten rid of à la Loar. The ultimate theory, if really successful, would be nothing but a (partial) scheme for a naturalized psychosemantics (e.g., in the tradition of two-factor semantic theories).[22]

8. IF COGNITION IS COMPUTATIONAL, HOW CAN PSYCHOLOGICAL LAWS BE INTENTIONAL?

This is what Fodor called the "Eponymous Question" in his (1994). As I said, this question has in fact been around, constantly popping up here and there, and haunting people working in the field, for more almost two decades now, mostly thanks to Stich and Fodor.[23]

This question is also related to certain puzzles computationalism has created *vis-à-vis* mental causation. According to the computational picture of mind (CRT, LOTH), mental processes are defined over mental symbols physically realized in the brain. But computationalism says that for these mental processes to qualify as computational, it is the non-semantic, in particular syntactic, properties of symbols that the processes must be causally sensitive to. In fact, given a physicalist framework, it is not even clear what it would be like for mental processes to be causally sensitive to the semantic properties of symbols, which are relational, i.e., hold between the symbol (or the organism) and environment. Given the locality of causation, thought processes can be causally sensitive to only syntactic (at any rate, non-semantic) properties of symbols that are implemented neurally. If so, even though mental symbols are causally efficacious in reasoning and causation of behavior, it is in virtue of having certain syntactic properties, but not in virtue of having semantic properties, that they are so. Thus, as far as the science of psychology is in the business of causal explanation, the relevant properties of mental states in virtue of which they are covered by causal psychological laws are all non-semantic, or so it seems on the face of it. This is another way of seeing Stich's motivation in arguing against content-based psychologies and promoting his STM over them. As we have seen, Stich calls the Narrow Functional Account of typing symbol tokens "syntactic" typing, presumably meaning just non-semantic and non-physical. And this sort of typing, on his view, is what the STM (or CTM for that matter) is committed to. He then claims that STM/CTM is all a scientific psychology needs; hence, *pace* Fodor, there is no need to appeal to semantic/intentional properties of syntactically structured brain symbols in stating the laws of psychology. He accuses Fodor of having it both ways.[24] We are now in a position to see how it is possible to have it both ways, i.e., to see what the answer is to the Eponymous Question.

Let us suppose that computational psychology is correct. Any scientific computational psychology needs to postulate states in terms of which it can explain (and predict) behavior (construed broadly—bodily, verbal, mental behavior). This seems to call for covering laws or generalizations that subsume those states under an appropriate description. This is at least the assumption shared by all parties in the debate, and I will not challenge it. This means that these states, under the relevant description, are projectable, i.e. natural kinds from the perspective of the theory. As such they must have identity conditions. Computational psychology characterizes these states as symbol tokens realized in the heads of cognitive organisms. *Qua* symbols they have both syntactic and semantic properties. OK then, how are we to type them to suit the psychological laws covering their tokens? We have seen that they cannot be typed, in the required sense, by their narrow functional properties: NFA is hopeless. The Physical Account (PA) of typing them is hopeless too. Stich and almost everybody in the field agrees. The PA seems to commit one to a very strong version of type-type identity theory for propositional attitudes with specific content (like the belief *that snow is white*) cast across people. In this form, the PA has no defenders as far as I can tell. Our only other option, then, the Semantic Account, is in fact mandatory if psychological processes are to be computational. In other words, *if Stich's original question, i.e. the question of what it is for two symbol tokens of Mentalese in different heads to be of the same type, has an answer*, it must be some version of the SA.[25] It must be on the basis of their semantic properties we type symbol tokens across systems.

Therefore, I conclude that computational psychology (CTM, for that matter) itself is essentially committed to semantic type individuation of symbol tokens across systems. And it is across systems that a scientific psychology casts its laws over. Hence, the necessity for intentional psychology whose laws advert to the semantic properties of representations. If mental representations can be typed interpersonally *only* on the basis of their semantic properties, CTM cannot be an alternative to replace intentional psychology. Hence the answer to Fodor's Eponymous Question.[26]

University of Florida, Department of Philosophy,
330 Griffin-Floyd Hall, P.O. Box 118545
Gainesville, FL 32611-8545, maydede@ufl.edu

9. NOTES

[1] In his (1991) Stich argues against Fodor that narrow content taxonomies will differ from the narrow causal taxonomies, which he calls "fat syntax" taxonomies. The problem, according to Stich, stems directly from the SA, narrow or wide.

[2] Generalizations detailing the causal relations between proximal input events and T-states (thought-like states), among T-states, and between T-states and proximal output events. See Devitt (1990).

[3] See also (1983: 78-9), where Stich writes: "mental sentence theorists typically leave the notion of an internalized sentence token as little more than a metaphor. And it may well turn out that when the metaphor has been unpacked, it claims no more than that beliefs are relations to complex internal states whose components can occur as parts of other beliefs." Here, it is not clear what the contrast Stich is trying to convey is supposed to be. Of course, this is what is literally intended by a LOT theorist and more: under a suitable mapping function internal states are literally interpretable as constituting a symbol *system* with a combinatorial syntax and semantics so that the processes defined over these are sensivitive only to their formal/syntactic properties. That is the essence of LOTH and computationalism, see my (1997a, and 1998).

[4] Indeed, this was the very point of Brian Loar in his polemical article (1982b) written against Fodor's LOTH. He says that from a philosophical point of view his non-committal content functionalism is weaker than the LOT version of it and thereby should be preferred. He does not reject the LOTH, but he claims that its motivation cannot be due to its having more explanatory and predictive power. For, with respect to these, his pure functionalism is equally good. Loar views the LOTH as a scientific hypothesis, and as such he leaves its truth as an open question.

[5] Here is a passage (among many others) from an early piece by Fodor: "For purposes of (narrow) ascriptions of content, the essential properties of a mental state are its functional properties (the ones it has in virtue of its causal role vis-à-vis behavior and other mental states). Since, as a matter of fact, there is a certain set of functional properties that mental states normally have when they are brought about by, for example, visual encounters with banks, we can specify a set of functionally equivalent experiences by reference to some such contingent fact as that they are like seeing a bank. Something could have these functional properties (hence this content) without being caused by a bank (cf. bank hallucinations). To this extent, the description 'prescinds from semantic relations'... But that's alright since [methodological solipsism] isn't the claim that we must pretend, in describing mental states, that the world does not exist; what it claims is that the properties of mental states which are essential for determining how mental operations apply to them are their functional (hence formally specifiable, hence nonsemantic) properties. A mental operation may apply to a mental state in virtue of its being like seeing a bank but not in virtue of its being seeing a bank" (1980b: 102). Here Fodor is taking formal/syntactic typing, narrow functional/causal typing, and narrow semantic typing to be all virtually equivalent.

[6] He makes the same point in his (1983): "The core idea of the STM—the idea that makes it syntactic—is that generalizations detailing causal relations among the hypothesized neurological states are to be specified indirectly via the formal relations among the syntactic objects to which the neurological state types are mapped" (151).

[7] For an extensive elaboration of how computation is to be understood in the role it plays in LOTH and modern cognitive psychology, see my (1997a) and (1998). In the former work, I also give an analysis of the notion of syntax as it is deployed by LOTH.

[8] Note that Stich's claim is stronger than merely saying that lexically different sentence tokens have the syntactic property of just *being different*. His claim is that they belong to different *specific* syntactic categories.

[9] See my (1997a). Stich, in his (1991) article, calls the type identity of sentences that gets fixed on the basis of their narrow causal profile their "fat syntactic" identity. This is supposed to be contrasted by their "skinny syntax". The latter is to be fixed by the T-T generalizations alone: no causal relations to proximal stimuli and behavioral events can be used in the individuation of sentences. Stich insists that it is the fat syntactic type identity that would do the work for STM-style theories. As I said, I will argue that the NCA cannot fix the type identity of mental sentence tokens whether or not what gets fixed is their (fat) "syntactic" type. Devitt (1990) has argued that even if their type identity can be so fixed, what gets fixed

would be their narrow semantics not their syntax. Devitt's discussion also contains a very helpful criticism of Stich's notion of syntax.

[10] For some others, see Stich (1983: 53-60 and 137-44).

[11] For a similar and more striking discussion of the commitments of the NCA of typing where Stich goes through a similar example, see (1983: 53-4). The generalizations (4)-(6) Stich mentions here are all what I will call below, L-generalizations. They advert to the logical form of the sentences, hence are blind to the non-logical primitives the theorist postulates.

[12] It is ironical, and in fact a bit puzzling, that Stich himself makes the parallel point in criticizing content functionalism: "There are literally infinitely many inferential paths leading both to and from every belief" (1983: 24). His point is that since every particular belief is potentially connected to every other, the generalizations detailing this potential will not be able to define beliefs with particular content.

[13] In fact, the situation is even more complicated given that there is already a build-in vagueness in the "syntactic" individuation of particular B-states: for Stich a sentence token to count as of a particular type, it must satisfy a substantial number of generalizations. Stich seems to propose a cluster theory of type individuating sentence tokens, and this, as Stich himself admits, brings with itself a certain amount of vagueness. See below.

[14] A parallel distinction is drawn by Loar (1982a) between "L-constraints" and "M-constraints".

[15] These are supposed to be "ceteris paribus" generalizations. I'll generally ignore this in what follows.

[16] In what follows, in order to avoid long and cumbersome ways of expressing the same thing, I will simply adopt the following convention: I will mark an intentional expression with a '*' to express whatever its syntactic parallel may be. Also, I will hedge a content sentence with '#'s in order to indicate that I intend its syntactic parallel, i.e., whatever syntactic object or sentence might go in its stead.

[17] There are many versions of this approach in AI. Frames, scripts, etc. are all versions of the same underlying idea. The tradition of "semantic representation" in linguistics again relies on the idea that lexical items can be semantically decomposed.

[18] Rules may or may not be explicitly represented. CTM is neutral on this. However, given that the rules that implement S-generalizations reflect important pieces of "semantic knowledge" they are unlikely to be hard-wired.

[19] Block (1993: 3-4) writes: "Fodor and Lepore seem to assume... that... the inferential role theorist has the option of appealing to analyticity as a way of discriminating the inferential liaisons that are in inferential roles from those that are out. But if we stick to traditional ideas about the extension of 'analytic', there aren't *enough* analyticities. Consider the putative analytic truths involving 'cat'—'Cats are animals', 'Cats are living beings', 'Cats are grown up kittens', etc. The problem is that abstracting from the *words* 'cat', 'kitten', etc., appearing in these sentences, there is nothing here to distinguish 'cat' from 'dog'. Corresponding to 'Cats are grown up kittens', we have 'Dogs are grown up puppies'. Sure, 'nothing is both a cat and a dog' can be used, but so can 'nothing is both a dog and a cat'. Even if 'Cats are feline', and 'Dogs are canine' are analytic, this is of no help without other analytic truths that distinguish 'feline' and 'canine'...". See also his (1986: 628-9). Cf. Loar (1982a: 81ff).

[20] It is very curious that more or less the same criticism is given by Stich himself for the claim made by content functionalists that there are such generalizations: "[t]here is generally no characteristic environmental stimulus which typically causes a belief. There is no bit of sensory stimulation which typically causes, say, the belief that the economy is in bad shape, or the belief that Mozart was a freemason... Nor do most beliefs have typical behavioral effects. My belief that Ouagadougou is the capital of Upper Volta does not cause me to do much of anything" (1983: 24). Later on, he argues (1983: 180-1), on familiar grounds, that there can be no principled distinction between beliefs whose content is "observational" and those whose content is "theoretical". So, according to Stich, even for allegedly "observational" beliefs there seems not to be any particular set of stimuli nomologically connected to them.

[21] This is in fact more or less acknowledged by leading functional role semanticists like Block. Hence the destructive holism to which they are said to be committed.

[22] On this last point, see also Higginbotham (1988) and Crane (1990).

[23] See, among many others, Stich (1983, 1991), Field (1978), Schiffer (1987), Fodor (1980a, 1989, 1994), Devitt (1991), Jacob (1997) who take issue with the EQ one way or other.

[24] Stich (1983). See also his (1991). Devitt (1991) joins Stich in accusing Fodor of trying to have it both ways but only with respect to processes governing thoughts without I/Os.

[25] It is of course possible that Stich's question doesn't have an answer. I surely haven't argued here independently for the truth of the SA. In other words, if Stich is right about the fate of the SA, and if I am right about the fate of the PA and NFA, then scientific cognitive psychology as we know it today is impossible. I can't take this option seriously, in particular I can't take seriously a priori arguments against the cogency of the foundations of what appears to be an enormously successful and fruitful scientific approach to cognition. Cognitive psychology seems to be into intentional talk up to its neck. I take it that there is an enormous *prima facie* evidence for the truth of the intentional assumptions of present day cognitive psychology. I take this to be the best argument for the SA albeit a non-demonstrative one. I left Stich's positive arguments against the SA aside in the beginning of the paper. What needs to be done, of course, is to address Stich's criticisms in order to begin to give an independent argument for the SA.

[26] There are, to be sure, problems with any version of the SA, as is well known. Suppose that the SA is broad as in (late) Fodor. Then we have problems with Frege cases as well as Twin-Earth cases. A narrow SA would be equally problematic, as we have seen, if it relies on narrow functional roles of vehicles as their narrow semantic content. On the other hand, a Fodor-style notion of narrow content as mapping from context to broad content can perhaps handle Twin-Earth cases at best, but not Frege cases (see my 1997b, 2000). But being problematic is one thing, being wrong is another: I think that a SA that works can after all be salvaged in the face of apparent difficulties.

10. REFERENCES

Aydede, Murat (1997a), "Language of Thought: The Connectionist Contribution," *Minds and Machines*, Vol. 7, No. 1, pp. 1-45.

Aydede, Murat (1997b). "Has Fodor Really Changed His Mind on Narrow Content?", *Mind and Language*, Vol. 12, No. 3/4, pp. 422-58.

Aydede, Murat (1998). "Language of Thought Hypothesis" in *Stanford Encyclopedia of Philosophy*, edited by Edward Zalta, Stanford, CA: CSLI E-Publications.

Aydede, Murat (2000). "On the Type/Token Relation of Mental Representations," *Facta Philosophica: International Journal of Contemporary Philosophy*, Vol. 2, No. 1, pp. 23-49.

Block, Ned (1986). "Advertisement for a Semantics for Psychology" in *Studies in the Philosophy of Mind: Midwest Studies in Philosophy*, Vol. 10, French, P., T. Euhling and H. Wettstein (Eds.), Minneapolis: University of Minnesota Press, 1986.

Block, Ned (1993). "Holism, Hyper-analyticity and Hyper-compositionality," *Mind and Language*, Vol. 8, No. 1, pp. 1-26.

Crane, Tim (1990). "The Language of Thought: No Syntax Without Semantics," *Mind and Language*, Vol. 5, No. 3, pp. 187-212.

Devitt, Michael (1990). "A Narrow Representational Theory of the Mind," in *Mind and Cognition*, W. G. Lycan (Ed.), Oxford, UK: Basil Blackwell, 1990.

Devitt, Michael (1991). "Why Fodor Can't Have It Both Ways" in *Meaning in Mind: Fodor and his Critics*, B. Loewer and G. Rey (Eds.), Oxford, UK: Basil Blackwell, 1991.

Devitt, Michael (1996). *Coming to Our Senses: A Program for Semantic Localism*, Cambridge, UK: Cambridge University Press.

Field, H. (1978). "Mental Representation," *Erkenntnis* 13, 1: 9-61.

Fodor, Jerry A. (1980a). "Methodological Solipsism Considered as a Research Strategy in Cognitive Psychology" in *RePresentations: Philosophical Essays on the Foundations of Cognitive Science*, J. Fodor, Cambridge, Massachusetts: MIT Press, 1981. (Originally appeared in *Behavioral and Brain Sciences* 3, 1, 1980.)

Fodor, Jerry A. (1980b). "Methodological Solipsism: Replies to Commentators", *Behavioral and Brain Sciences* 3, pp. 99-109.

Fodor, Jerry A. (1989). "Substitution Arguments and the Individuation of Belief" in *A Theory of Content and Other Essays*, J. Fodor, Cambridge, Massachusetts: MIT Press, 1990. (Originally appeared in *Method, Reason and Language*, G. Boolos (Ed.), Cambridge, UK: Cambridge University Press, 1989.)

Fodor, Jerry A. (1994). *The Elm and the Expert*, Cambridge, Massachusetts: MIT Press.

Higginbotham, J. (1988). "Is Semantics Necessary?", *Proceedings of the Aristotelian Society* 88, pp. 129-41.

Jacob, Pierre (1997). *What Minds Can Do: Intentionality in a Non-Intentional World*, Cambridge, UK: Cambridge University Press.

Loar, Brian F. (1982a) *Mind and Meaning*, Cambridge, UK: Cambridge University Press, 1982.

Loar, Brian F. (1982b). "Must Beliefs Be Sentences?" in *Proceedings of the Philosophy of Science Association for 1982*, Asquith, P. and T. Nickles (Eds.), East Lansing, Michigan, 1983.

Schiffer, Stephen (1987). *Remnants of Meaning*, Cambridge, Massachusetts: MIT Press.

Stich, Stephen P. (1978). "Autonomous Psychology and the Belief-Desire Thesis" in *Mind and Cognition*, W. G. Lycan (Ed.), Oxford, UK: Basil Blackwell, 1990. (Originally appeared in *The Monist* 61, pp. 573-591, 1978.)

Stich, Stephen P. (1983). *From Folk Psychology to Cognitive Science: The Case Against Belief*, Cambridge, Massachusetts: MIT Press.

Stich, Stephen P. (1991). "Narrow Content Meets Fat Syntax" in *Meaning in Mind: Fodor and his Critics*, B. Loewer and G. Rey (Eds.), Oxford, UK: Basil Blackwell, 1991.

PART IV
CAUSES AND ACTION

ARDA DENKEL

CAUSATION, PARTS AND PROPERTIES

1. The explanatory function of science hinges largely on causal relations. This explains why the philosophy of science has shown much interest in the nature of causation, investigating, among other things, whether causation has different types, whether it is necessarily connected to laws, whether its terms are events or facts and whether causes are powers or capacities. Not much work has been done, on the other hand, concerning the ontological presuppositions of causation. In this paper I wish, first, to indicate a certain ontic necessary condition of causation, and then to argue that what is regarded as one of the most promising physical ontologies of our time, *tropism*, cannot hope to make coherent sense of causation and thus be compatible with science, unless it abandons what I will call its 'standard' form, that is, its version refusing to recognize Aristotle's principle of inherence. Standard tropists have believed that adopting a particularistic ontology, or in other words, maintaining that anything that exists is a particular property or consists of properties that are particular, commits one to regarding properties (or tropes) as physically independent entities.[1] I will argue that this consequence does not follow. A particularistic ontology can be consistent with the principle of inherence and hence with scientific explanation.

Leaving aside the question whether or not causal relations entail generalities, I shall focus here on particular causes or effects. I wish to characterize the terms of a causal relation as 'property-occurrences'. By this notion I understand the existence of a particular property from a certain date onwards.[2] Thus a property-occurrence is a convenient specification for particular causes and effects, since it captures the essential nature of a cause as a property, entailing at the same time a change (the property's coming about). My characterization is agreeable to those who maintain that "When things act causally, they act in virtue of their properties"[3] and remains impartial to the dispute between interpretations of a cause as an event and as a fact. To cite a prominent representative of tropism, Keith Campbell says the following: "The causal agent is a state, or event, or process, always particular and always qualitative. It is not the stove, the whole stove, that burns you . . . It is the temperature that does the damage . . . Causes are always features . . . and every particular cause is a particular feature or constellation of features."[4] I agree completely.

G. Irzık & G. Güzeldere (eds.), Turkish Studies in the History and Philosophy of Science, 207-216.

Speaking thus, positively, causes and effects are no doubt property-occurrences. But, negatively, I must maintain that they cannot be simply that. It does not follow from the positive thesis that the causally efficacious qualities do what they do independently of the substances in which they inhere. On the contrary, if something were only a property-occurrence, in abstraction from an object, it would fail to be causally efficacious.[5] To be a cause or an effect, to engage in causal interaction, any property-occurrence must be property of some substance. The reason is that without the compresent support of determinates under every determinable necessary for objecthood,[6] property-occurrences could not bring about their usual effect; they would be simply inert.

Let me illustrate this. Consider a stone thrown at a window, which, upon hitting the glass, shatters it into fragments. We say that the stone's impact on a certain part of the stationary brittle glass caused it to break; that the cause of the breaking was that the glass was hit by a solid object with a certain mass and velocity. But if it is abstracted from the object, the mass would not have that effect.[7] The same thing applies to the velocity (and also to the solidity and shape) of the object, taken in abstraction. How can mass, without the object's solidity, surface boundary (shape), velocity, and so forth, break the glass; how could it have an impact on a particular part of the glass without such accompaniments? How could a free-floating velocity affect an object? Neither property, on its own, would yield the type of effect the object can have on the glass, in virtue of it. Not even the compresence of *some* of the select particular properties, which, as inhering in an object can cause a certain effect, will yield such a thing in abstraction. Nothing less than the full support of the mutually bonding compresence or bundle of determinates under all the determinables that stand for primary qualities will enable a property to have causal efficacy.

Let us take a second example. Imagine impressing a mark by a seal on a blob of wax. Without hesitation, we say that the cause of the mark is the seal's shape in relief being driven into the soft lump. Now if this shape were isolated from the seal, it would be devoid of causal relevance. In abstraction from the seal a shape could neither be visible or tangible, for it would fail to reflect light or to resist interpenetration. Unless it were compresent with the other properties amounting to those of an object, the effect we ordinarily expect from it would not ensue. Thirdly, think of a drop of acid, which, on a piece of chalk, causes the bubbly corrosive action. Similarly, the abstracted qualities of the acid would remain ineffective in the absence of the rest. The full support of weight, fluidity, and so forth, which enable the corrosive quality(ies) to spread on the surface of the chalk and to penetrate downwards is essential to the effect to obtain.

An ontology that permits physically independent and isolated existence for properties will render them causally inefficacious. Such an ontology yields a causally inert world to the extent that it regards properties as entities that can exist in abstraction. To the same extent, therefore, it fails to describe and to account for the actual world, depriving itself at the same time of the advantages of scientific explanation. In fact, it turns out to be incompatible with science. I will discuss how the commitment to the thesis of independence arises for standard tropism, and then will show that this is not a general consequence of ontologies that regard properties as particular entities.

2. Among other things, standard tropism is the conjunction of two of theses:

a) Any property that exists physically is a particular entity, that is, a trope, and,

b) Any particular property (trope) exists in its own right, independently of anything else. It is only a contingent fact that a large majority of properties, though not all, exist in inhering substances.

Tropism is naturally, but not necessarily, combined with the 'bundle thesis':

c) An object is a (complete) compresence or bundle of properties.

Standard tropists believe that conjoining (a) and (c) yields (b); that once one adopts a particularism of properties and accounts for objects in terms of bundles of properties, then one has to regard particular properties among the parts of objects. One basis for such an inference seems to be the worry that once, by (c), one lends tropes the most fundamental ontic status and accounts for objecthood in terms of them, it would be logically defective to make their existence, in turn, dependent upon that of objects. But such a worry is unfounded. If an object *is* a bundle of tropes, then suggesting that tropes exist only in compresences makes them dependent on other tropes only, and not necessarily on objects. All one needs to do in addition is to explain why tropes have to exist in space and time as linked with other tropes under different determinables.

Another reason motivating standard tropists to think that (a) and (c) yield (b) is based on the mereological assumption that

d) if bundles constituted out of tropes as *wholes*, then the constituent tropes are the parts of such wholes.

Then, granting (c), tropes should be among the parts of the *object*. Since parts are separable from the whole they constitute, it follows that tropes can exist in isolation from objects.

Tropism is not alone in maintaining that (a) and (c) have such an implication. Its arch-rival, the substance theory, is well-known for its adamant rejection of (b). Nevertheless, a prominent form of the substance theory agrees with standard tropism that (a) and (c) yield (b), and it is in order to block this consequence that it rejects the bundle thesis (i.e., [c]). The principle tenet of this specific position is that a bundle of properties forms an object only if its constituent properties are held together by a *substratum*, that is, by a non-qualitative particular entity which also individuates the object. The substratum version of the substance theory is, therefore, that

e) an object is the compresence or bundle of properties plus a substratum that holds the bundle together.

Concerning the nature of the relation of compresence, standard tropism sets off from the true premise that an isolated, free-floating trope is a logical possibility. Now what follows from this proposition is that properties do not hold in compresences by *logical* necessity. However, standard tropism makes more of the same thing. It goes as far as denying that there are bonds or (perhaps) (meta-)physically necessary links that hold particular properties in a bundle. According to it, properties are compresent only by accident. The world is made of compresent properties that form objects by a pure coincidence.

> If abstract particularism calls for a fundamental tie, the most natural candidate is *compresence*, that in virtue of which many tropes can combine to yield ordinary concrete particulars. But although a very widespread relation, compresence is not required by the very terms of the ontology to embrace all examples of the other category or categories... In trope theory, individual, isolated tropes, compresent with nothing, are admitted as possibilites.[8]

But if there are no bonds that secure compresences, the world, made of compresent tropes now, may not be so the next moment. Campbell argues that the relation of compresence is not internal or founded.[9] Moreover, one should deny that compresence is a trope. If it were affirmed, one would have to explain how, as a trope, *compresence* links other tropes, how it *is compresent* with them, and this paves the way for a regress. Admitting substrate, on the other hand, is self-defeating for the bundle theory.

Here is a conclusion I draw from all this: If bundles were accidental, in the sense that there are no bonds that would prevent their coming apart any moment, then the particular properties that constitute them could not support one another in the way causation requires. Thus without mutual bonds that hold bundles in one piece, what is true for free-floating tropes applies equally to tropes in (coincidental) compresences. It follows that for standard tropism *no* property can be causally efficacious. A very untenable consequence indeed!

I maintain that a particularistic ontology can consistently adopt (a) and (c) without accepting (b), and hence that it does not have to drop (c) in favour of (e). Before offering my reasons, however, I wish to demonstrate by means of quotations, that standard tropism commits itself to (b) quite explicitly. To begin with, let us read the major progenitor of contemporary tropism, Donald Williams:

> Besides... division into concreta, however, we are accustomed to think of abstraction as a mode of analysis and of analysis as also a mode of division, and we need only to take this literally to affirm that there are abstract parts and objects also—or, if we are squeamish about 'parts' in this connection, then abstract 'components', each of them as actual an entity as any concretum... Abstract entities differ from concreta in that many of them can and do occupy the same plime [spatiotemporal position]... Two tropes which are together in the sense that they occupy or pervade the same plime we call 'concurrent', and say that they are 'embraced in' or 'inhere in' and in a certain good sense are 'qualities of' the concretum which is the total occupant of the plime.[10]

Following the teacher's footsteps, Campbell declares that "...exist out there waiting to be recognized for the independent, individual items that they have been all along. For Williams and for us following his usage, abstract does not imply *indefinite, or purely theoretical.* Most importantly, it does not imply that what is abstract is *non-spatiotemporal*... Abstract here contrasts with *concrete*..."[11] Campbell continues: "Aristotle got us all off on the wrong foot when he treated qualities as existing only as inhering in a substance... It is a matter of fact, and not a metaphysical necessity, that tropes commonly occur in compresent groups... free-floating tropes are at least metaphysically possible... Individual isolated tropes, compresent with nothing are admitted as possibilities."[12] Campbell looks upon sounds as "prime candidates for items in our experience which *seem* to be qualitative yet without any substantial support. They support the idea that free-floating tropes are at least metaphysically possible."[13]

I find it difficult to understand how a 'metaphysical possibility' is being exemplified here, beyond ordinary logical possibility, which we have granted already. In the quoted material Campbell is talking about the *experience* of sound, and thus nothing ontologically objective (i.e., independent of the mind) is specified by his example. There existing no clearly demonstrable actual instantiation of a free-floating trope, Campbell's isolated properties remain as dim hopes, rather than genuine substantiations. It is interesting to observe that in Campbell's book explicit commitment to (b) is accompanied by unreserved statements that causation is a relation between tropes: "The terms of the *causal* relation are always tropes. It is the heat of this stove, here and now, that burns you, on the finger, here and now."[14] A cause (e.g., a chemical feature) is the basis on which a functional feature such as a certain power (e.g., a soporific power) of an object (e.g., a pill) supervenes.[15]

> Strictly speaking it is not the earth and the compass needle as entire complex wholes which are cause and effect in the compass' pointing north. Rather it is the *magnetic characters* of these complexes which do the work. Their other features have nothing to do with it. Events are changes of abstract particulars, in the typical case, where, for example, sunlight fades the drapery. Where conditions, rather then events, are involved in causal situations, they too will be tropes, usually tropes belonging to a compresent complex or concrete particular.[16]

What Campbell does not seem to realise is that, as *abstract* particular properties, tropes cannot have any causal powers. Something is seriously amiss in standard tropism, and I will try to expose it.

3. Consider a material object and contrast it with the bundle or compresence of particular properties (or tropes) it is, or possesses. For the sake of the argument we are not presupposing here (c), namely, that an object is analyzable as a bundle of particular properties; we allow an object to be a bundle of properties that are held together by (or inhere in) a non-qualitative substratum, and thus remain consistent with (e). A material object can be an articulated thing, with a functional structure essential to its identity. Thus we may suppose that an object is *at least* the sum of its parts, where the latter are all (potentially) material things. To many (though not to standard tropists), it seems relatively uncontroversial to suggest that the particular properties or features borne by the object are not among its parts. What can be said, in this respect, about the relationship linking the *bundle* with the properties? Is it analogous to the former (whole object-parts) or the latter (object-features)? In other words, is (d) true? Standard tropists view it as a whole-part relationship, and substance theorists, on the other hand, whether they commit themselves to particular properties or not, reject this view.[17] In fact it is among the arguments of the latter party that *unless substrata are assumed to exist* in objects as the bearers of these objects' properties (i.e., [e]), tropes will be recognised as entities that may exist independently of substances (i.e.,[b]). (According to them this is something that runs counter to both intuition and observation).[18] Standard tropists also assert directly that without a substratum, properties are conceived (implausibly) among objects' parts; that on a bundle analysis of objects, it is 'as if the shape of an object is a part of it like its top half".[19] Accordingly, the reason it is objectionable to explain a material object purely in terms of the bundle of its properties (i.e., in terms of [c]), is that it regards the relation in question as that existing between wholes and parts. So

far as the usual sense of 'parthood' is concerned, I shall deny this specific reason. The truth is that only *some* bundle theorists conceive of the relation in such a way. What we have here is not a commitment of the bundle theory; the commitment rather belongs to a certain version of it, namely to standard tropism.

I fully agree that the consequences pointed out by the substance theorists are undesirable. But the notion of an unknowable substratum as a bearer of properties is no less repulsive; if possible, an explanation of objecthood should be purified from it. We *can* analyze a material object as the compresence of its particular properties, without adopting the standard tropist's approach. With the purpose of showing how such a thing is possible, I have previously argued[20] that the bundle theory does not entail that the particular properties forming a compresence are the parts of this compresence. The *parts* of an object either overlap in space, in the sense of sharing their own parts, or else they are parts that are spatially discrete.[21] In contrast with discrete parts, many of the properties in a bundle exist in an interpenetrated state: as in the roundness, softness, redness and acid taste of a tomato, they occupy the same positions in space at the same time. Their individual presence in the bundle does not enlarge the volume or extension of the compresence. Whereas the bundling of properties does not involve the addition of their extension in a cumulative way, the extensions of non-overlapping parts in an object do add to one another, and thus yield the extension of the whole object. The spatial parts of the compresence are themselves compresences, however, and not simpler qualities. As a first step, I conclude that the properties making up the bundle of the object is considered to be, are not among the *discrete* parts of this object. Is there reason for supposing that they can be viewed among the object's *overlapping* parts? Now both the properties and the overlapping parts of an object occupy the same spatial positions together at the same time. Tibbles the cat, as a material object, for example, shared with its proper part Tib (Tibbles minus its tail) much of its spatial position at the same time. Similarly, the colour, warmth and softness of Tibbles' and Tib's bodies extend over exactly the same positions. There is a crucial difference, however. Tibbles and Tib share positions, or overlap, *by sharing their parts*, but properties under different determinables, which share positions, or overlap in the sense of interpenetration, do *not* share any of their parts. In fact, they cannot. Compresence is a relation that holds among entities of different kinds (different determinables).[22] In contrast, the parts of a whole do not have to be entities of different kinds.[23] Moreover, while it is inconceivable for the overlapping parts of a body, such as the top half and the top quarter of a pole to separate while remaining what they are, the separation of many overlapping properties, such as the acid taste and the roundness of a tomato, is easily imaginable. Secondly, therefore, I conclude that the properties making up the bundle of the object is considered to be are not overlapping parts of this object, either. It follows that, unless one stipulates a new sense for the word 'part', the properties making up the bundle that amounts to the object are not the parts of this bundle.

4. Joshua Hoffman and Gary Rosenkrantz have offered a criticism[24] of this argument, pointing out that *generally speaking*, the distinction I have used in my first premise does not exhaust the possible types of parts. They grant that in my application to material objects, the distinction *is* exhaustive, but restricting it *a priori*

to such a type of existence would beg the question. Taking into account a third type of parthood, on the other hand, will undermine the contrast I aim at drawing between objects' parts and properties of bundles: They say that in the *collection* of properties a bundle is, properties occupy the same positions *without having parts in common*; nevertheless, each such property is a *part of collection* (i.e., of the bundle). Thus, in their opinion, the contrast fails.

The idea of tropes as parts of a bundle is entailed by the construal of a bundle as a *collection* of properties. Many philosophers, including Hume, have adopted such a conception,[25] and no doubt the literal meaning of the word 'bundle' implies being a collection.[26] But *qua* philosophical terminology, 'bundle' does not have to be taken literally.[27] In the sense of a *compresence* it does not mean a collection at all, and I suggest that if we want to speak plausibly of a 'bundle' of *properties* we should use it in this non-collective sense. My view is that the alleged third type of parthood is an *ad hoc* extension of the usual notion of a part, and that the proper application of this notion is to the category of material objects. Now I will offer my evaluation of such an extension.

If a bundle were essentially a collection, there would be nothing, no linking principle, that would hold such a bundle of properties together; nothing would lend them a unity, apart from their being a collection.[28] But the parts of a collection do not have to share the same position, and if bundles are *mere* collections, it is a miracle that properties remain together so uniformly in compresences and that objects do not disperse into free-floating tropes. Precisely because it implies this, standard tropism offends common sense and fail to explain the way the world is. Given that only concrete objects are substances and that tropes never exist independently, there must be a much stronger link, a bond, that keeps properties in compresences. The substratum may be one such possibility, though many other bonding relations between particular properties can be entertained.[29] If, however, particular properties are held together by mutual bonding relations, their existing together in bundles is a derivative fact, and their togetherness is not due to the bundle itself. Rather the converse seems true: in such a case it is the bundling of the properties that owes its existence to the properties' mutually clinging together. As Hoffman and Rosenkrantz point out in a different but analogous context, the existence of such bonds will rule out the consideration of the bundle as essentially a collection.[30]

If objects are bundles of tropes, then granting that properties do not exist independently, bundles cannot be collections essentially, and thus tropes cannot be proclaimed to be the parts of bundles on these grounds. Since one is compelled to acknowledge that there are bonds holding bundles of properties integrally, the latter cannot be collections essentially: compresences have their tropes essentially in unity, but it is not essential to a collection to do so. We have seen that the ides of a bundle as (essentially) a collection of particular properties *entails* that such properties are parts of the bundle, but that idea is untenable. Now, let us reflect on the converse relation. Does *assuming* that tropes are the *parts* of bundles commit one to considering bundles as collections of tropes, essentially, and how plausible is this idea? No such commitment exists, and the reason is the same one we have just entertained above. Particular properties exist always in *concreta*, and for this reason, if objects are bundles of properties, then bundles are not collections essentially, whether or not the properties composing them are regarded as their parts.

Let us examine in greater detail the notion of particular properties as mutually linking parts of a bundle. Ontologically, this notion is quite benign. At any rate, the two objections made by the advocates of the substratum thesis do not undermine it. First, on such a conception properties are not physically separable from bundles.[31] Secondly, given the same conception, particular properties are not envisaged as entities such as the top half of an object, for while a top half is thinkable either as a discrete or overlapping part *of the object*, a property is neither; it is alleged to be a third and different type of part. What is more, this sense of parthood is even compatible with the substratum thesis: without inconsistency, particular properties seen as the parts of a bundle may be said to be bonded together by inhering in the same substratum.

As long as we understand 'part' in the latter sense, I do not object to the claim that particular properties are parts; moreover, I agree that my argument presents no obstacle to them. But my argument was not intended to reject such entities anyway: since they are not physically detachable from the whole, these entities are devoid of a principle characteristic of what it is to be a proper part. Hence construed in this sense, tropes will hardly serve the purpose of the standard tropist; their independent existence being excluded, tropes cannot be viewed among the physical units of existence.

5. I have argued that in the ordinary sense of 'parthood' particular properties cannot be conceived as parts of bundles. As for a stipulated sense according to which parts overlap without sharing their own parts, my judgment is as follows: either bundles are envisaged as collections essentially, and as a consequence of this, the thesis that the tropes are parts is committed to regarding the fully general fact of concreteness in the physical world as a miracle; or it is accepted that particular properties are bonded together, so that bundles are not collections essentially, and in such a case the thesis that properties are parts excludes standard tropism and becomes acceptable.

I conclude that whether we qualify particular properties as 'parts' of bundles or not, so long as the particular properties that are compresent in an object are seen as mutually linking, the prospects for a bundle analysis of objecthood remain high. Such a view avoids a collectionistic conception of bundles, and the implausibilities contained in it.[32] Moreover, we see that the inability to make consistent sense out of causation, and therefore of scientific explanation, is not the shortcoming of a particularistic bundle theory that combines the theses (a) and (c). The trouble lies in the misconception that these two theses logically lead to the idea of independently existing tropes.[33]

Department of Philosophy, Boğaziçi University, Istanbul

NOTES

[1] By 'physically independent' I mean the condition of anything that is capable of existing in physical space, by itself, without requiring the support of anything else.
[2] Presumably, this is to be conceived as the replacement of a determinate by another, under the same determinable.
[3] Armstrong (1989, p. 2).
[4] Campbell (1990, pp. 22-3).
[5] This would not necessarily be because such isolated existence is itself impossible, but rather because no property-occurrence in isolation could yield any effect.
[6] That is to say, under the determinables of the so-called 'primary qualities'.
[7] A task too hard to do, if mass is understood as the quantity of matter *in a body*!
[8] Campbell (1990, pp. 58-9).
[9] Campbell (1990, pp. 130-2).
[10] Williams (1986, pp. 3-4).
[11] Campbell (1990, p. 3).
[12] Campbell (1990, pp. 21, 55, 59).
[13] Campbell (1990, p. 55).
[14] Campbell (1990, p. 22).
[15] Campbell (1990, pp. 120-1).
[16] Campbell (1990, p. 122).
[17] As a realist of universals, Armstrong (1989, Chapter 6) is a substance theorist who does *not* commit himself to tropes, while Martin (1980) is one who *does*.
[18] To see the implausibility of this, all we need is to consider a relation-trope in isolation, that is, without its relata. See Armstrong (1989, pp. 114-5).
[19] Martin (1980, p. 7). For other arguments and counterarguments see LaBossiere (1994).
[20] Denkel (1992), and (1996, p. 40 ff).
[21] The latter are spatially adjacent, and in unity and continuity they make up the object.
[22] Properties under the same determinable are mutually exclusive occupants of their positions.
[23] Even if they are their difference is irrelevant to their parthood.
[24] Hoffman and Rosenkrantz (1994, p. 63).
[25] Hoffman and Rosenkrantz classify all bundle accounts of objecthood under "Collectionist theories of substance". See (1994, Chapter 3). This mistake is quite a widespread one, and is clearly discernible in Martin (1980), too.
[26] The entry for 'bundle' in the *Oxford English Dictionary* is "A collection of things bound or otherwise fastened together; a bunch; a package, parcel".
[27] Compare the concepts of 'matter' and 'form', which have also been transformed in philosophy.
[28] See LaBossiere (1994, pp. 363-4), for the implausibility of this conception.
[29] See for example Simons (1994) and LaBossiere (1994).
[30] "It might be objected that substance is a species of *Collection*, viz., that, necessarily, a substance is a collection of other substances (its parts). In reply, we would argue that it is impossible for a material substance to be a collection of this kind, since it is essential to a material substance that its parts have some principle of unity, e.g., physical bonding, whereas it is not essential to a collection that its parts have any such principle of unity." Hoffman and Rosenkrantz (1994, pp. 20-1).
[31] Accordingly, it is only a *logical* possibility that particular properties may exist apart from (and independently of) objects, and this is something even the substance theory admits.
[32] What of the possible objections that on the view I am promoting here, the bonds linking particular properties would be too many to be plausible (for one thing, they would be more than the number of the properties they link, or even infinitely many), and being particular properties they would themselves require bonding, thus leading to a Bradley-type regress? (See Simons [1994], LaBossiere [1994]) The thesis that the bonding of properties is by *internal* relations avoids this kind of objection, and one version of such a suggestion can be found in Husserl's notion of a 'founding relation'. (Husserl [1970, p. 478]. For discussions of internal bonding relations see Simons [1994] and Denkel [1997].)

[33] I wish to thank Stephen Voss for offering me useful criticism. I have read an earlier version of this paper (mainly sections 3-5) at the 2nd European Congress of Analytic Philosophy, held at Leeds, in September 1996. My work has been supported by the Turkish Academy of Sciences.

REFERENCES

Armstrong, David. *Universals: An Opinionated Introduction*, Boulder: Westview Press, 1989.

Campbell, Keith. *Abstract Particulars*, Oxford: Blackwell, 1990.

Denkel, Arda. "Substance without Substratum" *Philosophy and Phenomenological Research*, v. 52 (1992): 705-711.

Denkel, Arda. *Object and Property*, Cambridge: Cambridge University Press, 1996.

Denkel, Arda. "On the Compresence of Tropes" *Philosophy and Phenomenological Research*, v. 57 (1997): 599-606.

Hoffman, J. and Rosenkrantz, G. *Substance among other Categories*, Cambridge; New York: Cambridge University Press, 1994.

Husserl, Edmund. *Logical Investigations*, London: Routledge, 1970.

LaBossiere, Michael. "Substances and Substrata" *Australasian Journal of Philosophy*, v. 72 (1994): 360-370.

Martin, C. B. "Substance Substantiated" *Australasian Journal of Philosophy,* v. 58 (1980): 3-10.

Simons, Peter. "Particulars in Particular Clothing: Three Trope Theories of Substance" *Philosophy and Phenomenological Research*, v.54 (1994): 553-575.

Williams, D. C. "Universals and Existents" *Australasian Journal of Philosophy,* v. 64 (1986): 1-14.

SUN DEMIRLI

CAUSAL RELATIONS IN HUME

One old view is that Hume endorsed a reductive analysis of causation: all causation can be reduced to the regular succession of events, and there are no irreducible causal relations, connecting causes to their effects. Recently, however, this old interpretation has been challenged, and a new understanding of Hume's view on causation has been proposed. According to this new interpretation, Hume never intends to give a reductive analysis of causation; he never believes that causation is nothing more than regular succession.

The defenders of the new Hume draw attention to the language used in the pages of the *Treatise* and the *Enquiry* where the notion of an irreducible causal relation is discussed; in these pages, they maintain, Hume asserts merely that one can never conceive of "necessary connections" or "causal powers"; all of Hume's talk of inconceivability about irreducible causal relations, they argue, is used in making an epistemological point about our knowledge of causal relation and the ways in which we arrive at causal truths. Their view is that it is a distortion of Hume to see him as making a metaphysical point about the nature of reality.

The new Hume clearly holds that there is a discrepancy between what we know of causation and what causation in itself is; causal relations in so far as we know them can be analyzed in terms of regular successions of causes and effects; but evidently, he thinks, causation is more than that; there are "necessary connections" that must be taken into consideration.

Unfortunately, the issue is obscured by a lack of clarity on the part of Hume and his recent commentators. In both the *Treatise* and the *Enquiry*, Hume never addresses the question of whether there are irreducible causal relations head on; he never asserts unambiguously or argues clearly that there can be no "necessary connections" or "causal powers". What he asserts (instead) is that we can never observe such causal relations and that our causal inferences are never based on the supposition that there are such items in nature. Hume's text, therefore, seems to suggest that the metaphysical issue about the existence of causal relations is never uppermost in his thinking. That leaves two very interesting questions open: (i) Is there a case to be made for the claim that Hume accepts the existence of such relation that we can never know? And if not, then (ii) does he have anything against the existence of "necessary connections" and "causal powers".

G. Irzık & G. Güzeldere (eds.), Turkish Studies in the History and Philosophy of Science, 217-230.

In this paper, a return to the old Hume will be proposed. In defending this position, we will tackle these two questions. In connection with the question (i), I will try to show that Hume's text doesn't support the new Hume: there is no evidence that Hume presupposes the existence of irreducible causal connections. The discussion of the question (ii) will be more thorough. I will show that Hume has a case to be made against the existence of such causal connections; and this case will be formulated on behalf of Hume.

1. CAUSAL RELATIONS

Whether one considers the *Treatise* or the *Enquiry*, we can see Hume employing a number of referring expressions such as "necessary connexion", "causal power", "real power", "ultimate cause", "agency", "force", "efficacy", "energy", etc.; according to the defenders of the new Hume, these expressions are different characterizations of one and the same item which is supposed to connect causes with their effects. Leaving aside (for a moment) the question of whether those commentators are right, one needs to consider the question of how we must understand the notion of an irreducible causal relation; what kind of item can provide a tie between cause and effect.

The defenders of the new Hume do not give an ontology for causal relations and say precious little about them. Among those authors, Craig specifies them in an intuitive way. He says: when we observe a causal interaction, we believe that along with the two events which are said to be related as cause and effect, there is a third item that we try to capture with the phrases "causing", "producing", "bringing about" and so on; according to Craig, necessary connections or causal powers are nothing other than this additional third item which is thought to be involved in causal interactions.[1]

Craig's characterization reflects what we intuitively think about causal relations. Take a causal interaction between the blow of a hammer and the shattering in a coffee cup. We describe such an interaction by saying "the blow of the hammer causes or produces or brings about the shattering of the glass". No matter how we formulate the case, we express it by a sentence of the form "c causes e".

Now any causal statement of the form "c causes e" contains three syntactic components: two subject terms and one two-place predicate. But does this show that along with cause and effect there is a third item which corresponds to the relational predicate? But surely, though, if we follow Craig and read off our ontology from the structure of a sentence of that sort, we must acknowledge that along with the two items which are related as cause and effect, there is a *two-place entity*, tying these two items together. Now this two-place entity answers to the two-place predicate "... causes ..." in the same way as some items in the world answer to the subject expressions "the blow of the hammer" and "the shattering in the glass".

There is something annoying, however, in using a syntactical criterion for determining whether there are relational facts. Take the sentence "Sam is wise". There is nothing relational about the predicate "... is wise ...". And it seems to correspond to a completely intrinsic characteristic of the individual, Sam. Nonetheless, one can chose to translate this non-relational sentence into a relational sentence such as "Sam is wiser than Socrates", and decide to let Sam be wise just in case he is wiser than Socrates. Now should we say that the fact that one is wise entails a relational fact between this individual and Socrates? From the use of relational predicates, it doesn't seem to follow automatically that there are relational facts. Why should we stick with such a syntactic criterion in determining whether there are causal relations tying causes with their effects? Unless one thinks that how we speak provides a reliable guide into the nature of reality, there is room for denying that there are irreducible causal relations.

One doesn't need to press this objection any further. For the moment, we will look at the passages where Hume argues that we can never observe "necessary connections" or "causal powers".

2. ONE CAN NEVER EXPERIENCE NECESSARY CONNECTIONS

In both the *Treatise* and the *Enquiry*, Hume discusses the question whether we can observe causal relations. Let us first look at the discussion in the *Enquiry*. At the beginning of the section "Of the Idea of Necessary Connection", he writes:

> To be fully acquainted... with the idea of power or necessary connexion, let us examine its impression; and in order to find the impression with greater certainty, let us search for it in all the sources, from which it may possibly be derived. (E. 63)

Hume here does not mention just "necessary connections" but "causal powers" as well. It seems that he takes these two to be one and the same. Leaving this point aside for the moment, let us follow him in his search for causal connections.

Hume's first step is to note carefully what is perceived when we observe some causal interaction. He considers the example of colliding billiard balls. When we observe one billiard ball colliding with another there is nothing, apart from the motion of the first billiard ball and the subsequent motion of the second, of which we can have an impression. We never experience a necessary connection, along with our experiences of the cause and its effect. As he puts it, "we only find that the one does actually, in fact, follow the other. The impulse of one billiard ball is attended with the motion in the second. This is the whole that appears to the outward senses." (ibid.) This leads him to the conclusion that in our observation of a causal interaction, we never experience a necessary connection.

Having established that, Hume begins to pursue a different line of argument. He writes:

> From the first appearance of an object, we never can conjecture what effect will result from it. But were the power or energy of any cause discoverable by the mind, we could foresee the effect, even without experience; and might, at first, pronounce with certainty concerning it, by mere dint of thought or reasoning. (E. 63)

He here talks as if he has already abandoned the question whether there is a necessary connection which "appears to the outward senses". As will be noticed, he is no longer looking for something which is distinct from the cause and its effect but for something which can be discovered in the cause. Furthermore, just as he drops all talk of a separate item which we can associate with a necessary connection, he also drops the notion of necessary connection. Neither in this passage nor in the following twelve pages, does he mention this notion. He talks only about power or energy. This is strange when we recall that he seems to equate necessary connections with causal power and that he has used them together at the beginning of the section "Of the Idea of Necessary Connexion".

In the light of these two points, the following conjecture seems to be reasonable. Hume is no longer arguing against the view that we can *directly perceive* a necessary connection. He is certain that there is no separate item which can be taken as a necessary connection—a two-place entity tying cause and effect together. However, he supposes that there might be something in the cause which provides a necessary connection between this cause and its effect; and he explores the possibility that some property, or aspect, of the cause may imply that the cause is connected with its effect. Now since the term "necessary connection" does not apply to a property (or aspect) of a single object but to a relation which holds between two objects, this supposed property of the cause cannot be appropriately called a "necessary connection" (even though it is supposed to imply a connection between the cause and effect). So he tentatively calls it a "power" or "energy" and entertains the view that this supposed power can provide a tie between cause and effect. Accordingly, he begins to search for such a power or energy. But according to him, anything which counts as a power in the cause must be such that anyone who apprehends it has to be in a position to know *a priori* what the effect of that cause will be. So, he maintains that if we perceive this power we should then be able to tell by inspection of the cause what effect it would have without waiting for experience of that effect.

Hume then calls up his famous argument from the conceivability of the contrary. He writes:

> Motion in the first Billiard-ball is a quite distinct event from motion in the second;... when I see, for instance, a Billiard-ball moving in a straight line towards another; even suppose motion in the second ball should by accident be suggested to me, as the result of their contact or impulse; may I not conceive, that a hundred different events might as well follow from that cause? May not both these balls remain at absolute rest? May not the first ball return in a straight line, or leap off from the second in any line or direction? All these suppositions are consistent and conceivable. Why then should we give the preference to one, which is no more consistent or conceivable than the rest? All our reasonings a priori will never be able to show us any foundation for this preference. (E. 29–30)

Hume here first contends that the motion in the first ball and the one in the second, being related as cause and effect, are completely distinct events. Since they are distinct, he argues, we can conceive that one occurs without the other's occurring, and we can never tell *a priori* by observing just the cause what effect it would have. From this he concludes that we can never experience a necessary connection between the cause and its effect.

The discussion of the idea of necessary connection in the *Treatise* is similar to the discussion in the *Enquiry*. As he does in the *Enquiry*, he presents two different arguments: one argument against the view that there is a necessary connection existing as a separate item and a second argument against the view that there is a power in the cause which indicates a necessary connection between that cause and its effect. At the beginning of the section "Of the Necessary Connexion", he presents his first argument in the following way.

> [W]e must find some impression, that gives rise to [the] idea of necessity, if we assert we have really such an idea. In order to do this I consider, in what objects necessity is commonly suppos'd to lie; and finding that it is always ascrib'd to causes and effects, I turn my eye to two objects suppos'd to be plac'd in that relation; and examine them in all situations, of which they are susceptible. I immediately perceive, that they are contiguous in time and place, and that the object we call cause precedes the other we call effect. In no one instance can I go any farther, nor is it possible for me to discover any third relation betwixt these objects. (T. 155)

Hume here argues that when we observe two objects related as cause and effect, we perceive a relation of contiguity and precedence between them; but we can never perceive a necessary connection. Therefore, he concludes, there can be no necessary connection, as a separate item, combining the cause with its effect.

After denying that there is a necessary connection, as he does in the *Enquiry*, Hume considers the possibility that there is a power in the cause and this power may provide a connection between the cause and its effect. Accordingly, he writes:

> [W]e speak of a necessary connexion betwixt objects, and suppose, this connexion depends upon an efficacy or energy, with which any of these objects are endow'd. (T. 162)

But if we observe such a power in the cause which in fact provides a necessary connection, then,

> We must distinctly and particularly conceive the connexion betwixt the cause and effect, and be able to pronounce, from a simple view of the one, that it must be follow'd or preceded by the other. This is the true manner of conceiving a particular power in a particular body. (T. 161)

In this passage too, Hume claims that the true conception of a power is a conception of something which licenses us to know *a priori* that the cause will be followed by its effect: if we perceive a power in a cause which brings about a necessary connection between this cause and its effect, then we have to conceive this connection "distinctly and particularly" in the sense that from an observation of the cause we must know *a priori* what effect it would have. After making this claim, he denies that we can have such *a priori* knowledge. But in contrast to his discussion in the *Enquiry*, he does not appeal here to his argument from the conceivability to the contrary. He writes:

> [N]othing is more evident, than that the human mind cannot form such an idea of two objects, as to conceive any connexion betwixt them, or comprehend distinctly that power or efficacy, by which they are united. Such a connection wou'd amount to a demonstration, and wou'd imply the absolute impossibility for the one object not to follow upon the other. (T. 161–62)

3. HUME'S DRIFT TOWARDS EPISTEMOLOGY

According to Craig, neither in the *Enquiry* nor in the *Treatise* does Hume seem to be much interested in the metaphysical questions about the existence of necessary connections.[2] In Craig's view, we must distinguish between *Hume's original statement of his intention about the aim of his arguments*, and *what these arguments are actually aimed at*. Craig maintains that in both the *Enquiry* and the *Treatise*, Hume professes his intention to look for a separate item which may be associated with necessary connections. But, Craig contends, the passage in which Hume states that there is no such separate item that we experience in causal interactions is perfunctory, and this suggests that he is simply paying lip service to the question whether there are necessary connections. According to Craig, in Hume's real argument, he makes it clear that he is not so much concerned with questions about necessary connections, and drifts towards epistemology, where his true interests lie.

Craig draws attention to the contention that if we perceive any power or energy in the cause we should then be able to know the effect *a priori*. According to him, what Hume requires from causal powers is gratuitous. Craig believes that there is no apparent reason why he should insist that anything which counts as a power in a cause must enable us to predict the effect of that cause *a priori*. He says:

> What reason have we been given . . . to accept [Hume's] premise, that if there were a causal power we could predict the effect a priori? (1983, p. 93)

Craig maintains that Hume seems to dismiss an obvious possibility. There can be a power in the cause, which connects this cause with its effect, the experience of which does not enable us to know the effect *a priori*: and he does not argue that we cannot have *any* power but rather that we cannot have an experience of a power with a certain *epistemological property*—a power, the experience of which enables us to know the effect *a priori*. Craig argues that this "suggests that Hume's real interest is in the epistemological questions, how we know or come to believe truths about causes."[3]

Craig claims that Hume does not have any quarrel with the existence of a power in the cause. According to Craig, the argument in question is used only in making an epistemological point: given that there is a power in the cause which in fact connects this cause with its effect, from the experience of this power we can never know *a priori* that the effect will ensue.[4]

Let us consider the case of colliding billiard balls. The motion in the first ball and the subsequent motion in the second are related as cause and effect. There is an aspect of, or a power in, the motion of the first ball which connects this motion with that in the second ball. Nevertheless, according to Hume (Craig believes) from the experience of the power in the motion in the first ball, we can never predict *a priori* the motion of the second ball. So, Craig maintains, when Hume says that there is no causal power which provides a connection between cause and effect, the way in which he talks is deceptive; he does not argue for the (ontological) thesis that there is no causal power but for the epistemological thesis that given that there is a power in the cause, from the experience of this power, we can never know the effect of that cause *a priori*.[5] Therefore, according to Craig, such an epistemological point tells us little or nothing about Hume's ontological views, and in particular is not evidence against the view that he denies causal connections.

4. CRITICISM OF CRAIG'S INTERPRETATION

In Craig's interpretation, Hume sets off his discussion with the question of whether there is a causal power, which "appears to the outward senses"; nonetheless, he immediately abandons this line of investigation, and he focuses on the question of whether there is anything in the cause which enables us to know the effect *a priori*. For Craig, this is evidence for the contention that Hume is not interested in the ontological questions about the existence of necessary connections and causal powers but in the epistemological questions concerning how we acquire causal truths.

Craig overlooks one crucial aspect concerning Hume's discussion. He fails to see that Hume presents two distinct arguments, and that in each argument Hume denies the existence of a different type of item. According to Hume, one can speak of a necessary connection between cause and effect in two different cases: either (i) there is a necessary connection as a separate item—a two-place entity, as it were, tying cause and effect together, or (ii) there is a power—an aspect of the cause—which provides a necessary connection between the cause and its effect. Now (in Hume's view) if (i) is true then one must observe that separate item. On the other hand, if (ii) is true then the observation of the cause must enable one to know its effect *a priori*. In his first argument, he deals with the first kind of causal connection and shows that we can never observe a separate item along with the cause and its effect. After making sure that there is no such item, he begins his second argument and goes after a supposed power in the cause; accordingly, he shows that nothing in the cause enables us to know the effect *a priori*. Thus, he concludes, it is not only that there is no causal connection as described by (i) but also that there is no connection as specified by (ii).

Craig's second point is more interesting. He maintains that (in what we call Hume's second argument) Hume puts a gratuitously demanding requirement on causal powers. To Craig, it is not clear why Hume claims that a causal power in the cause must take us to the effect *a priori*. In Craig's view, what Hume requires for causal powers suggests that Hume does not deny the existence of any power but only a power of the sort which enables us to make an *a priori* inference. But then, according to Craig, this shows that Hume is not interested in the ontological questions but in the epistemological questions concerning our ways of arriving at causal truths.

It is true that Hume places a highly demanding condition on causal powers. In his second argument, he is in fact looking for nothing less than a power with an *epistemological property*—a power, the experience of which enables us to predict the effect of a cause *a priori*. The whole weight of his argument hangs on his denial that there is such a power with that epistemological property. Nevertheless, this should not lead us to the conclusion that he is merely making an epistemological point. Nor should this lead us to the conclusion that according to him, there is a causal power in the cause of a sort which does not allow us to predict the effect *a priori*. He is not concerned merely with the epistemological questions how we know or come to believe causal truths. Rather, he is concerned with the questions concerning the existence of necessary connections.

In the first place, we need to recall that Hume begins his discussion with the question whether we observe necessary connections and ends it by presenting his analysis of causation where he makes it clear that the notion of a necessary connection cannot be a part of the concept of causation. This strongly suggests that he is mainly interested in the ontological questions concerning the existence of necessary connections not in the questions of the epistemological sort.

Secondly, there is a more substantial objection against Craig's interpretation. In supporting his claim that the argument in question is used in making an epistemological point, he makes three different claims:

(C1) Hume does not deny that there is a causal power in the cause but just that we can observe a power with an epistemological property—a property which enables us to predict the effect of that cause *a priori*;

(C2) there is a power in the cause, the experience of which cannot enable us to know the effect *a priori*;

(C3) by denying just the existence of a power with the epistemological property, Hume does not argue for the ontological thesis that there is no power which can provide a necessary connection, but for the epistemological thesis that given that there is a power, nonetheless from the experience of this power we can never predict the effect *a priori*.

According to Craig, the claims (C1) and (C2) must take us to the claim (C3) which constitutes the gist of his position. Now (C1) is obviously true. Hume does not explicitly discuss whether or not there is a causal power of a kind which cannot allow us to know the effect of a cause *a priori*. He denies just that there is a power with the epistemological property. But (C2) is controversial, at best. Furthermore, even if we contend that (C2) is also true, (C3) may still be false. Unless we accept the additional claim that Hume himself believes (C2) to be true, there is no reason for us to accept Craig's conclusion (C3), that is: Hume is merely rejecting the epistemological thesis that given that there is a power, from the experience of this power, we can never predict the effect *a priori*.

In Criticizing Craig's position, we need to focus separately on two claims:

(C2*) Hume himself believes (C2) to be true.

and

(C3) by denying just the existence of a power with the epistemological property, Hume does not argue for the ontological thesis that there is no power which can provide a necessary connection, but for the epistemological thesis that given that there is a power, nonetheless from the experience of this power we can never predict the effect *a priori*.

When we look into Hume's argument carefully, we see that (C2*) is false. As will be recalled, in both the *Treatise* and the *Enquiry*, this argument has the following form.

(1) If there is a power which can provide a necessary connection, then this power must have an epistemological property—a property which enables us to know the effect *a priori*. "This is the true manner of conceiving a particular power in a particular body." (T. 161)

(2) Since we can never know the effect *a priori*, there can be no power with the epistemological property in question.

Therefore,

(3) There can be no power which can provide a necessary connection.

The first premise asserts that anything which counts as a power must have the epistemological property. Now this implies that Hume does not believe that there can be a power which does not enable us to know the effect *a priori*. Hence, (C2*) must be false.

Let us turn to (C3). Since (C2*) is false and thus Hume is denying the existence of a power that he thinks to be the only kind there is, it seems that (C3) is false. Is there any other way to support Craig's conclusion? Perhaps it might be said that in his argument Hume imposes an epistemological condition on the power. He accepts no power but one which satisfies an epistemological property—a power which enables us to predict the effect *a priori*. Couldn't this be regarded as an indication that he is more interested in epistemological issues?

It is true that Hume denies that there is a causal power which can provide a necessary connection between cause and effect on the grounds that we can have an experience of a power which can satisfy a certain epistemological condition; and he appeals to certain epistemological considerations in denying that there are powers as such. Nevertheless, all of these are used in the service of an ontological conclusion that there is no causal power. As we shall see in a moment, there are Humean reasons to believe that there cannot be any genuine power that lacks this epistemological property.

There is a further point to be made. In his arguments, Hume typically speaks of the experience of causal powers. So, he typically argues for the conclusion that there can be no power providing a necessary connection by denying that we can observe such a power. On these occasions, what he requires is that the experience of this power has the epistemological property to enable us to predict the effect *a priori*. Nonetheless, he sometimes speaks of *the power itself* (rather than the *experience* of this power). On these rare occasions, he puts a requirement upon the power itself, and demands that such a power be infallible and to necessitate with absolute necessity; and he reaches his conclusion about causal powers by claiming that there cannot be a power in the cause which operates infallibly and which brings about the effect with absolute necessity. Here is an example from the *Enquiry*.

> We call one object, Cause; the other, Effect. We suppose that there is some connexion between them; some power in the one, by which it infallibly produces the other, and operates with the greatest certainty and strongest necessity. (E. 75)

Hume here maintains that anything which counts as a power in the cause must operate infallibly and with greatest certainty and strongest necessity. In the *Treatise* also, there are similar passages.

> If we have any idea of power in general, ... we must be able to place this power in some particular being and conceive that being as endowed with a real force and energy by which such a particular effect results from its operation. We must distinctly and particularly conceive the connexion between the cause and effect, and be able to pronounce, from a simple view of the one, that it must be followed or preceded by the other. This is the true manner of conceiving a particular power in a particular body.... Such a connexion would amount to a demonstration, and would imply the absolute impossibility for the object not to follow or to be conceived not to follow upon the other: Which kind of connexion has already been rejected in all cases. If any one is of contrary opinion, and thinks that he has attain'd a notion of power in any particular object, I desire he may point out to me that object. (T. pp. 161–62)

He here declares that "the true manner of conceiving a power" which provides a necessary connection between cause and effect must be in terms of absolute necessity:

if there is a power which connects the cause with its effect, then this power must be contained in the cause and, in virtue of having this power, the cause must absolutely necessitate its effect. In his view, when we suppose that there is such a power, we must suppose it is inconceivable that the connection between cause and effect is fallible or is subject to change.

It is worthwhile to note that while what Hume requires from the experience of power is an epistemological property, what he requires from the power itself—infallibility and absolute necessitation—is not epistemological. So, it seems, the fact that he imposes an epistemological condition on the perception of power cannot be presented as evidence for the view that he is mainly interested in epistemological issues concerning the way in which we come to know the effect of a cause. According to him, if there are necessary connections, then we have to experience a power which provides a necessary connection between the cause and its effect and this necessary connection must be understood in the strongest sense as absolute necessity and infallibility. This is his reason for his familiar contention that the experience of such a power, because of what it is supposed to be an experience of, must have the epistemological property of enabling us to predict the effect of a cause *a priori*.

5. WHY SHOULD A POWER HAVE THE EPISTEMOLOGICAL PROPERTY?

As we have seen, Craig's claim that Hume uses his argument in making a merely epistemological point concerning our knowledge of causal truths is not tenable. Whether one considers the version of the *Treatise* or that of the *Enquiry*, there is strong evidence that this argument concerns the (ontological) thesis that there is power in the world which can provide necessary connection between cause and effect. The most Craig can hold is the conclusion that the argument in question does not succeed in establishing this (ontological) thesis.

As will be recalled, in the preceding section, we presented Craig's position in terms of the three claims that he makes; one of these claims was that (C2) there is a power in the cause, the experience of which cannot enable us to know the effect *a priori*. So far we have not said much about this claim except to make the following point. According to Hume, *anything which counts as a power in the cause* must provide a necessary connection and we must understand this necessary connection in the strongest sense, i.e., as infallible and absolutely necessary; therefore, the experience of such a power, because of what it is supposed to be an experience of, must enable us to know the effect of a cause *a priori*.

Why should Hume contend that there can be no power which is fallible and subject to change? In so far as Hume's discussion goes, he doesn't have an argument for the conclusion that there can't be a power which is fallible and subject to change and which necessitates in a weaker sense than absolute necessity. Furthermore, even if one were to grant this point to Hume, a different question could be raised: why should Hume contend that a power which is infallible and not subject to change must be such that its experience must enable us to know which effect will follow *a priori*? One does not need to concede that the experience of an infallible and absolutely necessitating power must have this epistemological property. Now if he does not have a legitimate reason for his claims that anything which counts as a power must be infallible and must necessitate in the sense of absolute necessity and that the experience of such a power must enable us to know the effect a priori, then since he does not argue against the existence of such a power, we have to conclude that his argument fails to establish the conclusion that there cannot be a necessary connection.

We can summarize what Hume thinks of causal connections in terms of two qualifications that he makes: (1) a power must be infallible and not subject to change, and (2) the experience of such a power in the cause must enable us to know the effect *a priori*. Hume's view is that anything which counts as a power must satisfy the restrictions (1) and (2). He says: "This is the true manner of conceiving a particular power in a particular body." (T. 161) Why? The answer seems to be that Hume understands the notion of a causal connection in accordance with a rationalistic conception of such connections. Let me explain.

Bennett writes that some commentators accuse Hume of "willfully restricting reasons to deductive reasons". He quotes Craig speaking of such criticisms: "Hume's arguments are easily overcome: just deny the dogma that that all reasons are deductive and sit back."[6] Hume construes causal connections in accordance with the rationalist dogma that was common currency in his time. On this rationalistic view, there is no distinction to be made between causal and logical connections. Effects follow from the causes in the way the conclusion follows from the premises in a deductively valid argument. That was Spinoza's view. Someone who understands causation as absolute necessity must hold that causal laws are of the same kind as the laws of logic; that is: causal necessity is nothing other than logical necessity. According to a rationalist such as Spinoza, saying that the cause occurs without being followed by its effect is nothing but a contradiction. Another important feature of the rationalistic view of causation is that causal connections are completely intelligible: If there is a causal connection between two events, then, since this causal connection is of a logical sort, a complete understanding of the cause must take us in principle to the effect *a priori*.

When Hume puts the restriction (1) and (2) on the notion of a causal connection, he seems to understand these connections in the same way as a rationalist such as Spinoza does. Therefore, his arguments against the existence of causal connections seem to be an easy victory: just deny that rationalistic conception of causation can be true and sit back.

But there is evidence for saying that Hume does not rely merely on a rationalistic conception of causation in denying the existence of necessary connections. There are some passages in both the *Treatise* and the *Enquiry* which suggest that Hume considers the possibility that there is some kind of necessity which necessitates in a less strong sense than absolute necessity.

6. A WEAKER CAUSAL NECESSITY

We observe many regularities in the world. I take a sleeping pill and fall asleep in the next hour. There is an observed regularity between these two events: taking the pill and falling asleep. Every time one takes a pill, he or she falls asleep. Locke and many other philosophers think that this regularity cannot be taken as a brute fact and that such regularities cry for an explanation. As Strawson puts it, if there were no underlying facts which explained such observed regularities, then that would be a coincidence of gigantic proportions.[7]

Locke thinks that the observation of regularities in the world must provide us with good reasons for supposing that there are underlying facts which sustain and explain such regularities. This is what is commonly known as the "Lockean inference to causal powers": when we observe a causal interaction, we must suppose that there is a power in the cause, and in virtue of having this causal power the cause brings about its effect, and this is what explains the causal interaction in question and also all other interactions of the same type.

Hume is familiar with this conception of Lockean causal power. He mentions it in the *Treatise* and summarizes it as follows:

> Such an object is always found to produce another. It is impossible it could have this effect if it was not endowed with a power of production. The power necessarily implies the effect; and therefore there is just a foundation for drawing a conclusion from the existence of one object to that of its usual attendant. The past production implies a power; the power implies a new production; and the new production is what we infer from the power and the past production. (T. 90)

He next presents an argument against the Lockean conception of powers. Whatever the merits of supposing that there are such powers, Hume argues, such powers cannot do the job that we expect them to accomplish. The reasoning behind Hume's argument is quite intuitive. Why do we need to suppose that there are causal powers operating in nature? Lockean powers are theoretical entities. They are introduced to solve the 'inference problem'. We observe a regular pattern: the cause is regularly followed by its effect. Now we believe that this regular pattern will continue to hold in the future; whenever we observe the cause, we infer that the effect will ensue as it did in the past. What is our justification for this inference? For those who deny causal powers, the justification is the belief that the future will resemble the past. For those who appeal to the notion of causal powers, things must look much better. "The production of one object by another in any one instance implies a power, and this power is connected with its effect." (T. 91) This power, connecting the cause with its effect, must provide us with reason (which goes beyond the inductive belief that the future will be like the past) for our inferences. Otherwise, why should we stick our neck out and suppose that there are causal powers, operating in nature? At this point, Hume shows that there is nothing we can gain by supposing the existence of such powers.

> But it having been already proved that the power lies not in the sensible qualities of the cause, and there being nothing but the sensible qualities present to us, I ask why in other instances you presume that the same power still exists, merely upon the appearance of these qualities? Your appeal to past experience decides nothing in the present case, and at the utmost can prove that the very object which produced any other was at that very instant endowed with such a power; but [it] can never prove that the same power must continue in the same object or collection of sensible qualities, much less that a like power is always conjoined with like sensible qualities. (T. 91)

For those who deny causal powers, what was problematic is to extrapolate the observed conjunction between cause and effect into the future; such an extrapolation could only be based on our inductive belief that the future will be like the past. Hume's point is that when we suppose that there are causal powers, we face a similar problem. Lockean powers are not observed, but they are inferred on the basis of certain sensible qualities of the cause; we believe that such powers were conjoined with these sensible qualities in the past, we suppose that the same power will continue to be conjoined with the same sensible qualities next time. But how can we extrapolate this conjunction into the future? In other words, how can we know that this conjunction will be satisfied at the very moment where we make a causal inference? The only justification is that causal powers were always conjoined with the sensible qualities of the cause. Why should they fail to be there in the future? But then the supposition that there are Lockean powers which sustain causal interactions does not take us beyond inductive reasons.[8]

A similar line of argument is present in the *Enquiry*.

> Should it be said that, from a number of uniform experiments, we infer a connexion between the sensible qualities and the secret powers: this, I must confess, seems the same difficulty couched in different terms... When a man says, I have found, in all past instances, such sensible qualities conjoined with similar secret powers. And when he says, similar sensible qualities will always be conjoined with similar secret powers; he is not guilty of tautology, nor are these propositions in any respect the same. You say that the one proposition is an inference from the other. But you must confess that the inference is not intuitive; neither is it demonstrative. Of what nature is it then? To say it is experimental, is begging the question. For all inferences from experience suppose, as their foundation, that the future will resemble the past, and that similar powers will be conjoined with similar sensible qualities. (E. 36–7)

It is important to note that the success of Hume's argument does not rest upon the point that the powers are not observed but inferred. Suppose that one associates Lockean power with sensible qualities of the cause. We observe a regularity between the cause and its effect and maintain that some sensible quality of the cause is a causal power—and thus, is responsible for this observed regularity. But how can we know that this power will be effective and will be successful in bringing about the effect in the future? The only justification is the belief that the future will resemble the past: the power was effective and brought about the effect in the past, and it will continue to be effective and will bring about the effect in the future. That means: even an 'observed power' does not provide us with additional reasons for expecting the future will be like the past.

It seems that Lockean powers cannot do the job that we expect from genuine causal powers. When we suppose that there are such powers, we are not in a better position than someone who denies that there are such powers. Now it is obvious that only a power which satisfies the restrictions (1) and (2) can assure us that the regular succession between the cause and the effect will continue to hold in future; and thus, only a power which is understood in accordance with the rationalistic conception of causation can take us beyond our inductive belief that the future will be like the past. In Hume's view, therefore, if we call something which is understood in a less robust sense a 'causal power', we have to keep in mind that such a thing cannot be the kind of thing "on which the regular course and succession of objects totally depend." (E. 55)

Towards the end of his discussion of causation in the *Treatise*, Hume says that he is tolerant of a view according to which an insensible quality of an object is called a 'power'. But after saying this, he adds that "it will be of little consequence to the world" should someone insist to call such an insensible quality a 'power'. (T. 168) He is sure that such powers cannot play any significant role in our understanding of how causes are connected with their effect

7. NOTES

[1] Craig (1987), p. 93
[2] Craig (1987), pp. 93–7
[3] Craig (1987), p. 94
[4] Craig (1987), pp. 98–9
[5] Craig (1987), p. 99
[6] Bennett (2001), p. 265
[7] Strawson (1989), pp. 22–31
[8] A similar point is made in Bennett (2001, pp. 264) and in Blackburn (1990)

8. REFERENCES

Bennet, J. F., *Learning from Six Philosophers* v. II, Oxford Clarendon Press, 2001

Blackburn, S., "Hume and Thick Connections", *Philosophy and Phenomenological Research* 1990, Volume 50 suppl., pp. 237–50

Craig, E. G., *The Mind of God and the Works of Man*, Oxford: Clarendon Press, 1987

Hume, D., *A Treatise of Human Nature*, L. A. Selby-Bigge and P. H. Nidditch (ed.), Second Edition, Oxford: The Clarendon Press, 1978

Hume, D., *An Enquiry Concerning Human Understanding*, L. A. Selby-Bigge and P. H. Nidditch (ed.), Third Edition, Oxford: The Clarendon Press, 1975

Strawson, G., *The Secret Connexion: Causation, Realism, and David Hume*, Oxford Clarendon Press, 1989

BERENT ENÇ

HOW CAUSES CAN RATIONALIZE: BELIEF-DESIRE EXPLANATIONS OF ACTION†

1. INTRODUCTION

On one model of rational action, when an agent acts on the reasons she has, these reasons have a dual role: they *cause* the bodily behavior that constitutes the action, and their contents stand in some logical relation to some appropriate description of the action in question, thereby *rationalizing* the action. On this model reasons consist of two categories of mental states: desires, which define the objectives of the actions, and instrumental beliefs which lay out the means for attaining the objectives.[1] This model, variously known as "the Causal Theory of Action", or "the Standard View", can also be used to generate a criterion for distinguishing (rational) action from "mere" (non-intentional, non-voluntary or arational) behavior.[2]

This essay will take the above model for granted. It will be devoted to an examination of the relation between the causal antecedents of behavior in general and rationality. The objective of the examination will be to find some of the conditions under which the *causes* of behavior constitute the *reasons* for that behavior. The question to be pursued may be put this way: "On the assumption that all behavior is caused by the internal states of an organism, what sorts of internal states, and what sorts of causal pathways are needed to give rise to *rational* action?" Or alternatively, "What sorts of causal relations must obtain between motivational states on the one hand, and behavioral output on the other, for these motivational states to constitute the agent's reasons for the resultant behavior, and for them to render the behavior rational?"

The strategy to be employed in pursuing the answers to these questions is to examine motivational states of simpler organisms, starting from pathways laid out by natural selection, moving to learning by operant conditioning, and to argue that the causal relations that hold between motivational states with this kind of etiology and behavior are not the right kind needed for rational action–that none of these states are candidates for *reasons*. The argument will take its force from one central thought: the instrumental beliefs that make up part of one's reasons for one's actions have conditional content, and unless the representation of this conditional plays a causal role, the resultant behavior will not be rational behavior. It will be shown that

G. Irzik & G. Güzeldere (eds.), Turkish Studies in the History and Philosophy of Science, 231-248.

the cognitive representations of simpler organisms, even if they can be assigned conditional content, act as "straight wires". It will be suggested at the end that the essential element in rational action is a computation that involves deliberation, the weighing of pros and cons of the consequences of one's prospective actions.

Simply put, I want to try to answer the following question: What sorts of things are these reasons, which cause an agent to act a certain way, and thereby render her action rational? In pursuing this task, I will be ignoring several problems that may be quite central to the understanding of the nature of reasons or rationality. The first problem I will ignore is the so-called Kripke-Wittgenstein paradox, formulated in the context of rule following behavior, that involves specifying the determinate content of the reasons for which one acts. A second problem, which will also be left outside the focus of this paper, is what sense is to be made of situations in which the reasons an agent acts on are deemed to be bad, or "irrational". In other words, I will not be interested in figuring out how the *quality* of the reasons one has is associated with rationality. I will also blithely pretend that one can inquire into the source of the rationality of action without having an adequate theory of how the contents of the agent's reasons for that action are determined.

2. GENETICALLY HARD-WIRED BEHAVIOR

One obvious thought that might get the project started is that a minimum condition on rational behavior is that the behavior may be the result of *following a rule*, as opposed being the enactment of a mere disposition. In the recent years, two notable examples that develop this thought are to be found in Dennis Stampe's and Ruth Millikan's essays.[3] Both of these philosophers quite rightly observe that following a rule is distinct from conforming to a disposition. Millikan speaks of unexpressed biological purposes as determining the "proximal" and "distal" rules that a male hoverfly follows when he darts after a sighted female, launching himself on a course that is determined by the angular velocity of the image of the female across his retina. "To say that a given male hoverfly has a biological purpose to conform to the proximal hoverfly rule is very different from saying that he has a disposition to conform to it. The normal hoverfly has ... a disposition to squash when stepped on, but [this] disposition [does] not correspond to biological purposes or to competences" (p.220) Stampe, too, points out that when I want to eat the peach in my hand, I am disposed to take a bite out of it. But my desire, in addition, gives me a reason to take a bite out of it, whereas a mere disposition to do something, like the disposition I have to fall when tripped, or to blush when embarrassed cannot constitute *a reason for* doing those things. Stampe's answer to the question, "why *should* the desire cause the behavior?" takes him to the ideal causes of desire, and to the observation that the state of the body that under ideal conditions causes the desire is the state in which the organism benefits from the behavior that is the effect of the desire. The two authors have different objectives: Millikan, seeking to bring determinate content to the rules which rational agents follow, and Stampe, seeking to understand why desires should have the functional role they are seen to have. But both turn to the same form of teleological consideration involving normal functional explanations, or what, in my view amounts to the same thing, to a notion involving what would happen if the state in question were produced under a set of conditions of well-functioning.

Both Stampe and Millikan are correct in their observation that mere functional characterization of beliefs and desires, the identification of their causal role, cannot supply the answers to their respective questions. And I for one believe that teleological considerations they offer get us started in the right direction.

A second example that has exactly the same structure as that of the hoverfly is found in the behavior of the moth. When the moth's ears register a sound wave of a frequency between 20-100 kHz, the moth takes a dive. Again, we can ask: "why should the state in which the moth's receptors are registering a sound-wave of high frequency cause such behavior?" The answer is: because under ideal conditions, such a state is caused by the sonar mechanisms of the predator bats who will momentarily descend on the present location of the moth. And under those conditions, taking a dive benefits the moth.

The point can be made explicit in the following way: Suppose we take *S* to be a state of the agent, which is to comprise the agent's reasons for doing something. And let us label the proximal thing done *B*. The following schema is satisfied by the inner state of the moth that causes it to dive or by the inner state of the hoverfly that causes it to dart off at an angle to the direction of the present location of the sighted female:

(i) When a set of specifiable well-functioning conditions, *C*, obtain, *S* is triggered by an input, *I*, (the high-pitched sound, in the case of the moth) and *S* triggers *B* (the dive).

(ii) When a set of specifiable "normal" conditions, C', obtain, *I*, the input, is caused by environmental (or internal physiological) facts, *F* (the presence of the bat).

(iii) When a set of specifiable "normal" conditions, C'' obtain, *B* has the consequence *Q* (escaping the predator).

(iv) An essential part of the explanation of why *S* causes *B* when *F* is that normally *B* causes *Q*.

This schema makes clear the type of explanation we seek in order to answer the preliminary question, why the state we have labeled *S* *should* cause the relevant behavior. But the examination of the schema leads me to suspect that we do not yet have a complete account of how certain motives, i.e., the reasons one has, can render an action rational. For in the moth and the hoverfly we do understand why the state *should* cause the output. This understanding is brought to us by first finding out what the far reaching goals of these systems are,–their unexpressed distal purposes, as Millikan puts it (survival or mating), and then by seeing that, normally, proximal effects of these states best serve these goals, or purposes, when the normal cause of the state obtains. But in such cases the system's being in that state fails to comprise the system's having a reason for behaving a certain way. In other words, the features of a system that yield an explanation of why the system's state *should* cause a certain type of behavior do not thereby make the system a rational agent. This suggests the conclusion that the necessary condition for having rule-following behavior, as opposed to behavior that is governed by mere disposition, which is implied by such an explanation, is too minimal to yield the source of rationality in behavior. The point may be best defended by considering an objection to this conclusion.

It might be argued in this objection that we could think of the moth's state *S* that causes its dive as a state in which a belief-desire pair is scrunched together to form a proto-belief-desire, so to speak. After all, the moth has the goal *Q* of escaping predation. The sound it registers is a representation of *F*, the presence of a predator. It is also equipped with a means, *B*, of achieving the goal. So the internal wiring of the moth may be taken to represent the fact that when a predator is present, and a dive is taken, then predation will be avoided:" If *F* and *B*, then *Q*".[4] The causal role played by these representations, the objection might continue, is what confers the rationalizing role to the causes of behavior. We could represent the causal relations among the structures that embody these representational contents in the following way.

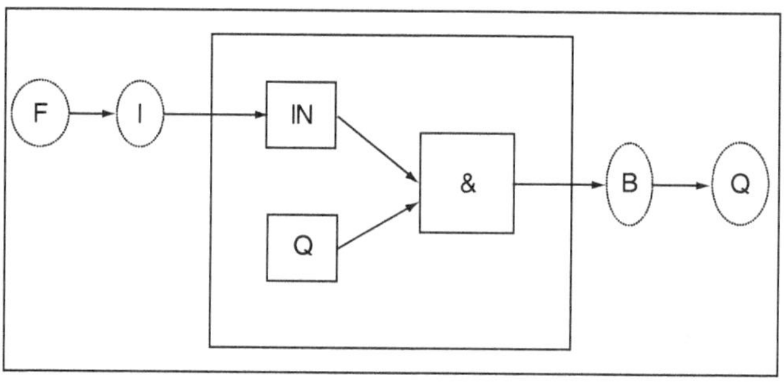

Figure 1. *An organism's internal wiring formed by Natural Selection*

Here *I* is the input which causes the auditory perceptual state *I′*, and *I′* represents *F*, the presence of a predator, and *Q* is the content of the omnipresent motivational state: avoid predation. A natural selection story makes it plausible to assert that the wiring of the &-gate, i.e., the way *I′* is hooked up to *B*, the dive, represents the present conditions as being one in which taking a diving action results in avoiding predation. It is this representational content that makes the moth's dive something more than an outcome of its mere disposition, something with a purpose. (It has a proto-desire to escape predation and a proto-belief that when bats are nearby if it takes a dive, it will escape predation.) This story is told elegantly by Dretske. Dretske calls the causal role attributed to the fact that in the past when predators were present, the dive was correlated with successful escape, "a structuring cause": it causes the structure of the network wherein *I′* has been recruited to be cause of *B*. The structuring cause explains why the &-gate is the way it is. This is in contrast to the causal role attributed to *I′*, which Dretske calls "a triggering cause". The wiring of the &-gate acts as, what we might call, a standing causal condition, one that (causally) explains, as Dretske will have it, why *I′* causes *B*. Why isn't this picture sufficient to give the moth something like a proto-reason for taking the diving action? After all, there is an important difference between systems, like a simple thermodynamic system, in which the links between inputs and outputs are supported by straightforward causal laws, and teleological systems, in which the links that determine what output a given input is to cause are themselves determined by past consequences of that type of output. It seems clear to me that this is a valuable insight which forms a good starting point for a naturalistic account of behavior. And I also agree that the difference between purely causal systems and teleological systems may exhaust the difference between stumbling when tripped or blushing when embarrassed on the one hand, and pulling one's hand away by reflex when one touches a hot surface, on the other. But the kind of answer that an appeal to the intentional content of the &-gate provides does not advert to any *reason* that the moth might be said to have for diving, rather than, say, staying in a holding pattern. Simply put, I reject the suggestion that the internal state of the moth can be thought of as a proto-reason for the moth's behavior. It is gratuitous to identify this state as a desire state because it does no work in determining the character of the moth's behavior: the input *always* causes the dive, regardless of whether the moth "wants" to avoid predation or not. To look for the difference between the alleged proto-belief-desire complex of the moth's system and the reasons on which an agent acts, we might examine the following questions:

Question 1. Why should one in general act in ways that one believes will satisfy one's desires?

Question 2. What is it about the desires and beliefs that cause one's behavior that renders them constitutive of one's *reasons* for one's behavior?

It is true that mere motives that, when triggered by incoming information, produce behavior, and desires which, when coupled with beliefs, give rise to rational action, share at bottom a feature that is accurately captured by the answer to Question 1. There is a sense of "should" in which it is literally true that given that I want a peach, I *should* do whatever it is that I think will satisfy that want, *and* that the moth, when it registers the tell-tale high pitched sound, *should* take a dive. Both of these uses of "should" take their force from the fact that if these systems were to fail to do what they *should* do, they would not be fulfilling that which is their natural function to fulfill. The systems are natural teleological systems, and as such, they are expected to conform to certain norms that are dictated by their function *and* their *design*. It is these norms that legitimize the application of "should" to both of these systems. The moth's system is "designed" to make it dive when certain conditions are present because doing so under those conditions is "good" for the organism. The rational agent's system is "designed" to make her do whatever she thinks will satisfy her desires under the circumstances because doing so is in the long run "good" for her.[5]

On the other hand, identifying what is common to these systems does not establish that the system of the moth contains beliefs and desires, or that what constitutes *reasons* is confined to the fact that one is designed to act so as to satisfy one's objectives. In spite of one underlying similarity between the moth's system and the system that governs rational behavior, there is an important difference between them. And the correct answer to Question 1 does not help us locate the source of this difference.

If the reader finds I have been belaboring the obvious in arguing that the moth does not act on reasons, I can put the argument on its head: When one asks, "What are the conditions that a set of states needs to satisfy if it is to comprise the reasons on which the agent acts?" it is not enough to propose as an answer, "whatever explains why one should act in ways that one believes will satisfy one's desires" because we can then take the set of states in the moth, call them proto-belief-desire states, and show that one can explain why the moth should act in ways that it "thinks" will satisfy its "desires". But if it is obvious that the moth does not have anything that vaguely resembles reasons, it is then equally obvious that the proposed answer provides at best a very non-informative minimal necessary condition for being a reason.[6]

3. REASONS AND LEARNED BEHAVIOR

It is maintained that there is an important difference between the structuring causes that arise in Natural Selection and the structuring causes that operate on individual organisms during their lifetime.[7] Taking the &-gate of the moth, depicted in Figure 1, one can express the difference in question in the following way: The &-gate that has been shaped by natural selection fails to explain why the *individual* moth dives, rather than gets into a holding pattern, because all that can be explained by natural selection is why this type of behavior is prevalent among moths, not why individual moths act the way they do. This admission would make sense of why we may think Question 1 finds an answer in an appeal to the selectionist story, for example, showing how, during evolution, acting in ways one "thinks" would satisfy one's "desires" has contributed to the fitness of past individuals, and how this trait has become prevalent in the species. However, in order to explain the behavior of an individual organism now, one has to find contents of representations that have acted as causes during the lifetime of the individual. Such causes are best located in learning histories–it is only in organisms with such histories that we can hope to find the first emergence of anything that comes close to beliefs and desires. Hence if we want to the answer Question 2, we should look to structuring causes which have acted during the lifetime of the individual agent through learning. Only states that have been acquired through such learning processes can explain the actions of an individual agent (Dretske).

Correct as the observation of the difference between the scope and the respective explananda of selectionist explanations and explanations due to learning may be, I am skeptical that the difference captures what is essential to reasons for action.

What follows is a series of considerations that are designed to ground this skepticism, and to steer us to a positive thesis about the constitutive source of reasons.

First we should note that there is a type of learning that is actually said to hold in the case of rats and some birds. When these creatures eat something that smells or tastes a certain way and thereupon get sick, they learn not to eat food that smells or tastes that way ever again. In these cases, one might be persuaded that the rat's reason for not eating soap, for example, was that the previous time it ate soap, it got sick. This is certainly *the reason* why it is not eating it now. In a similar mood, I may have as my reason for not eating fish in a particular restaurant the fact that when I ate fish in that restaurant in the past, I got sick. A Skinnerian would find no difference between me and the rat and would insist on reducing my talk of *my reasons* for an act to *the reasons* that are grounded in some version of the Law of Effect. But the key question here is: are the respective causal roles played by the representational structures in the rat and in me the same? And I think there is reason to doubt that they are the same.

When the rat has learned to avoid eating soap, we may presume that a neural connection has been wired in the individual rat that transforms the input of a certain taste or smell into the output of not eating. The structure involved here is the same as that involved in the moth of Figure 1. In contrast, we like to think of ourselves when we act on reasons as being *moved* by those reasons, that is in a Davidsonian spirit, we tend to think of our reasons as being the triggering causes of our actions, as opposed to assuming that these reasons are represented by ruts that have been formed, or by channels that have been carved, by our past experience, and that some perceptual input flows through them to trigger muscle contractions of a specific kind. Perhaps Figure 2 depicts the story we would like to think separates us from the rat.

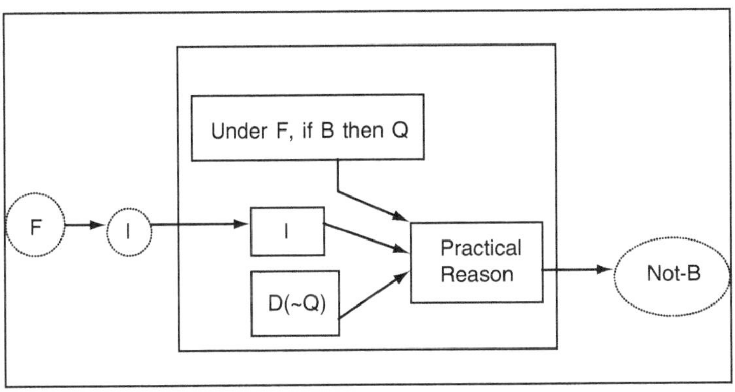

Figure 2. *A rational agent's internal wiring*

Here my belief that if I eat fish at Slimy Joe's, I will get sick as well as my belief that I am at Slimy Joe's, together with my ever-present desire not to get sick, get fed into my reasoning faculty, and as a consequence I choose not to order fish. Must I admit that this picture of my presumed "Freedom and Dignity" may be just wishful thinking on my part, and that if we go "Beyond" these presumptions, we will find the same kind of &-gate in my neural wiring as I claimed must exist in the rat?

4. REASONS AND UNCONSCIOUS BEHAVIOR

Psychologists have recently been studying cases where, due to localized brain damages, the patients manifest behavioral deficiencies. A type of case reported by Penfield involves epileptics who can be struck with seizure while they are walking, driving, or playing the piano, and who are able to continue their activities while they are in seizure. But these *petit mal* patients, as they are called, perform these activities, during their seizure, mechanically, without the flexibility, adaptability, and creativity they are capable of when they are normal. Penfield uses the label "mindless automata" for them. Searle in his discussion of them insists that they are acting in goal-directed ways without any consciousness. The descriptions of the behavior of these patients remind one of the egg-rolling behavior of the greylag goose, made famous by Lorenz and Tinbergen. (See Gould.) An object of a certain size near its nest initiates the fixed-action pattern of rolling the egg up to the nest and continues the motions in air even if the object is moved out of its path. Now, even if Skinner can persuade some of us that I am just like the rat in my fish avoidance behavior, it is harder to deny the difference between a normal piano player who has reasons for continuing to play the tune she has started and the petit mal patient who continues to play the tune during an epileptic seizure. It seems literally correct to say of the normal person that given her reasons and her perceptions, she *should* continue to play, whereas the *petit mal* patient seems clearly to lack any reasons to continue to play–at least, he lacks the kinds of reasons that render his action of continuing to play that of a rational agent.[8] What is the basis then for the difference in our judgments in these two cases? Here is one tentative suggestion: with the normal piano player, we presume that inputs are being routinely monitored and a set of desires and a set of beliefs are being constantly kept in sight, so to speak, and these are potentially capable of causing a change in the behavior in question. On the other hand, in the petit mal patient, the input is being channeled into the output in a way that makes talk of the agent's reasons for continuing to play totally inapplicable.

A related contrast can be made between the way we normally allow incoming information to influence our behavior and the way information that is sometimes described as being *unconsciously* processed influences the behavior of certain types of patients. People who report to have no sight in one half of their field of vision apparently have some unconscious awareness of the objects in their blind half.[9] In one experiment, patients who are shown a river in their blind half, overwhelmingly disambiguate the word "bank" shown in their sighted half as "river bank".[10] It seems again that it would be a mistake to ask for the *reasons* the blind sighted people have for their choice of meaning in such experiments.[11]

What these examples, and many others that can be cited from similar studies, suggest is that the causal structure that underlies the behavior of these patients is different from that of normal agents. If we combine this conjecture with the intuition that I have been urging above, namely that goal directed behavior is not always behavior done for a reason, we arrive at the result that the causal role of reasons that are involved in deliberative action is different from the causal role played by the motivational and cognitive states of these patients. In fact, on the basis of such results some philosophers have come to defend the view that the function of consciousness is that of enabling the information represented in our brains to be used, among other things, in rationally guiding action.[12] If this line of thought is on the right path, then we should look to the causal structure in question for the normative power of reasons.

5. REASONS CONTRASTED WITH MOTIVES

In drawing the contrast between these behavioral deficiencies and normal agency, I have alluded to the way reasons seem to operate in normal behavior. It is perhaps possible to describe acting on one's reasons by contrasting such action with acting in conformity with some character trait. When an agent finds herself in a situation where she wants something and decides on the means of satisfying that want, she typically will have evaluated the different routes that seemed to be available to her. The evaluation will have included estimating the consequences of the different means and ordering them according to some "desirability quotient". The decision, the choice of the means, then, is the result of this process of evaluation. In other words, the choice is determined (i) by what the agent thinks is available to her in terms of proximal behavior, (ii) by how likely she takes these behaviors to yield the object of her desire, and (iii) by how "costly" each of these available behaviors seem to be. Furthermore, acting for a reason also presupposes that these choices, once made, are revisable after the project of pursuing the object of the desire is launched: discovering unforeseen obstacles, or finding better or easier means to the goal, or stumbling on competing projects that appear overall more desirable than the one in progress might make the agent change course. Not all of these factors need to be operative in everything one does do for a reason, but most of the time, most of them seem to be.

This process requires a particular type of interaction between the different things that are desired and the contents of instrumental beliefs. Some philosophers have suggested that the Aristotelian Practical Syllogism should be viewed as a piece of abductive reasoning.[13] Suppose I want a car. This is my desire. My instrumental belief is that if I get a Ford, I will have a car. So I go and get a Ford. It is hard to design something similar to the &-gate of Figure 1 to perform the reasoning involved here because such an &-gate would make me go for a Ford whenever I want a car. If, on the other hand, I have performed a piece of practical reasoning, I would be expected to have detached the consequent of the conditional belief and identified it as the object of a desire. It is only then that the two premises can be separate but joint causes of my action. It is easier to see this feature if we include in the reasoning some simple deliberation: We start with: "If I get a Ford, I will have a car." We add the following: "If I get a Toyota, I will have a car," "If I get a Ford, I will spend less money now and more on repairs later," "I prefer spending less money now and more on repairs in the future to spending a lot of money now." So, perhaps foolishly, but perfectly rationally, I go and get a Ford. If these instrumental beliefs are to be the causes of my action, their consequents need to be matched up with what I want and what I don't want. And that operation does not seem to be possible within the model of structuring causes, which is exhausted by the operation of &-gates.

To contrast acting for a reason with acting in response to an inner force, we might look at Hume's theory of motivation. Hume thought, for example, that kindness to children, together with hunger, lust, love of life, was one of the primary passions we are possessed of. When an agent acts in a way that constitutes acting kindly toward children, and being kind to children is an overriding character trait for the agent, it seems wrong to say that the reason she had for her action was that she was kind to children. Here I do not want to imply that one cannot have the desire to be kind to some child. It is just that when one acts *out of kindness*, where being kind is a built in trait, as opposed to acting *because one wants to be kind*, one does not have the burden of considering alternatives and evaluating their costs and benefits and comparing them to those of acting kindly.[14]

I want to suggest here that the difference between acting on the trait of kindness on the one hand and acting because I want to buy a car, on the other, lies in the fact that in the former, the trait dictates only one course of action and does not permit of any deliberative choice among alternatives. My desire for a car becomes my reason in so far as that desire is just one among other incompatible desires I may be presumed to have, like the desire to take a long trip, or desire to save money. Furthermore, the action chosen as a way of satisfying the desire is again only one among several actions that are available at the time. Change the presumptions, and block the alternatives, by, for example, making me a compulsive car buyer, who is incurably addicted to the feel of Fords, you deprive me of all my reasons for buying the Ford–it becomes just as incorrect to say of me that, given my compulsions and addictions, I had a reason for buying a Ford, as it was of the woman who is driven by kindness to children, of the soap avoiding rats, or of the hoverfly that they had reasons for their actions.

6. REASONS AND VON NEUMANN MACHINES

The point of my remarks of the previous section may be illustrated perhaps more accurately with an example from Artificial Intelligence.

For obvious reasons, there is some interest in AI in distinguishing between systems that are merely rule conforming and systems the behavior of which can be described as *following* a rule. Presumably, the windshield wipers in my car merely conform to the rule, "If the switch is on, the wipers move." They do not *follow* that rule. D. Lloyd writes,

> A system is a Rule Conforming System if ["If *X* then *Y*" describes the system such that when the system is in state *X*, it tends to move into state *Y*]. A system is a Rule Following System when "If *X* then *Y*" causes (as well as describes) [the system to be such that when the system is in state *X* it tends to move into state *Y*]...A computer exemplifies a Rule Following System. Somewhere in the computer...is an explicit rule, a representation, or a set of representations, which can be interpreted (by programmers) as meaning that when Control-N is typed, execute a line feed on the screen. Because that rule is there, inside the computer, the computer has the disposition to respond to Control-N with a line feed. Given a different program, the behavior of the computer would have been different. In contrast, the planets are familiar examples of a Rule Conforming System. The laws of gravity describe their behavior, but they do not move because those laws are inscribed somewhere. (Lloyd, 123-4)

What Lloyd describes here is very much like what classical AI envisaged. In a Von Neumann machine, the Central Processing Unit is presented with two types of inputs, first, the Rule that it accesses, i.e., the rule that determines what function is to be performed, and second, the input that constitutes the argument for that function. Presumably, in Lloyd's example, the rule, "If Control-N, then line feed" is called in, and the rule operates on some previously recorded parameter that

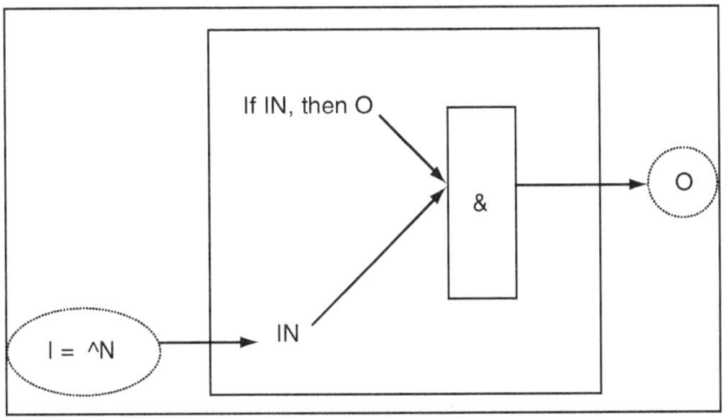

Figure 3. *Internal wiring of a Von Neumann machine*

determines the value of the line feed. Suppose that we are willing to admit here that the rule, the conditional, "If Control-N, then line feed" is functioning as a cause of the output. What is important to realize though is that the causal role of this conditional can be captured by an &-gate similar to that of Figure 1.

The only difference between Figure 1 and Figure 3 is that in Figure 3 we attribute to some state of the system a representational content, and the content is the conditional, "If Control-N, then *O*". But in this system, the input is the *antecedent* of the conditional rule. This is what enables us to represent the wiring as an &-gate. The inner structure is similar to the "rule" embodied in the wiring of the moth: If high-pitched sound, then dive. The conditional rule in the Central Processing Unit does no more causal work than the "rule" by which the moth acts does causal work in the moth. This is in contrast to what I maintain takes place in practical reasoning. There, the input provided by the desire is the *consequent* of the conditional. And if the conditional and its consequent are to act as causes, the wiring cannot be represented as an &-gate. The wiring has to be such as to allow for detaching the consequent and entering it into the computation. I suggest that it is this kind of detachability that constitutes the normative force of reasons.

7. REASONS AND DELIBERATIVE COMPUTATION

What is, then, the connection between this somewhat mysterious process, which I have called detaching the consequent, and the normative force of reasons?

So far I have been promoting the intuition that for something to constitute my reasons for acting the way I do in a given situation, the reason needs to be reached as a result of a deliberation–a deliberation that involves surveying several alternatives; where each alternative is a potential reason, which, if chosen, would lead to a different course of action. Each of these alternatives picks some relevant proximal behavior that is available under the circumstances, and computes what consequences would follow if the behavior were produced under the present conditions. Presumably, the agent already possesses a representation of the present conditions, *F*. We could conceive of each of these alternatives as being represented as, "Under *F*, if B_i, then Q_i." Here B_i is one of the many basic behaviors that are available to the agent under the present conditions[15], and Q_i refers to the (anticipated) distal outcome of B_i under the same conditions. In addition, the agent also has some ordering of the different *Q*s determined by her desire profile, *Q*s here standing for the contents of her desires. So there are three elements in one's acting for a reason:

(i) the presence of alternatives represented as conditionals, the consequent of which are the estimated outcomes of the actions referred to in the antecedents,

(ii) a process of deliberation that picks one of these alternatives by computing which alternative had the consequent that represents the most attractive outcome, and

(iii) the actual course of action that is launched as a direct causal result of (ii).

If these elements are constitutive of one's reasons for action, then it should be clear that the Von Neumann model of Figure 3, as it stands, does not realize a system which incorporates these three elements. If we think of a Central Processing Unit, in addition to a transducer which generates representations, *I′*, of the present conditions, *F*, we need a scanner that finds the best *Q* in an ordered list of *Q*s, analyzes each of the conditionals to locate the one with the best *Q* as its consequent, and passes the corresponding *B* of this conditional to an executive subsystem. A block diagram that schematizes these operations is given in Figure 4.

The operations involved in analyzing the conditionals to find the one with the preferred *Q*, and identifying its antecedent, though, presuppose representational structures which are so articulated as to embody a complex representational content, where the individual elements of the content, *F*, *B*s and *Q*s are themselves represented in the structure. Only if the neural wiring that represents "Under *F*, if B_i, then Q_i will

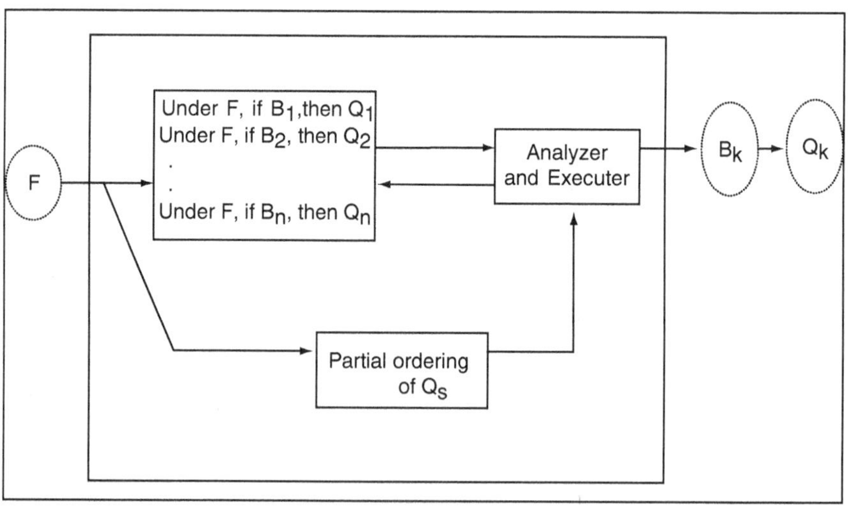

Figure 4. *Internal wiring of a rational agent whose action is the result of deliberation*

occur" is such that distinct parts of the wiring contain representations of B_i and Q_i, can the analyzer make the computations it needs to make. That is the only way it can "choose" the course of action which, under the circumstances, and given what is wanted and what is not, it *should* choose. In other words, for an agent to act for a reason, there is an important requirement that needs to be satisfied. The requirement is that the representations of the conditionals that express the agent's reasons must function as triggering causes. This requirement is satisfied in the operations I have been describing, and it seems to me that the only way it can be satisfied is to have a complex representation, the components of which are themselves representations.

In this model there is no presumption that the agent is the "ideally rational agent" sometimes assumed in Decision Theory. In other words, not all the possible courses of action need to be considered, nor is it assumed that all the foreseeable consequences of the actions actually considered are anticipated. So I am willing to admit that in the simplest case where the only alternatives considered are merely doing something and not doing it, if the consequence of each alternative is taken into account and the scanner identifies a preference as existing between these two, then the agent is acting as a rational agent–albeit perhaps not a very wise or an effective one.

Indeed Figure 4 simplifies the process involved in deliberative action perhaps too severely to capture the full force of action rationalization in Folk Psychology, which typically takes the following form: "*S* did B_k because she wanted Q^*, and she thought B_k would be a 'good' way (the 'best' way (!)) of getting Q^*." In the schema depicted in Figure 4, there is no explicit reference to Q^*, the object of one's desire, and as such the conditionals represented do not explicitly mention the conative force that initiates and fuels the deliberative process. Suppose I am compelled toward obtaining food. Here my goal, food, (Q^*) is the principle which the Central Processing Unit (i.e. the scanner) uses in selecting the conditionals. The consequents of the conditionals that are made available to the scanner, the *Q*s, share Q^* among them–each *Q* is the anticipated consequence of a specific way of getting food, each "way" corresponding to a different *B*, for example, stuffing the doughnut on the counter in my mouth; or making an omelette with the eggs in the refrigerator and eating it; or going to the store, buying a chicken breast, lighting the charcoal, grilling the chicken, and eating it with mushroom sauce, and so on. As long as at least two alternatives are available, and there is some preference ordering for the consequences of these alternatives, and that ordering is implicated in the selection of the proximal behavior B_k, I am judged to have reasons for doing B_k, as opposed to any of the other *B*s.

So, for example, my reason for making an omelette was that making an omelette was computed as being the least effortful way of getting least harmful food. (This computation is the result of comparing the consequents of the following three conditionals: (i) If I eat the doughnut (B_1), I will get food and it will be bad for my health (Q_1); (ii) If I make an omelette (B_2), I will get food, and I will spend some effort, and it will increase my cholesterol a bit (Q_2); (iii) If I go and get a chicken breast to grill (B_3), I will get food, and it will be a lot of effort, but it will be good for my health (Q_3). And the comparison favors Q_2 over the other two *Q*s.) But did I have a reason for *getting food*, simpliciter? The answer to that question depends on whether I considered any other current desires, which generated alternatives to eating, during my deliberation, and on whether the set of conditionals selected for comparison to the scanner included conditionals with consequents that represented those alternatives. For example, if my desire to lose weight made me consider not eating and taking a pill to stop the hunger pangs, or being stoic and doing nothing, as opposed to eating something, and the anticipated consequences of the former were less favorable than those of making and eating an omelette, then not only did I have a reason for making and eating an omelette, as opposed to, say, eating the doughnut, I also had a reason for eating, as opposed to not eating. However, if my "desire" for food has a built-in strength that always overrides every other alternative and blocks them from consideration, like Hume's overriding passion of kindness to children, then I did not have a reason for eating, versus not eating–not eating was not something that was presented to me as an option.

Naturally there are intermediate cases. For example, I may have standing desire to lose weight, and at some time in the past, I may have examined and (foolishly) found all the alternatives that would satisfy that desire less favorable than eating something high in calorie whenever I feel hungry. The memory of this deliberation in the distant past may not even be accessible to me now. But still, it is arguably correct to say that in choosing the omelette now, even though I do not consider not eating as an alternative, I had a reason for eating. I will resist the temptation to incorporate these interesting complications to the simple model of Figure 4. I would be happy if one feature of that model gives us a sufficient condition for having reasons in the form of desires (i.e., the existence of alternative courses of action, the preference ordering over the consequences of which contribute to the choice of action), and a second respect of that model provides a necessary condition for the same thing (i.e. the existence of representations of conditionals that have compositional complexity so that their consequents can play a causal role). I do not have the more ambitious intention of giving a necessary and sufficient condition for having reasons for having acted in some way.

Furthermore, I should make it explicit that I am assuming that the ordering of the *Q*s may be partly determined by the nature of the conditions *F* that are detected. So certain kinds of consequences that may be desirable under one set of conditions may cease to be so when the conditions change. And there may be some one ultimate principle that determines the ordering. Hedonism offers a candidate for such a principle. But whatever explanation the ordering might have, if *Q*s are to constitute the contents of desires, then these explanations must be constrained by the correct answer to the question, "why should such systems in general act in ways that are sensitive to the ordering of their *Q*s?" I suggested earlier that the requirement that there be a correct answer to this question, Question 1, is a minimal necessary condition on desires.

I proposed above a second requirement: the representation of the means-end conditionals must have compositional complexity, i.e., the components of these representations themselves must be representations of basic behaviors and representations of expected outcomes of such behaviors under the circumstances. This requirement places a further constraint on the *Q*s if they are to be associated with reason constituting preferences. In other words, if these preferences are *desires*, then they have to be representational; they cannot be the non-representational passions that Hume spoke of.[16] For if they are non-representational, then their strength will be sufficient to force the choice of action by an &-gate like that depicted in Figure 1. If they are non-representational, then they will lack contents (at best they will only have intentional objects), and consequently the *Q*s would not be separately represented in the representations of the conditionals and hence will not be able to enter into the computations required by the model I propose for rational action.[17]

It is clear that this model is in the spirit of a compatibilist view of deliberation and of one's reasons for action–whatever element of indeterminacy is allowed for in this model would have to come from the weak ordering of the *Q*s and from the role of randomizers that may be switched in if some Qs are tied for top, or if inconsistencies are revealed in the ordering. There is no place in the model of a libertarian vision of an uncaused agent who "authors" its own actions.

The model can be complicated by adding whistles and bells, for example, modules that monitor the output and its immediate consequences, and feed them back to the system to check to see if the originally projected consequences are likely to come about given the way the chosen course of action is progressing, and if not, to alter the course in accordance with the new estimates. This model would incorporate the flexibility and the creativity that is supposed to inform rational behavior. But on the view I have defended here, these features are not essential to the nature of reasons; they just render the rational agent a more effective rational agent.

Department of Philosophy
University of Wisconsin, Madison

8. NOTES

[†] I have presented versions this essay at various philosophy departments, including University of Wisconsin-Madison, University of Alberta-Edmonton, University of Calgary, Warwick University, Kings College, London, and I have benefited from the comments I received from the audience as well as written comments that were sent to me. I am especially grateful to Greg Mougin, Fred Dretske, Terry Penner, Mohan Matthen, and David Papineau.

[1] For the purposes of this essay, I will ignore the question whether *intentions*, as mental states that are not reducible to beliefs and desires, are also to be included among the immediate causal antecedents of action. I tend to think they are. As a result, the account given here, in so far as it ignores the role of intentions, is incomplete. But I am hoping that this move, made with the aim of simplicity of presentation, will not invalidate the main thesis of this essay.

[2] R. Hursthouse provides a very sensitive and cogent discussion of actions taken in the grip of emotions, and calls such actions "arational".

[3] Stampe ("Defining..."), and Millikan.

[4] The assignment of such a representational content to the wiring is not too far-fetched. It could be defended by a view which extracts a causal theory of representation from the teleosemantic approach outlined in the schema above (conditions i–iv). On this view, the content of a representational state is determined by what can be identified as the function of the state. In the case of the moth, it is arguable that the function of the state is to help the moth escape predation. And it serves this function by eliciting a diving behavior when a bat is present. So the fact responsible for the formation of the mechanism that realizes the function, i.e., the fact that in the past, diving when bats were present resulted in escaping

predation, becomes the representational content of the mechanism. The objection that is being considered here would be a non-starter if such a view of representation were totally implausible.

[5] The idea that satisfying one's desires is in the long run good for the organism is premised on assumptions that take desires to be representations (or misrepresentations) of what would be good for the organism. *cf.* Dennis Stampe ("Authority . . . "). See also David Papineau for a different approach to a similar idea. I am not interested in pursuing this idea any further because the correct answer to Question 1 is not my concern here.

[6] The same point can be made by examining agents capable of rational behavior, and by contrasting their conditioned or unconditioned reflexes with their reasoned actions. When one reflexively pulls away one's hand from a hot surface, one is doing what one "should" do, but the fact that the surface is hot is hardly the *reason one has* for doing so.

[7] This difference was originally introduced by Sober (*Nature*), and since has become a source of controversy. See, for example, Neander, Sober ("Natural Selection . . . "), and Walsh. It was used to good effect by Dretske.

[8] I do not mean to deny here that in so far as the action of *playing* the tune is concerned, the *petit mal* patient may be doing something as a fully rational agent; his playing the tune may be no different from my stopping at a red light when I am driving. The action I mean to focus on by the description, "continuing to play" is one that is in contrast to stopping in the middle of the tune.

[9] Weiskrantz ("*Neurophysiology . . .* ")

[10] Weiskrantz (*Blindsight*).

[11] The contrast conscious/unconscious plays no important role in my arguments in this paper. That is why I am deliberately ignoring the controversy that exists both about the details of the methodology used in the setup of the experiments, and as to what exactly these findings show about consciousness.

[12] Umiltà cites several experimental results all of which suggest that there is a central conscious process that exercises "strategic control over lower order mental operations".

[13] See Kenny. The similarity between the structure of practical reasoning and that of abductive reasoning has also been noted by R.M. Hare and H. Simon.

[14] Some of Hursthouse's examples are all actions that are correctly explained by the agent by saying, "I did it because I just wanted to, or I felt I had to." Among the cases she cites are behavior explained by joy: running, jumping, leaping up reaching for leaves on trees; behavior explained by horror: covering one's eyes when they are already shut; explained by fear: hiding one's face, burrowing under the bed clothes. In each of these cases I would maintain, and Hursthouse would mostly agree, that it is hard to find *reasons* for which the agent does these things.

[15] For a defense of a the need for an appropriately defined category of basic behaviors, see B. Enç.

[16] A passion is an original existence, or, if you will, modification of existence, and contains not any representative quality, which renders it a copy of any other existence or modification. When I am angry, I am actually possessed with the passion, and in that emotion have no more a reference to any other object, than when I am thirsty, or sick, or more than five foot high." (Hume, 415)

[17] It is interesting that Hume's theory of action renders all acts of agency into behavior that is modeled on acting out of love for children, not acting for the reasons one has. This point was brought home to me by Terry Penner during discussions we had on Socratic Egoism.

9. REFERENCES

Dretske, F. *Explaining Behavior: Reasons in a World of Causes*. Cambridge, MA: MIT Press, 1988.

Enç, B. "Units of Behavior." *Philosophy of Science* lxii (1995): 523-542.

Gould, J. *Ethology: The Mechanisms and Evolution of Behavior*. New York: W.W. Norton, 1982

Hume, D.*Treatise*. Selby Bigge, ed., (revised by P.H.Nidditch) Oxford: Oxford UP, 1980

Hursthouse, R. "Arational Actions." *The Journal of Philosophy* lxxxviii (1991): 57-68.

Kenny, A. "Practical Inference." *Analysis* xxvi (1966): 65-75.

Lloyd, D. *Simple Minds*. Cambridge, MA: MIT Press, 1989.

Millikan, R. "Truth Rules, Hoverflies, and the Kripke-Wittgenstein Paradox." *Philosophical Review* xcix (1990): 323-353.

Neander, K. "What does Natural Selection Explain? Correction to Sober." *Philosophy of Science* lv (1988): 422-426.
Papineau, D. *Philosophical Naturalism*. Oxford: Blackwell, 1988.
Penfield, W. *The Mystery of the Mind: A Critical Study of Consciousness and the Human Brain*. Princeton: Princeton UP, 1975.
Searle, J. *Intentionality*. Cambridge: Cambridge UP, 1983.
Sober, E. *The Nature of Selection*. Cambridge, MA: MIT Press, 1984.
Sober, E. "Natural Selection and Distributive Explanation: A Reply to Neander." *British Journal for the Philosophy of Science* xlvi (1995): 384-397.
Stampe, D. "Defining Desire" in Joel Marks, ed., *The Ways of Desire: New Essays in Philosophical Psychology on the Concept of Wanting*. Chicago: Precedent Publishing, Inc., 1986: 149-173.
Stampe, D. "Authority of Desire." *The Philosophical Review* xcvi (1987): 335-381.
Umiltà, C. "The Control Operations of Consciousness" in A. J. Marcel and E. Bisiach, eds., *Consciousness in Contemporary Science*. Oxford: Oxford UP, 1988: 334-356.
Walsh, D.M. "The Scope of Selection: Neander on What Selection Explains." *The Australasian Journal of Philosophy* (forthcoming).
Weiskrantz, L. *Blindsight: A Case Study and Implications*. Oxford: Oxford UP, 1986: 139, 149.
Weiskrantz, L. "Neurophysiology of Vision and Memory" in A. J. Marcel and E. Bisiach, eds., *Consciousness in Contemporary Science*. Oxford: Oxford UP, 1988: 183-199.

PART V
OTTOMAN SCIENCE STUDIES

BERNA KILINÇ

OTTOMAN SCIENCE STUDIES–A REVIEW

1. INTRODUCTION

A bibliography of publications in Ottoman science studies to date would not be a thick volume. There is always the skeptic with the question: Was there really an Ottoman science? Doubts of this sort may be the grounds for the lack of interest in this topic. I believe however that these doubts are unfounded. They stem from a wrong-headed conception of science and history of science, the presuppositions of which must be superseded. To this end, I promote a philosophical vision of science which I believe must guide any historical research into the earlier forms of knowledge. This vision comes from historical epistemology. Shifting the focus of attention from individual discoveries or discoverers modern sciences take pride in to forms of validity of knowledge, historical epistemology is a call to historicize the epistemological concepts and practices, with the sensibilities of an anthropologist who is able to savor "strange" cultures.

With this perspective in mind, I present in this paper an assessment of some recent historical studies of Ottoman science and technology. I claim that many (but not all) of those studies suffer from a lack of historical perspective: They take for granted concepts such as "experience", "experiment", "nature", "rational justification", "evidence", "objectivity", and even the very notion of "science", as if these were available to historical agents with their modern connotations, no matter where and when. As such, these studies conflict with the basic tenet of historical epistemology. It need not be the case that all of our epistemological concepts were of recent origin; however, one cannot simply assume that any such concept was an invariant of history and geography. Historical epistemology obliges us to understand how different cultures produced or reproduced their specific epistemological categories.

Science studies in the periphery, however, are not only part of an anthropology of peripheral cultures. The very term periphery requires an attention to its complement, the center. Scientific cultures in the periphery were not entirely insulated formations. In ways that need to be made more precise, patterns of center-periphery interaction permeated the styles of learning and research at the fringes of Europe.[1] Challenging overly global visions of a universalist scientific culture, spatialization of reason threatens facile generalizations about the fusion or fissure of traditions, which may mistake amalgams for compounds. However, spatialization of reason is not tanta-

G. Irzık & G. Güzeldere (eds.), Turkish Studies in the History and Philosophy of Science, 251-264.

mount to confinement of reason to local settings. That is to say, scientific cultures are not necessarily isolated formations, separated from each other in conformity with a topology that marks off human settlements on a map. As several anthropologists have pointed out recently, one has to take into account regional and global forms of connectedness, and be wary of segmenting the world into discrete self-contained cultures and peoples.

What are the boundaries of the object of Ottoman science studies? In order to sketch out a preliminary delimitation, I focus in my review specifically on accounts of Ottoman scientific interactions with European centers. What these accounts indicate is that transmission of science was not simply a gradual diffusion of learning but rather active acquisitions and reconstructions thereof, which can be characterized in terms of forms or strategies of reception. Proposing two such schemes, what I call the instrumentalist and positivist strategies of reception, I highlight in this paper the epistemological underpinnings of these forms of selectivity. My discussion is canvassed against an overview of the Ottoman efforts at modernization, which undergirded both forms of reception strategies. In this connection, I focus particularly on the philosophy of science of an important ideologue of the Turkish Republic, Ziya Gökalp, in the writings of whom science played a crucial role to integrate the ideals of Turkism, Islamism and modernism. I end by considering an issue that pertains to both Turkish and Ottoman identities, as well as to doing history of science under any such identity, namely, nationalism.

2. HISTORICAL EPISTEMOLOGY

Many historians of Ottoman science dwell in an eternal present in matters of epistemology. For instance, the historian Ekmelettin İhsanoğlu brings to our attention the work of a native Anatolian mystic, İbrahim Hakkı of Erzurum, who wrote in a 1757 manuscript *Marifetname* approvingly of the heliocentric theory of the solar system. The treatise is curious for its combination at one stroke of astronomical views of European origin with legends of eastern provenance. In place of exploiting something like the genre of dialogue, whereby different viewpoints or audiences can be addressed simultaneously in one continuous narrative, İbrahim Hakkı seems to have stacked two distinct narrative styles one after the other. İhsanoğlu evaluates this transition from astronomy to fiction as a "regression", and describes it repeatedly as an "unscientific" and inappropriate move (İhsanoğlu, 1992). İbrahim Hakkı appears as an anomaly in İhsanoğlu's selection of "scientific" works from this period. I believe, in contrast, the very textual style of İbrahim Hakkı is a topic for dissertations. Literary properties of a text are usually linked to its epistemological claims. İbrahim Hakkı's literary style can be studied, for instance, with a goal to understand the absence or limited presence of the genre of dialogue in the Ottoman philosophical or scientific writings. As is well known, authors such as Galileo Galilei or Marin Mersenne had exploited that genre in order to evaluate side-by-side different worldviews of the seventeenth century. In contrast, İbrahim Hakkı alternated between the astronomical and the legendary, framing the novel with the traditional, thereby enhancing or preempting the claims of the one or the other.

Historical epistemology teaches us one way to resist this atrophy of historical sensitivity. This perspective derives from the French tradition–in particular from the writings of Gaston Bachelard, Georges Canguilhem and Michel Foucault.[2] Its primary injunction is to historicize not only concepts peculiar to a given science but also epistemological categories and values which sustain sciences across disciplines, with a particular tenderness towards their variant or extinct forms. Scientists were not always equipped with the current values, logics or methods of inquiry, nor, *pace* Kant, did they always utilize the same categories of understanding. Epistemology, whether implicitly or explicitly practiced, has a history that can be revealed in the very historical processes of knowing. From this vantage point, an historical epistemological approach does not take epistemology historically prior to sciences. A theory of knowledge is not necessarily anterior or exterior to sciences–according to the historical epistemological insight, the former must be extracted from the latter.

Historical epistemology thus bids us to achieve a symmetry *vis-à-vis* what we currently take to be truth and falsity: Not only what has come to be established as true science is worthy of historical scrutiny; also, former scientific claims which are currently discarded reveal epistemological truths. One of the tasks of historical epistemology is, in my opinion, a philosophically sound evaluation of the reasons for why former scientific certainties are current falsities. There is no commitment here to either epistemological relativism or universalism–I believe historical epistemology can be retooled in ways compatible with either (or with neither). This kind of a study can be normative by providing a comparative evaluation, in which the virtues and vices of two epistemic styles are put under purview.

Historical epistemology is at odds with a historiographical viewpoint–preformationism as I will refer to it–which was widespread in the older historiography and is also commonly presupposed in many works on Ottoman science.[3] Typically, this attitude surfaces in the attributions of a sudden birth of sciences, or *the* scientific method, either in the Greek antiquity or during the scientific revolution. It would be claimed, for instance, that the scientific revolution of the seventeenth century contained, perhaps implicitly and potentially, the whole set of categories and values which made modern sciences possible. Accordingly, the experimental methods, quantification techniques, patterns of explanation, notions of evidence and the like emerged all during the seventeenth century, and rapidly permeated all areas of natural and moral philosophy. This view is challenged in several recent studies, which attend more carefully to the works and deeds of not only what are considered to be the geniuses and giants, but also to scientific workers of lesser stature or products of lesser ambitions, like textbooks or handbooks. The resulting historiography, with its emphasis on practices of reading as well as writing, usually reveals that ingredients of distinct forms of knowledge were gradually constructed rather than instantly generated as if they already preexisted in some seminal historical event.[4] An analogy with epigenesis promises to deliver better histories of sciences in the periphery.[5] In particular, this analogy should give pause to those historians who search for a miniature European science in the midst of Ottoman cultures. It is also an antidote to the thesis that European sciences unfolded from the Greek origins which were preserved and transmitted by the Arabic scholars, as if the latter contained the blueprint of a development hindered by the slumber and ignorance of centuries of intermediary traditions, and as if such a blueprint–had it existed–could serve to legitimize or sanctify an otherwise unacceptable European science.[6]

3. OTTOMAN SCIENCE

Refraining from a unitary theory of epistemology in science studies, I propose these rough guidelines of historical epistemology in order to achieve an understanding of the distinctness and diversity of past scientific cultures. Pluralism of this sort requires a delimitation of the discrete elements it aims to reveal. How do we circumscribe the subject matter of Ottoman science studies that we conventionally refer to as "Ottoman science"? To this end, I focus in this part primarily on the issue of autonomy and insulation. The very question of whether Ottoman-European scientific interaction before the nineteenth century can be described through center-periphery models is a topic of debate. One of the earliest accounts of Ottoman science, Adnan Adıvar's *Osmanlı Türklerinde Ilım*, claimed that the Ottoman educated class was by and large indifferent to the Western traditions in science, at least until the nineteenth century. According to Adıvar, only with the reform movements of the 1830s was there a radical shift in view, although even in this period there was no massive importation of western scientific products.[7] This view is challenged by some recent historians, especially by İhsanoğlu, who can document at least some translation projects beginning in the seventeenth century (İhsanoğlu, 1996b). The issue, however, remains as one of emphasis until further studies are carried out: there were not many translations of European scientific treatises, and the question of how significant those few were in shaping or contributing to the extant traditions can still not be assessed with confidence. While more readily interacting with Arabic and Persian traditions, it is plausibly the case that the Ottoman intellectual realm was relatively impermeable on its western borders.

Neither before nor after the nineteenth century, the transmission of Western science to the Ottoman realm was a random affair. Curiously, it was also not a selection of what we would nowadays consider the classics of Western science, at least not until the nineteenth century. It is possible here to distinguish between at least two strategies of reception. The first can be called an instrumentalist strategy. Particularly operative in the choice of works translated within astronomy, this strategy appears as a favorable disposition towards scientific works of practical significance rather than towards those providing theoretical frameworks.

It was thus that one of the first translations within astronomy was parts of the French astronomer Noel Durret's work *Novae Motuum Caelestium Ephemerides Richelinanae* of 1641 by Tezkereci Köse İbrahim of Szigetvar in the years 1660-1664.[8] This was followed by translations of some atlases, including Janszoon Blaeu's *Atlas Major* (completed by al-Dimashki in 1685) and Andreas Cellarius's *Atlas Coelestis* (translated in 1733 by İbrahim Muteferrika), aiming to apply the practical knowledge of heavens to the realities of navigation. The translations of Alexis-Claude Clairaut's and Jacques Cassini's astronomical tables in 1768 and 1772, and later in 1826 those of Lalande, testify once again to the practical and "factual" orientation of Ottoman learned class regarding the knowledge of the heavens. What the Ottomans translated in this period, they used mainly for navigation, timekeeping and calendar making.

Conspicuously absent in this list of translations are the canonical works of Copernicus, Kepler, Galileo or Newton. There is, furthermore, no evidence that the controversy over the heliocentric versus the geocentric views of the solar system occupied a prominent place among Ottoman scholars. When the Copernican hypothesis was accepted eventually, it was done so without much ado, without any major controversy, apparently because the Ottoman scholars did not have any firm theoretical commitments countering that reception. What the Ottoman scholars took from European science was primarily practical rather than theoretical knowledge. It is plausible that in this context, theoretical knowledge, especially when there are several competing alternatives such as the Ptolemaic, Copernican and the Tychonic systems, would not be considered scientific knowledge.[9] It is also possible that the religious outlooks with which astronomy in particular and European sciences in general were deeply intertwined from Renaissance to Enlightenment were repugnant to the Muslim scholars of the Ottoman empire.

The reception of European astronomy is typical of the larger efforts by the Ottomans to import western technology, especially military and engineering technology, during the eighteenth and nineteenth centuries. These efforts gained momentum with the establishment of military and technical schools.[10] The gradual recognition of the technological weakness of the Ottoman armies in comparison with the European ones prompted many perceptive Ottomans to publicize the need to adopt European military arts.[11] Surprisingly, the Europeans were not in general reluctant to disclose military and engineering technology to what they considered to be their traditional enemy. There is evidence documenting that the Ottoman elite could easily procure the military technology from the European pool, along with the experts who could implement them, possibly because of the way they exploited the rivalries among European states in order to obtain assistance. They received assistance variously and severally from France, Great Britain and later in the nineteenth century from the German Reich. In the nineteenth century, they began adopting several industrial technologies, shortly after their development in western Europe. Although not widespread, there was nonetheless a considerable technology transfer involving the railroad, the steamship and textiles.[12]

Another ideal about science seems to have shaped the reception of European sciences and technology in the nineteenth century, when the impact of this importation on traditional values and life styles came to be felt more intensely. This was the period of a rapid dissolution of Ottoman dominion, marked by endless military defeats on European front, and an almost permanent fiscal crisis. Military weakness and retreat before the Europeans in the two centuries of decline, beginning late seventeenth century, was the main stimulus to what has been called *defensive modernization* (Parla). Modernization, whether it can be called Ottoman Enlightenment or not, was thus a byproduct of military and economic decline.[13] Unlike many other European contexts, modernization attempts among the Ottomans did not fuel or crown imperial power.

Modernization was tantamount to Westernization, for the Ottoman ruling elite chose to emulate the strong in order to resist the strong. Attributing the military and economic power of Europe to its technological and scientific superiority, the Ottoman rulers initiated a series of reforms to update the state institutions, facilitating the transfer of sciences and technology. In the speeches and reports of many reformers of the period, an equation between science and power was the leitmotif. In a report of the Board of Useful Affairs dating from 1838 is stated:

> All arts and trades are products of science. Religious knowledge serves salvation in the world to come, but science serves perfection of man in this world. Astronomy, for example, serves the progress of navigation and the development of commerce. The mathematical sciences lead to the orderly conduct of warfare as well as military administration.... Through science one man can now do the work of a hundred." (Quoted in Berkes, 105).

The first phase of modernization (1718-1839) began with the adoption of Western military technology, training and organization. This was followed by a second phase (1839-1876) when more extensive reforms in the administrative and educational institutions were undertaken.[14] It was in this period that a new system of schools for the education of Westernized bureaucrats was introduced, which led to the eventual eclipse of the traditional religious school system (*medrese*) along with the diminution of the political influence of the learned clergy (*ulema*).

While modernization was invariably taken to be westernization, this should not hide the fact that there were alternative renderings of the "west" that provided the models to imitate. The "west" of the eighteenth century was primarily France, whereas from the second quarter of the nineteenth century on, the "west" came to include Great Britain. By the late nineteenth century and well into the twentieth century, German influences trafficked in, along with the new conceptions of the "west".[15] The receptivity to these specific cultural models was conditioned by constructions of the "west", and needs to be studied more extensively.[16] The crossing of cultures was neither simply chosen by the Ottomans–recall the trade agreements–nor was it entirely imposed by the European powers.

The reform movements of the nineteenth century posed problems for many Ottoman intellectuals. One of the chief concerns of intellectuals like Namik Kemal, a leader of the Young Ottomans, or Ziya Gökalp, a prominent political thinker of the period, was the preservation or cultivation of traditional Islamic values in the face of the trend towards westernization. The leaders of the young Ottomans generally sought ways to reorganize the society by western style liberal ideals, while remaining pious and increasingly more patriotic.

Ziya Gökalp (1876-1924) was one of the most influential ideologues of the reform movement as well as of the Turkish republic that was established in 1923. Born in an eastern Ottoman province, Gökalp's upbringing involved a combination of western sciences and eastern learning.[17] Throughout his career as poet, journalist and teacher, Gökalp was at pains to combine Turkish, Islamic and European forms of life into a viable synthesis. He based this synthesis upon a distinction between the notions of culture and civilization. He maintained that Turkism, Islam and modernism were not irreconcilable ideals, since each pertained to different aspects of life. Modernism signified the pursuit of the scientific, technological and industrial achievements of the West–its civilization. Gökalp meant by civilization the more or less similar social institutions, actively developed and explicitly adopted by different nations.[18] Modernism did not mean the embrace of European moral values or life styles in toto, but rather "the acceptance of the theoretical and practical sciences and techniques from Europe".[19] Culture, in contrast, encompassed collective values, aesthetic judgments and customs implicitly shared by the groups in a nation. Civilization, according to Gökalp, was transnational and transferable, whereas cultures were anchored in local traditions, hence immobile. Rational inquiry and sciences were part and parcel of civilization, and could be grafted onto national or religious cultures without deforming them.[20] Hence the possibility, in his vision, of Modern Muslim Turkism.

Implicit in Gökalp's synthesis is a fact-value distinction. Does not "science" comprehend an intrinsic set of values which may conflict with the indigenous ones? Affirmative or negative answers to this question lead to different conceptions of science. For most of the Ottoman elite intent upon importing European scientific goods–be they textbooks, technologies or experts–sciences, particularly natural sciences, were value neutral.[21] They could flourish in the same way in different cultural milieus. In particular, while promising to enhance the economic power of the Empire, production and dissemination of scientific beliefs did not pose any threat to Islamic values and life styles.

Gökalp's conception of science aligns readily with positivism, as advanced by August Comte in the first quarter of the nineteenth century. Gökalp took science to be an activity that aimed to represent the "outside" of nature or society. Based solely on observations and experiments, Gökalp maintained, scientific knowledge could not represent the "inside"–that task was left to metaphysics, with its investigations into the nature of mind and consciousness.[22] Philosophy, in contrast, was concerned mainly with the formulation and assessment of values, i.e., with ethics (ibid., 49-50). Philosophy and metaphysics did not overlap with science, for the study of interiors and values of human life had nothing to do with the study of the exteriors, of nature or of society. This conception of science is intimately connected to what I call the positivistic reception strategy: this is characterized by an openness towards what are allegedly value-free scientific theories, especially those in natural sciences, which do not occasion any need to worry about the transfer of values.[23] How can sciences occasion frictions with moral economies, ingrained in the mental and ethical life of a nation, if sciences are concerned solely with the representation of surfaces? Thus could Gökalp proclaim "our first objective, as individuals and as a nation, is science".[24] This objective coincided with the means and the ends of human progress conceived by Comtean positivism, readily conformable to non-Christian cultures because of its secular, or better, areligious proclamation.

4. IMAGINED COMMUNITIES

With his project of interweaving the national and the transnational, Gökalp was a chief contributor to the imagination of the modern Turkish Republic. Whether combined in a viable synthesis or not, visions of Turkism or Islamism continue to shape the imagination of some Turkish intellectuals. Some historians of Ottoman science, in particular those promoting traditional Islamist values, are prone to construct a holistic account of Ottoman-Islam science. Overlooking the long span of time and the vast range of cultures comprised under the Ottoman identity, those historians delight in discovering European scientific ideas presumably antecedently or contemporaneously in the oriental realm. Implicit in this delight are some ideological commitments. Nationalism or religious chauvinism is a most injurious element in Ottoman science studies, and has to be distinguished carefully from the recent emphases on local knowledge within historiography.

A powerful analysis of the often elusive and sinister ways in which we are surrounded and divided off by nationalist sentiments is provided by the political scientist Benedict Anderson in his *Imagined Communities*. Paying special attention to the way print-capitalism (in particular novels and newspapers) provided coercive fields replacing the earlier forms of unity based on sacred values, Anderson reveals a variety of ideological strategies which have shaped modern feelings of nationalism.[25] A nation, for Anderson, is "an imagined political community" (Anderson, p. 6), unlike the real communities based on actual proximity and communion such as a family or a village population could be. It is, furthermore, imagined to extend in history, like "a sociological organism moving calendrically through homogeneous, empty time" (ibid., 26). Memories of a nation and its histories, written in books or displayed in museums, tend to magnify this age. Cults of nationalism have ever since the nineteenth century relished in this world of imaginations.

Not only historical continuities, but also the spatial ones are being questioned in the recent investigations of some anthropologists. More precisely, the notion of a natural identity rooted in locality and community is subjected to closer scrutiny.[26] The assumption that cultures are territorialized native formations is increasingly disconcerting when transnational culture flows, whether through mass media or through mass movements, crisscross homelands and geographic regions in ever expanding waves. Analogous to the manner in which some historians of science, such as Bachelard, Foucault and Kuhn, had questioned the premise of historical continuity, social scientists began examining the assumption of spatial connectivity implicit in the concepts of "culture", "society", "community", "nation", etc. with a goal to develop a finer conceptual apparatus to delimit collective identities. Not only nations but also spatial connectedness may be imagined or enforced without necessarily being reflected in our representations of local cultures. Spatialization of reason must be sensitivized to these constructions of the places, to the mapping of the globe as a constellation of nations, on which "units" we find superimposed the subnational or supranational organizations as spatial sites. Local histories of science, knowingly or unawares, contribute to the making of places as well as of place-bound histories. Reflection on these spatio-temporal assumptions of historiography is a corrective to an overhasty localization, and promises to deliver better representations of the gradations of the difference and identity of communities.

Little is known about the local and translocal interactions weaving the intellectual life of the Ottomans, its alleged cosmopolitan networks comprising Jewish, Christian and Muslim creeds as well as Greek, Armenian and other ethnic groups. Ottoman territories were scaringly vast and variegated–even linguistic versatility showed considerable variation from borderlands to inlands. Ottomans were not one society, one ethnic group, one culture–only in 1923 the multi-ethnic and multi-religious Ottoman Empire was transformed into the nation-state of the Turkish Republic. Ottoman intellectual historians need to examine those hybrid, cosmopolitan encounters as much as the possibly rooted native ones in order to understand better the co-productions in cultural history. Take, for instance, the Jewish presence in the Ottoman lands. Two waves of Jewish immigration from Spain, one in 1492 and the other in 1536, brought a number of scholars from the Andalusian Arab-Jewish tradition. The intellectual life of the Ottomans was enriched by the migration of Jewish people from Spain, but we do not have detailed studies of this intellectual diaspora. "Intermediary" groups, such as the Greeks, the Jews, the Armenians and the converts to Muslim, capable of communicating easily with the eastern and western cultures, played an important role in the cultural life of the Ottoman empire (Inalcik). Only in the nineteenth century did the populations under Ottoman imperialism begin separatists movements organized along ethnic lines and inspired by nationalist ideologies. The multi-cultural social life of the Ottomans is in stark contrast with the official monocultural outlook of modern Turkish Republic. Playing down the importance of this social factor would be another form of presentism.

Unless histories are written with the explicit and wrong aim to contribute to the imagining of communities, they must avoid identification with and glorification of one group of people on the basis of religious, national or ethnical affiliations. While sciences may have differentiated along national divisions, at least since the nineteenth century, there is no reason why the historians cannot participate in a supranational Republic of Letters.[27] Minimally, this means impartial, disinterested, independent assessments of cultural products, be they from Turkey or Greece, Russia or the U.S, with the goal of fostering meritocracy over and above national, religious or ethnic favoritism.

5. CONCLUSION

A monument to local and translocal identities of a cultural object is the magnificent sixth century sanctuary of Hagia Sophia in Istanbul. The former patriarchal seat of Eastern Christendom, Hagia Sophia was transformed in 1453 into the imperial mosque of Ottoman Empire, and since 1934 it has become a national museum. Witnessing the massive cultural transformations wrought by the successive dominations of pagan, Christian, Muslim and secular states, the building is an archeological site, welding layers of decorations, renovations and at the same time meanings.

Despite the wishes of the Ottoman sultan Mehmed II, the "conqueror", Hagia Sophia's Ottomanization and Islamization was not an instant matter.[28] The figural imagery of its decorations had to be refashioned, albeit somewhat reluctantly, in conformity with the aniconic Islamic tradition, an aesthetic style reflecting the dedication of the temple to a nonanthropomorphic Muslim God. The reappropriation of the building by the Ottomans went hand in hand with the creation of mythical histories which reconstituted the former imperial Byzantian church as a symbol of universal cosmopolitan rule. One of the histories stated, for instance, that its rare stones were sent from India, Arabia, Persia, China, Turkistan and Europe (Necipoglu, p. 200). These strategies to universalize a cultural product have close parallels to the way a positivistic conception of science would dislodge it from its context of production. Of course, architecture and sciences are differently localized in their sites of production–one can talk about the history of architecture in Istanbul but not about a history of science in Istanbul, at least not in a fruitful way. Yet the ever imposing physical presence of Hagia Sophia can be taken to be a symbol of resilience to cultural appropriations, allegorizing in this way the career of sciences in Ottoman lands and Turkey.

This career should be studied with historiographical considerations I have presented in this paper. We need micro-histories of Ottoman science but with macro-perspectives. Most studies of Ottoman science are written backwards–backwards from the perspective of the present sciences. Historical epistemology promises to deliver perspectives through which they can be written forwards, as if from the past. That perspective is complemented by considerations of place. Margins need not be conceptualized as new centers, but as sites of local and global historical encounters. Anderson's work is a powerful antidote against imagined communities, be they religious, national or ethnical. Several historians of Ottoman science have yet to learn to resist the language of possessive pronouns. Who are "our" ancestors? Once rid of imagined communities, it would be easier to avoid imagined continuities in history of science.

For historians of local cultures, localism is not tantamount to loyalty to the imagined or real communities of the region. Localities for research usually create abstract spaces of interaction, especially since late twentieth century. To borrow an expression from Foucault, these are discursive spaces–international, regional or joint congresses, journals, proceedings, electronic networks. The new information processing and communication technologies have changed even more drastically the scope of localities. While globalization in this way is not necessarily an instant route to an equitable cosmopolitanism, it nonetheless prepares a groundwork for such an ideal. These intellectual spaces challenge the tendency towards excessive localism of particularist cultural relativism.

Local knowledge has to be reunderstood in this manifold space of abstract localities. Both spatial and temporal unities of conventional analyses need to be reassessed. My very historiographical premises are permeated with "non-local"–if not universal–principles, deriving from other local traditions. To wit, the historical epistemological approach that I am advocating is of European provenance. Although one can easily establish that different cultures have different historical sensibilities, this is no good reason for the conservation of the parochialism of intellectual hinterlands.

Boğaziçi Üniversitesi, Felsefe Bölümü

6. NOTES

[1] On the center-periphery modes of interaction, see Gavroglu ed. *The Sciences in the European Periphery*.

[2] Among the numerous works in which these authors presented the historical epistemological outlook can be cited Bachelard's *The New Scientific Spirit*, Canguilhem's *The Normal and the Pathological*, Foucault's *The Archeology of Knowledge* and *The Word and the Thing*. Despite the sympathies of the advocates of this approach, historical epistemology should not be taken as an exclusive commitment to cultural histories of science. I believe this approach bears equally on intellectual histories as well as on histories of scientific ideas. This is because epistemologies are present in the practices of both writing and reading scientific works, and hence can be revealed in studies of both the production and interpretation of sciences.

[3] For a diagnosis of this historiographical approach, see Daston (1998).

[4] For instance, see Daston and Park, *Wonders and the Order of Nature*, and Wise ed. *The Values of Precision*.

[5] Of course, a developmental account of science which the epigenetic analogy may suggest is also anathema to historical epistemology. I do not believe there is a royal road to science, recapitulating the European historical transformations. It may seem, *prima facie*, that one of the most venerated accounts of science, that of Thomas Kuhn's, is preformationist in the sense that paradigms as exemplars are the seeds from which normal research flourishes. This is, however, an incorrect interpretation of Kuhn's account, for in the latter, the exemplars are actively interpreted, creatively refashioned and generalized in ways possibly transgressing the original intentions of the revolutionary authors. This emphasis on the normal science *activity* aligns Kuhn's account with epigenetical models in history of science.

[6] References to the Arabic roots of modern European sciences for the purposes of the legitimization of the latter is a common strategy, employed early on by the Ottoman reformers. For the early currency of this viewpoint, see, for instance, Sultan Mahmud's opening speech at the Mekteb-i Tibbiye-i şahane (Imperial School of Physical and Medical Sciences) in 1838. See Berkes, p. 113.

[7] Largely based on Adivar's account, Lewis also adopts this viewpoint.

[8] My account of these translation efforts relies on that of İhsanoğlu in İhsanoğlu, (1996b).

[9] Such an attitude is part of what is alluded to as oriental wisdom in Voltaire's *Candide*. But a similar attitude has pervaded European scientific societies in the seventeenth century, when sociable discourse favored in scholarly meetings a focus on factual findings over theoretical speculations for the regulation of academic manners. For the latter, see Daston (1992).

[10] The first military institution modeled on European ones was opened in 1735 with the assistance of a general of French origin, Claude-Alexander, Comte de Bonneval, who had earlier converted to Islam and had become an Ottoman. On the career of Comte de Bonneval (1675-1747), later known as Humbaraci Ahmed Paşa, see Berkes, p. 47. The second one, the *Hendesehane* (École de Théorie et de Mathématiques), was created in 1775 under the supervision of Baron de Tott, in order to train officers for the imperial fleet. A larger institution, Mühendishane-i Cedide (Imperial School of Military Engineering), was established in 1793. See İhsanoğlu (1996a).

[11] See, for instance, Ibrahim Müteferrika's *Traité de la Tactique* (1769). On the latter's viewpoint, see Berkes, p.45.

[12] See Quataert (1996). While the Ottomans could easily and speedily access the European ideas and technologies because of geographical proximity, the same factor can also be responsible for impeding economic growth and industrial development in the Ottoman lands. European manufacturers began selling their goods on Ottoman shores beginning the sixteenth and seventeenth centuries, and in ever expanding quantities until the twentieth century. On this history, concise accounts can be found in Zürcher, Shaw and Shaw.

[13] On the issue of Ottoman Enlightenment, see Faroqhi, p. 270.

[14] This second phase is called the *Tanzimat* period, the word "tanzimat" meaning reordering or reorganization. The proclamation of Tanzimat was inaugurated by the Gülhane Hatt-i Hümayunu (Imperial Rescript of Gülhane), the official document written in 1839 which declared to reform taxation and conscription, and to guarantee the property and inheritance rights of all Ottoman subjects regardless of their creed. The reforms were enacted by Ottoman upper bureaucracy, often in conjunction with European

powers. Great Britain already secured a vast market in the Ottoman realm through the Commercial Treaty of 1838 (Trade Convention of Balta Limani), which removed previous Ottoman trade restrictions and tariff walls. See Parla for more details.

[15] These influences can be tracked to some extent through a study of the loan words Ottoman and modern Turkish contain. See Lewis, p. 85 for some examples.

[16] Before the eighteenth century, the "west" was primarily the entire block of Christendom confronting the Empire of Islam expanded by the Ottomans. Only with the defeats on the European front did the Ottomans come to see themselves and the Christians as a network of states among which there might be allies and enemies (Lewis). For a study of the images of the west in Turkey, focused largely on the twentieth century, see Metin Heper et al eds. *Turkey and the West.*

[17] Although Gökalp absorbed nationalistic sentiments from a variety of sources during his youth, a single most important figure who inculcated a nationalistic ideology to Gökalp was ironically a Greek physician by the name of Yorgi, whom he respectfully referred to as "my teacher" (Gökalp, p. 38).

[18] See Gökalp's "Hars ve Medeniyet" from 1923, translated in Gökalp.

[19] Translated from Gökalp's "Üç cereyan" of 1913 in Gökalp, p. 76.

[20] These views are expressed in many of Gökalp's writings, for instance, in "Cemaat Medeniyeti, Cemiyet Medeniyeti" of 1913 and in "Üç cereyan" of 1913. See Gökalp, p. 71-6; 102, 104.

[21] See Hanioğlu for the controversies at the end of the nineteenth century over the transfer of science and the delineation of "positive" sciences in Mekteb-i Tibbiye, the medical school of Istanbul.

[22] See Gökalp, p. 47 for the translation of these views expressed in his "Bugünkü Felsefe" of 1911.

[23] On this aspect of Ziya Gökalp's approach to science see Irzık. Irzık also presents an incisive analysis of how modern radical Islamist intellectuals capitalize on post-positivist philosophies of science to legitimize their reaction against western science.

[24] See Gökalp, p. 280 for the translation of the views voiced in his "İlme doğru" from 1922.

[25] That newspapers play a crucial role in the construction of national identity is an observation also made by Gökalp in 1913, and is attributed to the social scientist Gabriel Tarde (Gökalp, p. 71). Jean-Gabriel de Tarde (1843-1904) was a well-known French sociologist and criminologist of his time.

[26] See Clifford's *Routes* as well as the collection of articles in Gupta and Ferguson eds. *Culture, Power, Place.*

[27] On the eighteenth century ideal of the Republic of Letters, see Daston (1991). Daston (1990) provides a detailed study of how at the end of the eighteenth century Napoleonic nationalism transformed the prevailing scientific cosmopolitanism into scientific nationalism.

[28] See Necipoğlu. The concept of conquering, especially when it bears on cultural confrontations, creates some difficulties. As Necipoğlu notes, regarding the history of Istanbul, "the conquerors chose to define their self-identity in terms of the conquered, while simultaneously remaining meaningful to their own past". (Necipoğlu, p. 225).

7. REFERENCES

Adıvar, Adnan A. *Osmanlı Turklerinde ILim.* 1st ed. in 1939. Istanbul: Remzi Kitabevi, 1970.

Anderson, Benedict. *Imagined Communities.* 2d ed. London: Verso, 1991.

Bachelard, Gaston. *The New Scientific Spirit.* Trans. by Arthur Goldhammer from *Le nouvel esprit scientifique*, 1934. Boston: Beacon Press, 1984.

Berkes, Niyazi. *The Development of Secularism in Turkey.* London: Hurst & Com., 1998 (facsimile of 1964 first edition).

Canguilhem, Georges. *The Normal and the Pathological.* Trans. from *Le normal et le pathologique*, 1966. New York: Zone Books, 1991.

Clifford, James. *Routes: Travel and Translation in the Late Twentieth Century.* Cambridge and London: Harvard University Press, 1997.

Daston, Lorraine. "Nationalism and Scientific Neutrality under Napoleon". In Tore Frangsmyr ed. *Solomon's House Revisited: The Organization and Institutionalization of Science.* Science History Publications, U.S.A., 1990, 95-119.

——"The Ideal and Reality of the Republic of Letters in the Enlightenment." *Science in Context* 4.2 (1991): 367-386.
——"Baconian Facts, Academic Civility, and the Prehistory of Objectivity." *Annals of Scholarship* 8 (1992): 337-363.
——"The Nature of Nature in Early Modern Europe". *Configurations, 6*: 149-172.
Daston, Lorraine and Katharine Park. *Wonders and the Order of Nature, 1150-1750*. New York: Zone Books, 1998.
Faroqhi, Suraiya. *Osmanli Kültürü ve Gündelik Yaşam: Ortaçağdan Yirminci Yüzyila*. trans by Elif Kilic from *Kunst und alltagsleben im Osmanischen Reich*, 1995. Tarih Vakfi Yurt Yayinlari 48, 1998.
Foucault, Michel. *The Archeology of Knowledge*. Trans. by A.M. S. Smith from *L'Archeologie du Savoir*, 1969. New York: Harper & Row Publishers, 1972.
——*The Order of Things: An Archaeology of the Human Sciences*. Trans. from *Les mots et les choses*, 1966. New York: Vintage Books, 1973.
Gavroglu, Kostas ed. *The Sciences in the European Periphery during the Enlightenment*. Dordrecht: Kluwer, 1999.
Gökalp, Ziya. *Turkish Nationalism and Western Civilization: Selected Essays of Ziya Gökalp*. Trans. and ed. by Niyazi Berkes. London: Ruskin House, 1959.
Günergun, Feza (ed). *Osmanli Bilimi Araştirmalari*. Istanbul: Istanbul Üniversitesi Basimevi, 1995.
Günergun, Feza and Shigehisa Kuriyama. *The Introduction of Modern Science and Technology to Turkey and Japan*. Istanbul: International Research Center for Japanese Studies, 1996.
Gupta, Akhil and James Ferguson (eds). *Culture, Power, Place: Explorations in Critical Anthropology*. Durham and London: Duke University Press, 1997.
Hanioğlu, şükrü M. *Doktor Abdullah Cevdet ve Dönemi*. Istanbul: Üçdal Neşriyat, 1981.
Heper, Metin, Ayşe Öncü and Heinz Kramer eds. *Turkey and the West: Changing Political and Cultural Identities*. London: Tauris, 1993.
Hunt, Lynn. *The New Cultural History*. University of California Press, 1989.
Ihsanoğlu, Ekmeleddin. "Introduction of Western Science to the Ottoman World: A Case Study of Modern Astronomy". In İhsanoğlu ed. 1992, 67-120.
——(ed). *Transfer of Modern Science and Technology to the Muslim World*. Istanbul: Ircica, 1992.
——"Changes in Ottoman educational life and efforts towards modernization in the 18th-19th centuries". In Günergun and Kuriyama 1996.
——*Büyük Cihad'dan Frenk fodulluğuna*. Istanbul: Iletisim, 1996b.
Inalcik, Halil. "Some Remarks on the Ottoman Turkey's Modernization Process". In İhsanoğlu ed. 1992.
Irzık, Gürol. "Philosophy of Science and Radical Intellectual Islam in Turkey". In W.W. Cobern ed. *Socio-Cultural Perspectives on Science Education*, 1998, 163-179.
Irzık, Gürol and Sibel Irzık. "Which Multiculturalism?" , *Science and Education* 11, 2002: 393-403.
Karpat, Kemal H."The Ottoman adoption of statistics from the west in the 19th century". In İhsanoğlu ed. 1992.
Lewis, Bernard. *The Muslim Discovery of Europe*. W.W. Norton & Company, New York and London, 1982.
Mark, Robert and Ahmet S.Çakmak. *Hagia Sophia from the age of Justinian to the present*. Cambridge: Cambridge University Press, 1992.
Necipoğlu, Gülru. "The Life of an Imperial Monument: Hagia Sophia after Byzantium". In R. Mark and A. Çakmak ed. *Hagia Sophia from the age of Justinian to the present*, 1992, 195-225.
Parla, Taha. *The Social and Political Thought of Ziya Gökalp 1876-1924*. Leiden: E.J.Brill, 1985.
Quataert, Donald. "The introduction of modern technology in Ottoman industry during the 18th and 19th centuries" in Günergun et al eds. 1996.
Shaw, Stanford J. and E.K. Shaw. *History of the Ottoman Empire and Modern Turkey*. 2 vols. Cambridge: Cambridge University Press, 1977.
Wise, Norton ed. *The Values of Precision*. Princeton: Princeton University Press, 1995.
Zürcher, Erik J. *Turkey: A Modern History*. London: I.B. Tauris, 1993.

EKMELEDDIN İHSANOĞLU

INSTITUTIONALISATION OF SCIENCE IN THE *MEDRESES* OF PRE-OTTOMAN AND OTTOMAN TURKEY

This study aims to show the process by which the teaching of sciences that were originally translated from pre-Islamic scientific legacies, particularly from Greek, were integrated into the formal teaching programs of the *medreses* – the most indigenous institutions of learning in Islam. This article being part of a wider study on the development of these institutions shows the transformation from the personal teacher-disciple relationship to the institutional model in Turkish *medreses* of the pre-Ottoman and Ottoman era until the sixteenth century. The process of the institutionalisation of science during this period is one of the most interesting and intricate subjects in the history of sciences in Islamic civilisation.[1]

Islam and the Islamic culture penetrated Anatolia or Asia Minor-the greater part of Turkey- in the first century of the Hegira. This epoch coincided with the reign of the second Caliph Umar and with the conquests that put an end to the Byzantine rule in Syria. After the conquest of the southern Anatolian regions in 640, the old settlements of Amid (Diyarbakir), Mardin, Ruha (Edessa-Urfa), Harran, Hisnkeyfa, and Meyyafarikin were won over to the new religion and culture. Some of these cities were instrumental in the transmission of the Hellenistic cultural and scientific heritage to Islam. The fortified cities of Adana, Tarsus and Antakya (al-Avâsim) and the vanguard cities of Malatya, Marash and Erzurum (al-Shugur) successively came under the rule of the first Caliphs, the Umayyads and the Abbasids.

The first Turkish conquests in Anatolia began during the reign of the Great Seljuk Sultan Tugrul Beg (1042-1045). The Malazgirt victory of Sultan Alp Arslan against the Byzantine Empire in 1071 opened the gates of Anatolia to the Turks. Within two centuries Anatolia acquired its Turkish and Islamic identity. These territories, which had belonged to the old Roman Empire, had been called the Roman lands (*Bilād al-Rum*) since the first conquests by the Arabs and continued to be called so during Ottoman times.

Towards the beginning of the twelfth century, the Seljuk Turks became the leading political and military power, and their language became the *lingua franca* in the area. Europeans referred to these lands as Turkey. In 1242-3, the Mongols destroyed the politically and militarily weakened Seljuk power. The Mongols made the Seljuks their vassals and the Seljuk state broke into several smaller Turkish principalities.

G. Irzık & G. Güzeldere (eds.), Turkish Studies in the History and Philosophy of Science, 265-284.

It was during this period, towards the end of the 13th century, in 1299, that the Ottomans first emerged as a small principality under the Seljuk Sultanate in Konya. They rapidly became a dominant state, advanced and spread into the Byzantine lands of Anatolia and the Balkans, and conquered Istanbul in 1453. Half a century later, Sultan Selim I (1512-1520) brought all the Anatolian principalities under one rule and united the area.

1. SOCIAL AND CULTURAL LIFE IN ANATOLIA DURING PRE-OTTOMAN TIMES

Before we turn to the study of the scientific and cultural activities in Anatolia during the Seljuk period, we should investigate social and cultural life during the 11th-14th centuries.[2] Under the Seljuks, the Anatolian people were divided into two basic religious groups, namely the Muslims and the Christians, and into three social classes, that is, the nomads, the peasants, and the city dwellers. The Seljuk city dwellers were the military and the civil servants, the learned men, sheikhs, Seyyids, dervishes, preachers, poets, physicians, craftsmen, and merchants. The sultans, begs, and distinguished people supported cultural activities by founding mosques, *medreses*, public kitchens, dervish lodges and hospitals, many of which were endowed as pious foundations or *waqfs*.[3]

Cultural development in Turkish Anatolia blossomed in the second half of the 12th century, and important cultural and artistic works were produced during the 13th century. The Mongol invasion must have had an unsettling influence on Anatolian urban life, and yet despite the unsettling consequences this political turbulence must have had, great progress is observed in every field in the 13th century in Seljuk cities.

Simon of Saint Quentin reports that there were 100 cities in the Seljuk State towards the middle of the 13th century. Ibn Sa'id al-Maghribi states that there were 24 provincial cities in the Seljuk State, each administered by an official governor, with one judge (*qadi*), mosque, baths, and cloth merchants. The Seljuk administrators were lenient towards non-Muslim communities and gave them the opportunity to live and practice their own religions freely in their cities.[4] Contemporary Western and Islamic sources acknowledge the fact that urban life in the main Anatolian cities of the 13th century was more developed than life in the cities under Byzantine administration. The Seljuks "built up the whole infrastructure of Sunni Islam" in Anatolia.[5] They appointed the Iranians, whose culture was more developed at the time, to the various bureaucratic posts in their administration and invited their scholars to come and settle in Anatolia. Thus, in many of the cities where the Seljuks had settled, Iranian culture became dominant. Arabic and Persian were used as the official languages of the Seljuks. As Cahen remarks, during the Seljuk era in Anatolia, the most favourable integration, from the viewpoint of political administration and cultural progress, was not between the local people and the Turks but rather between the Iranians and the Turks. The main cause behind the ensuing conflicts in Seljuk Turkey was the deterioration of the relations between the Turkmens who lived in rural areas and the city dwellers who had been deeply influenced by Iranian culture. The Turkish begs were also influenced by the Byzantine and Ilkhanid culture. Despite the fact that Turkish literature had progressed in the Turkoman principalities (*Begliks*), it had not improved sufficiently to replace Persian and Arabic literature. Nevertheless, a new culture which was instrumental in the impressive advance of the Turks in Anatolia arose and spread through all the Anatolian principalities.[6] The Turkish principalities also came under the influence of Syria and Egypt, areas that had belonged to the Mamluks.

The active relationship between Anatolia and the lands in its immediate surroundings, Iraq, Syria, Iran, Transoxania and Central Asia, i.e. lands which had been integrated at an earlier stage into the Islamic world, was instrumental in the diffusion of the various scholarly traditions and the establishment of institutions for learning in Anatolia. The spread of *medrese*s in the Seljuk lands and the countries under their rule increased literacy, while the spread of hospitals bearing names such as the *Dar al-Shifa*, *Dar al-Sihha*, *Shifahane*, *Bimarhane* or *Maristan* brought health care services, until then unavailable, to the populations in these areas. The spread of the educational and health facilities contributed to affluence and prosperity in these areas, and this in turn helped promote literature, art, and scientific activities.

The sultans of the Great Seljuk Empire, the Anatolian Seljuks and some begs supported the scholars involved in the study of these disciplines. Moreover, sources indicate that some of the sultans and chieftains, or begs, were personally engaged in these activities. Around that time, we start noticing some changes in the *medrese*s that started to incorporate some of the rational sciences in their teaching activities. In this paper we shall confine ourselves to certain aspects of this patronage and related activities in Anatolia, an area which is less studied than the Great Seljuk Empire.

There are historical reports that describe the involvement of the Anatolian Seljuk sultans in astronomy. The most notable is the one made by the historian Ibn al-Athir (1160-1233). He wrote that Kutalmish Beg, the father of the founder of the Anatolian Seljuk State, Rukneddin I Suleyman Shah (1075-1086), knew the sciences of the stars and other sciences too. According to him, the rulers from this lineage also knew about the sciences of the pre-Islamic period and the Hellenistic heritage, including mathematics, astronomy and medicine. He added that they protected the people who were involved in these sciences.

Al-Jazari is the author of a famous book on automata written for the Artuk Beg, Mahmûd b. Muhammed bin Kara Arslan (1200-1222). The interest which this major contribution aroused is a reflection of the place held by these sciences in Anatolia. A famous man of learning with diverse interests, Abd al-Latif al-Baghdadi (1220-1229), who left us about 170 works on the religious sciences, philosophy, medicine, mathematics, and literature, wrote some of his important works under the patronage of Alâ al-Din Dawud b. Behram, the ruler of Erzincan, who was a descendent of the Mengucek Ogullari. Although there are many other patrons and scholars who would be worth mentioning, we shall now focus on institutional developments as such.

2. HISTORY OF THE ANATOLIAN *MEDRESE*S

A comprehensive study of the Anatolian *medreses* and the educational life evolving around them has not yet been undertaken. Yet the surveys made by A. Kuran (1969) and M. Sözen (1972) allow us to make a partial analytical evaluation of the history of the *medreses* in Anatolia despite the fact that they were studied and classified as regards to their construction and architectural features. According to the given data, the earliest *medreses* in Anatolia were built under the Artuk rule in the cities of South East Anatolia. According to the existing information, the Eminuddin *Medrese* founded in Mardin in 1108-1123 is the oldest Anatolian *medrese*.[7] Unlike the Seljuk *medreses*, which were influenced by the Iranian architecture, the *medreses* built in the Diyarbakir, Urfa and Gaziantep regions, as of the 12th century, were built in the Syrian style.[8]

The first *medreses* built in the regions mostly populated by Turks were located in the relatively peaceful lands of the Danishmendogullari in the 12th century. This area was not under the domination of Anatolian Seljuks. During this period, the Anatolian Seljuks were in constant struggle with the Byzantines and the Crusaders in the West. In Turkey today, the two oldest remaining *medreses* in Niksar and Tokat respectively are, built by the Danishmend sovereign Nizameddin Yagi-basan (1142-1164) and known by his name. The central domes of these two *medreses* played a great role in the development of the Anatolian *medrese* architecture. The Danishmend architecture inspired many builders who built several *medreses*, *dar al-shifa*, and dervish lodges with central domes in the lands of the Seljuks, the Ilkhanids, the Begliks and the Ottomans.[9]

In the 12th century, the Anatolian Seljuks, who were involved in the wars against the Byzantines and the Crusaders, did not have the means to establish *medreses* or other institutions. Once peace was secured and order was established in the cities, Seljuk *medreses* began appearing toward the beginning of the 13th century.[10] As the Seljuks became politically stronger, they founded *medreses* which, unlike those in Syria, Egypt or North Africa, were evenly distributed in numerous large cities and also in smaller towns in Anatolia. This "remarkably even spread of facilities throughout the land may be best explained by the interaction of two complementary trends: a centralised building programme and -though probably to a lesser degree -a popular fashion for the *medrese* as an institution."[11] These *medreses* were not thought of as institutions reserved for an urban elite. Rather, they had deep popular roots and served the communities with which they kept close contacts. The Seljuks established the *waqf* institutions and their multipurpose facilities such as the medical care institutions, the mental hospitals, and the public kitchens (*imaret*), as well as the mausoleums that were often included in the building complexes.

In order to understand the shift in the methods of transmission of the rational sciences from the personal way of teaching between master and disciple, from the urban elite cultural milieu to the public institutions of learning during this period, it is necessary to keep in mind the basic stages of the emergence of the *medrese* institution in Islam, its spread, and its transfer to Anatolia.

Medreses came into being as organised educational institutions in the city of Nishapour during the Ghaznavid period in the 10th century and became widespread in the lands ruled by the Great Seljuk Empire. The *medreses*, known as Nizamiya *Medreses* built by Nizam al-Mulk, the grand vizier of the two Seljuk sultans Alp Arslan and Malikshah, spread over a wide area and became models for the others.

Medreses as developed educational institutions appeared during this period. Academic and political/religious factors were effective in their establishment.

Let us give a brief aperçu of one of the most prevalent theories about teaching in the institutions of learning in the Muslim world. Makdisi (1981) writes that the *medrese* institution came into being after a long evolution. He sees it as the outcome of a long development in the context of the struggle between what he calls the traditionalist and the rationalist forces, a struggle which ended with the triumph of the former. He traces the origin of educational tradition back to the second half of the 8th century when the personal schools of law were formed. Masjids for the study of law emerged in the 9th and 10th centuries and *medreses,* subsequently, combined the functions of the masjid and the adjoining inn, where the students were lodged, in the 10th and 11th centuries. Makdisi prefers to use the European word 'college' as an equivalent for '*medrese*' and maintains that the structure of the collegiate system rested on a legal basis defined, interpreted and maintained by the lawyers. He claims that collegiate learning was so organised as to give primacy to legal studies over all other fields and that all rationalistic studies were excluded from the regular curriculum.[12]

2.1. 'Exclusion' or gradual 'inclusion'?

We will argue, on the other hand, that an initial 'non-inclusion' stage was followed by a gradual 'inclusion' of these sciences into the teaching of the Islamic educational institutions. The *medrese* institution came into being as a result of the development of Islamic religious sciences and the long-standing academic and educational traditions. Concomitantly, the *medrese* was the reflection of a need for institutionalisation in order to counteract the Shiite propaganda pursued against Sunni Islam. It is worth noting that these institutions were established within the framework of the educational tradition for the teaching of jurisprudence (*fiqh*) in the training of jurists (*faqihs*). However, it is possible to surmise that the jurists did not allow the teaching of the sciences and subjects towards which they felt antagonistic, since their opponents had used these very sciences for their own doctrine and polemic.

The Abbasid Caliph al-Mamun lived through the *Mihna*, a time of religious oppression and violence that was the result of a conflict between the Mutazilite school and the *ulema* about dogmatic questions, especially the question of the creation of the Quran. Mutazilism became the official doctrine upheld by the Abbassid caliphs and was imposed on the Sunni majority. Rightly or wrongly, it is often argued that this doctrine was used for political aims and propaganda. The Mutazilites used philosophy as a doctrinal weapon to win the intellectuals and the public to their side.[13] The doctrine of the marginal Ismaili group of Alamut, which rested on the rational sciences and particularly philosophy and alchemy, was used in the same way.[14] This use of the foreign heritage by the Mutazilis and the Ismailis set up a reaction among some scholars against the sciences that had been transmitted from the pre-Islamic, mainly Hellenistic, heritage; these sciences were referred to as "the sciences of the ancients" or "the sciences of antiquity" (*ulum al-awāil*).[15]

Medreses founded within this framework spread in Anatolia and other places. The fact that the *medreses* did not formally include these sciences as part of their curriculum does not imply, however, that scholars did not pursue them privately. The teaching of these sciences was based on the traditional master-disciple transmission system. The biographical works on Muslim scientists, especially the biographical dictionaries of physicians, offer ample evidence of the extent to which the *ulum al-awāil* were studied and also transmitted through non-official or private channels. After the establishment of the Nizamiya *Medreses*, the integration process followed a gradual curve reflecting a change of outlook towards the *ulum al-awāil*. The pre-Ottoman Anatolian *medreses*, as well as the Ottoman *medreses*, and more particularly the ones built during and after the reign of Sultan Mehmet II the Conqueror, played a pivotal role in this process. We will come back to this point later.

As expounded in a previous study, which focused on the deeds of trust of the various foundations, the primary objective of the pre-Ottoman *medreses* founded in Anatolia was, above all, the teaching of Islamic jurisprudence and secondarily the religious sciences and related subjects, namely Arabic language and literature.[16] However, when sources other than the foundation deeds are examined for the study of educational activities in the pre-Ottoman Anatolian *medreses*, there is some evidence that the rational sciences such as mathematics and astronomy were not left out of the *medrese* education and that the theoretical *medrese* model expounded by Makdisi does not completely fit when historical facts are taken into consideration. The examples we shall provide here indicate that his model is in need of some modification.

Qutb al-Din al-Shirazi (d. 1311) was one of the most famous students of Nasir al-Din al-Tusi (d.1274), the great astronomer, mathematician and founder of the Maragha observatory, who lived in the last quarter of the 13th century. Al-Shirazi in his search for learning came from Maragha and studied the sciences in the city of Konya in central Anatolia. He was appointed judge (*qadi*) to the cities of Sivas and Malatya. While he was teaching at the Gokmedrese in Sivas, which was also known as the Sahibiye *medrese*, he wrote his important book on astronomy *Nihayât al-Idrâk fî Dirâyat al-Aflâk* in 1282[17] and taught this work to his students. We have clear evidence that two years after he had written his book, he was still giving lectures based on the same book in the same *medrese*. One of his students reported that in 1284, while al-Shirazi was still teaching in the *medrese*, he was taught from this book and even made a copy of his teacher's manuscript for himself.[18] Thus, there is little doubt that al-Shirazi taught astronomy in the Gokmedrese. Nonetheless, when one examines the deed of trust of this *medrese*, it is clear that clauses referring to teaching activities are rather similar to the standard phrases in other deeds of trust, with no specific mention of the teaching of any rational sciences.[19]

Like the Sivas *medrese* in Eastern Anatolia, the Jaja Beg *Medrese* of Kirshehir and the Wajidiyya *Medrese* of Kutahya in Central and Western Anatolia are two examples which show that the rational sciences were part of the institutional teaching. They are documented by Sayili in his well-known study on the *Observatory in Islam*.[20] The *medrese* of Kirshehir was built by Nur al-Din Jibril ibn Jaja in 1272. Ibn Jaja was the governor of Kirshehir during the time of the Seljuk ruler Giyath al-Din Kaykhusraw ibn Kilij Arslan. The deed of trust of the *medrese* does not contain any statement to the effect that astronomy was to be included in the official teaching. Nevertheless, there are some historical accounts to the effect that the minaret of this *medrese* was originally an observation tower used for astronomical purposes.[21] Likewise, there are several reports on the existence of "observation wells" at Jaja Beg *Medrese* in Kirshehir and Wajidiyya *Medrese* in Kutahya.[22]

We shall now dwell on a second example from Kutahya in Western Anatolia. Germiyanogullari (1300-1429) was one of the principalities (*Begliks*) in Anatolia that coincided in time with the rule of the Ottomans. During this period Umur b. Savji built a *medrese* in Kutahya in 1314. One of its teachers was Abd al-Wajid b. Muhammed al-Mashed al-Kutahi (d. 1435). Later this *medrese* became known by his name, and was referred to as the Wajidiyya *Medrese*. While he was a teacher in the *medrese*, he wrote three works on astronomy [23] and made astronomical observations, as shown by Aydin Sayili.[24] In all likelihood, the content of this deed of trust does not differ in any way from the other deeds extant.

It is clear that the deed of trust of these two *medreses*, the Jaja Beg of Kirshehir and the Wajidiyya of Kutahya, did not prohibit the teaching of astronomy. This was also the case with the *medrese* in which al-Shirazi taught. From this we can infer that when a particular science is not explicitly mentioned in a *waqfiyya*, this does not mean that the science is not taught there.

When we take into consideration the examples given above and the explicit historical evidence concerning the teaching of the rational sciences (including mathematics and astronomy) in Anatolian *medreses*, we can reach the following conclusion: Initially, as in the case of Nizamiya *Medreses*, there was indeed a reluctance to teach these sciences in the *medreses* as part of their regular curriculum, which included primarily religious sciences and secondarily complementary topics such as Arabic language and literature. However, because of the encouragement and support of the Seljuk sultans, Ilkhanid rulers and Turkish begs, the rational sciences were not categorically 'excluded' from the Anatolian *medreses* altogether. They were taught either because the teachers chose to teach them or because the students requested. Thus a new tradition began to emerge in the Anatolian *medreses*.

One of the important issues concerning education under the roof of the *medreses* would be the payment made to the teacher who taught the rational sciences. Would he be paid from the revenues of the foundation? The foundation deeds of all the *medreses* built within this period explicitly state the conditions of expenditure and this is in conformity with the tradition of the *waqfiyya*. The teachers would receive payment for teaching religious subjects from the revenues and the students were also given stipends. It is to be supposed that after the teacher was remunerated according to the conditions of teaching written in the deed, the teaching of subjects not listed therein would be a subject for discretion. This would be a plausible explanation for the teaching of rational sciences during the teacher's free time.

It is only natural that Qutb al-Din al-Shirazi, who had studied under Nasir al-Din al-Tusi, absorbed some of the teaching methods and contents of the teaching in the Maragha observatory, where besides the regular practical functions of the astronomical observatory, a new dimension had been added: teaching. Teaching was then transposed to the Sivas *medrese* where al-Shirazi was appointed as a professor. In this way a new domain is included in the activities of the *medrese*, the Islamic teaching institution *par excellence*, namely teaching of the rational sciences in addition to the regular religious subjects listed in the deeds of trust of all such foundations.

The examples which we give, that of the Sivas Gokmedrese being an obviously clear-cut case, and A. Sayili's reports on the Kutahya Wajidiyya *Medrese*, and the Jaja Beg *Medrese* in Kirshehir, as well as many others may be cited to substantiate this point. A study of Table 1 in the *Osmanli Astronomi Literaturu Tarihi/ History of Astronomy Literature During the Ottoman Period LXXVIII–LXXXV*, which contains numerous records of scientific manuscripts copied during pre-Ottoman and Ottoman times will surely provide ample evidence for the hypotheses proposed above and give us a picture of two aspects of the education carried out in some Anatolian *medreses*.

The first aspect is the formal education conforming to the objectives written in the foundation deed, consisting basically of religious sciences and the supplementary Arabic language and literature lessons. The teachers were appointed and were paid for pursuing this objective in education, and the students were given stipends for this same reason. The second aspect, i.e. the teaching of rational sciences, was not included in this formal requirement but took place under the roof of the *medrese*. It consisted of lessons in rational sciences given by teachers who were knowledgeable in these subjects to students who showed interest in these very topics. The rational branches of learning like logic (*mantiq*) had been naturally accepted and integrated into the teaching of the *medreses*. As to *'ilm al-kalam*, it was recognised as being the expression of theology or philosophy in its Islamic form. From the biographies of the learned men who were trained in pre-Ottoman Anatolia, it is assumed that the inclusion of courses on other rational sciences did not bring undue expense to the foundation and attracted much interest from the students. However, it cannot be said that this kind of education found favour everywhere and was overwhelmingly adopted in all the *medreses*. Indeed, according to the tradition that had been followed since the foundation of the Nizamiya *Medreses*, this kind of education was to be imparted outside the *medreses*. The Anatolian *medreses* fostered and contributed to the new tradition in teaching by including some of the rational sciences in their activities.

3. OTTOMAN *MEDRESES* AND INSTITUTIONS

The *medrese* system inherited from the Seljuk Turks was adopted and enriched by the Ottomans. The construction of a *medrese* by a mosque became a tradition among the Ottomans and an integral part of their policy of conquest. This tradition was geared both to the provision of the necessary religious, scientific and educational services for the society, and particularly to the training of the administrative and legal personnel for the state. It was in this fashion that the Ottoman state was able to provide itself with knowledgeable individuals according to the requirements of Islamic jurisprudence and customary practice.[25] Let me now give some examples.

3.1. Examples of 'Exclusion'

The first Ottoman *medrese* was built in Iznik in 1331 by the second Ottoman monarch Gazi Orhan Beg just after he conquered the city in 1330-1. Gazi Orhan Beg established many foundations in order to meet the financial needs of the *medrese*. The Iznik *medrese* trained the student in religious sciences (*al-'ulum al-diniyya*) in their totality, and famous religious scholars such as Dawud al-Kaysari, Taj al-Din al-Kurdi and Ala al-Din Aswad taught in this *medrese*. There is only one sentence in the Iznik *medrese*'s *waqfiyya* about tutorial activities: "the student in search of knowledge (*talib al-ilm*) would attend classes in the *medrese* every day of the week". It seems that the education offered in the *medreses* was left entirely up to the initiative of the professors, provided that it was within the framework of the conditions laid down by the *waqf*, and in conformity with the long established academic traditions.

Umur Beg *Medrese* is an example of the *medreses* that followed the Nizamiya tradition. This *medrese* was founded by Timurtash Pasha's son Umur Beg (d.1434) in Bergama. Nothing is left of it except its deed of trust, where the objective of the education is clearly stipulated as the teaching of *tafsir*, *hadith*, methodology of jurisprudence (*usul al-fiqh*) and the branches of jurisprudence (*furu' al-fiqh*). The deed clearly stipulates that the 'philosophical sciences' would not be taught.[26]

Dar al-Hadith founded by Murad II in Edirne in 1435 focused on the study of *hadith*. It explicitly forbade in its deed of trust any involvement in the study of philosophical sciences. The founder of the *waqf* stated the following: "The *mudarris efendi*, or professor, will teach the students the sciences prescribed by the religious law and the literary arts (*sher'i ilimler ve edebî fenler*). This is my stipulation that the professor will under no circumstance teach philosophical sciences (*al-funun al-falsafiyya*). There, every day, *hadith* and the subjects related to it will be taught by the professor".[27]

The two examples mentioned above support Makdisi's view that the rational sciences are 'excluded' from the *medrese* education. Yet, even the mere fact that 'exclusion' is stipulated in the deed suggests that some teaching of the philosophical and/or rational sciences had occurred beforehand. To understand better this negative attitude towards the teaching of the 'philosophical sciences', it is important to look at this issue in term of the fact that philosophy was used by heterodox religious groups and marginal political movements against mainstream Islamic thought, in the framework of the relationship between philosophy and *ilm al-kalām* in the 14th century and afterwards.

Sabra provides a new perspective on the subject. He argues that *kalam* should not be seen as apologetics, as is often claimed, but rather as a rational enquiry into revealed truth. It is "an argumentative approach to religion, which sought, through discussion and discursive thought, to interpret and transform the content of the Islamic revelation into a rationally-based doctrine".[28] As such, *kalam* would be a science parallel to philosophy, "an attempt to offer an alternative philosophy to *falsafa*".[29] The foreign *falsafa* inherited from the pre-Islamic cultures and the native Islamic *Kalam* were 'competitors'. This does not mean that *falsafa* (by which Sabra means the foreign body of philosophies, mainly the Hellenistic ones) was rejected or 'excluded' by Islam. On the contrary, the Muslim philosophers (*falasifa*) tried to adapt this foreign philosophical body and integrate it in their own Islamic culture and faith despite the suspicion of some. Nevertheless the attitude toward the foreign sciences other than *falsafa*, i.e. arithmetic, geometry, astronomy and music was different, perhaps because, as Sabra emphasizes, "science was not a direct competitor of *kalam* the way that *falsafa* was, and generally the specialised scientific disciplines were not perceived as sciences that posed a threat to religion".[30]

It is also important to establish whether or not the prohibition of the 'philosophical sciences' found in the deeds of trust refers to philosophy in general or to heterodox ideas like those of the Ismailis who threatened Sunni Islam in the 10th-12th centuries by targeting certain religious beliefs with philosophical arguments and by maintaining other philosophical doctrines that led to the denial of the existence of God. This can only be done by looking into the intellectual history of Anatolia during that period and by paying careful attention to the meaning of 'philosophy' and 'philosophical sciences', which seem to mean doctrines or sciences that put forward arguments that went against the mainstream religious beliefs and supported heterodox trends.

If we were to assume that the 'philosophical sciences' as such were rejected, it is important to determine whether these sciences were banned as a whole or not, and whether this ban included mathematical sciences (*al-'ulum al-ta'limiyya*) or not. The search for the answers to these questions will help us draw a picture of the scientific life prevalent in Anatolia during this period and of the development of sciences in the Islamic world. Whatever the outcome of the above debates is, the stipulations found in the deeds of trust of the two above mentioned *medreses* prohibiting the teaching of 'philosophy' show that there was a conservative attitude vis-à-vis the inclusion of different sciences in the teaching activities of *medreses*. The conservatives, labelled 'traditionalists' by Makdisi, were definitely against the teaching of subjects other than religious sciences and the related fields. Founders like Umur Beg and Murad II, or rather their scholarly entourage, were under the influence of conservative scholars who did not wish to have these subjects taught in the context of their *medreses*. These scholars were the legal authorities who would draw up the deed and had the jurisdiction over the activities proposed in it. The explicit restrictions included in the deeds actually provide us with clear evidence that the philosophical sciences, at least some of them, were -or had been-taught in some *medreses*. As a consequence, their explicit prohibition in the stipulations in the deeds had become necessary for founders who wished to exclude them from the teaching activities of their *medreses*.

3.2. Examples of 'Inclusion'

As seen through the examples given above, the founders of the Seljuk *medreses* and the early Ottoman *medreses* founded before Fatih Sultan Mehmet II followed the tradition of the Nizamiya and aimed at spreading religious education in general and jurisprudence (*fiqh*) in particular. However, the fact that we find in some of the Seljuk *medreses* some observation wells and hospitals (*shifahanes*) built close by the *medreses*, reveals a definite interest in teaching medicine and astronomy. During the era of the Seljuks and that of the early Ottomans (again before the reign of Mehmet II), the *medrese* curriculum did not include philosophical, mathematical and natural sciences. These were not part of the religious sciences *per se*. They were taught, as had been done in the past, outside the *medreses*, or in the *medreses* without having been explicitly mentioned in the deeds of trust, or in the hospitals (*shifahanes*), or in the houses of the learned men.

The turning point in the history of *medrese* teaching and the shift from the traditional Nizamiya system towards a more comprehensive institutional model took place during the time of Mehmet II. There are several reasons for this. First, there is the personal interest of Sultan Mehmet II in the rational sciences and his patronage of the scholars; second, this new tradition seems to draw directly from the Ilkhanid and Timurid institutions of learning, which included the teaching of the rational sciences. Finally, we must mention the role of Ali Kushju (d. 1474), a famous mathematician and astronomer who served both the Timurid Ulug Beg and the Ottoman Mehmet II. Kushju was brought up in Samarkand, where the tradition for teaching astronomy and mathematics had already been established by Kadizade al-Rumi (d. after 1427). Kushju was instrumental in the importation of the Timurid Samarkand tradition to the Ottomans.

The traditional system went through an important change in a very short time. We believe that a study of two *medreses*, namely the Eyup Sultan and the Fatih complexes, established by Mehmet II after the conquest of Istanbul in 1453, is pivotal in understanding this change.

The Eyup Sultan Kulliyesi was founded by Sultan Mehmet II in 1458-9. We learn from the deed of trust of this *kulliye* (complex) that the education in this *medrese* concerned those who wished to "pursue the fundamental sciences prescribed by the religious law and their ancillaries, as well as all the other noble transmitted sciences".[31] This implies that the goal of formal education was to teach all of the religious sciences including the transmitted (*naqli*) sciences. Thus the teaching objective of this *medrese* was in conformity with the established pattern.

Concomitantly, the Fatih Kulliyesi was founded between 1463-1470, shortly after the Eyup Kulliyesi, by Sultan Mehmet II and bore his name. It included eight intermediate *medreses* called '*tetimme*' and eight other high *medreses* called '*sahn*' (literally, "courtyard"). The deed of trust of the latter shows a very different educational goal and reflects a major change in the Ottoman *medrese* system. For the first time among the Turkish *medreses* recorded up to the time of Mehmet II, we see that the trust deed requires that the employment of professors is to depend on their knowledge, not only in the transmitted sciences, but also in the rational ones. The *waqfiyya* lists the qualifications of the professor appointed to teach in the higher colleges or *sahn*s: he must be "knowledgeable in the principles and basics of both transmitted and rational sciences."[32] However, we do not find any indication in the trust deed of the Fatih *Medrese* to the effect that the professor was required to teach both transmitted and rational sciences. There is but a clause stating that he should teach '*anwā'i-'ulum ve ma'arif*', that is 'the various sciences and learning', while his helper, the assistant (*mu'id*), has to be skilled in the 'various arts' (*funun-i shatta*) and teach the students the art of critical discussion under the direction of the professor (*mubahatha* and *mukataba*). Another implicit reference to rational sciences is included in the introductory descriptive statement about the qualifications of the teachers in the *sahns*. In it, *'hikma'*, a term used to refer to philosophy, geometry, and other rational sciences, indicates that these were indeed part of the basic education in this *medrese*: "The basic foundation of the studies in high colleges, is based on philosophical precepts and geometrical rules".[33] These clearly show that the goal of collegiate education had changed. *Fiqh* was no longer the exclusive subject taught; rather, many different sciences were introduced into the curriculum without any restriction being imposed on them.

This change becomes much more explicit with the Suleymaniye *Medreses*. In one of the clauses of the deed of trust of this *medrese*, it is clearly stipulated that the task of the *mudarris* should be "to teach the student religious sciences and to enlighten him with true knowledge." The deed also clearly stated for the first time that the professor should teach "the seekers of knowledge in classrooms, on school days, from the textbooks in use during that time. Students were to be taught both transmitted and rational arts through the questioning of one another and through the discussion of the topics."[34] In this sentence, the use of the terms *'tadris'* and *'mudhakara'* unequivocally indicates that the *medrese* formal education during the reign of Suleyman the Magnificent was widened to officially integrate the so-called rational sciences as a subject to be studied in the classroom. The *'mudhakara'* of all the rational (*maqul*) and transmitted (*manqul*) sciences was thus legalized, and the professor was formally asked to impart them to his students.

History of mathematical Literature during the Ottoman Period and *History of astronomy Literature during the Ottoman Period* provide the records of numerous copies of astronomical and mathematical works produced in the *medreses* during the Ottoman period. They make it clear that from the 16th century onwards there is an increase in their number until the 19th century.[35] Without doubt Bursali Kadızâde's (d.1432) two works on astronomy and mathematics, *Sharh al-Mulakhkhas fi'l-Hay'a* and *Tuhfat al-Ra'is fî Sharh Ashkāl al-Ta'sîs,* were two basic textbooks for students who wished to study these subjects in the *medreses*. There are more than three hundred extant copies of the former and approximately two hundred copies of the latter. Among these copies there are a considerable number of reproductions which were copied in the Anatolian and Istanbul *medreses*.[36]

4. CHANGE AS REFLECTED IN THE TERMINOLOGY AND EPISTEMOLOGY

The changes from the time of the pre-Ottoman Seljuk *medreses* until that of the Suleymaniye also display a shift in the terminology describing the student. The terms referring to the students, *'faqih'* or *'mutafaqqih'* which, as Makdisi stresses, were part of *medrese* terminology, were also current in the descriptive deeds of the Anatolian Seljuk *medreses*. The beginners were called *'mubtadi'*, the intermediates *'mutawassit'*, and finally those who had become skilful enough to reason or reflect upon legal matters, that is those who had mastered the *istidlāl*, were called '*mustadill*'. As we look at the early Ottoman *medreses*, we see that the same terminology is used. However, as the scope of education widened, the nomenclature changed also, and the subsequent Ottoman texts generally refer to the students as '*talebe-i ulum*' or '*tullab-i ulum*' (which means 'seeker(s) of knowledge, or sciences'). Those who were at the beginning level were called '*suhte*' and those at the advanced level were called '*danismend*'. The terminology used in the codes of law of the later *medreses* also corroborates the shift in the aims of collegiate education. Thus, such education was no longer exclusively limited to *fiqh* and had acquired a much broader scope.

The changes recorded concerning the formal objective pursued in collegiate education can be linked to the development of '*ilm*'. In order to understand these changes within a historical perspective, we have to survey the development of the concept of *ilm* preceding the foundation of the Ottoman state until the reign of Kanuni (i.e. Suleyman the Magnificent), which sets the time limits for our paper. We selected three scholars and their three epistemological systems to outline the progression in the understanding of knowledge and its repercussions on *medreses* and collegiate education: al-Ghazali, a major figure from the eleventh-twelfth century, who left an indelible mark on the history of Islamic thought; Ibn Khaldun, the renowned fourteenth century historian, who is most famous for his *Muqaddima;* and finally Tashkopruzade, one of the finest Ottoman scholars of the sixteenth century.

Al-Ghazali (1058-1111), in *Ihya' Ulum al-Din*, sees knowledge as being divided into two main groups: the rational sciences (*al-'ulum al-'aqliyya*) and the religious ones (*al-'ulum al-shar'iyya wa'l-diniyya*). The first group can be apprehended through reason and is often innate and natural to man or necessary (*daruriyya*) and acquirable (*muktasaba*) through effort and learning. The rational sciences are classified as praiseworthy, if useful to the community (medicine and calculus, for instance), and blameworthy, if not (namely magic, learning about talismans, etc.). Among the religious sciences are included the following: *'usul al-fiqh* (methodology of jurisprudence), *furu' al-fiqh* (branches of jurisprudence), the propaedeutic sciences or *al-muqaddimat* including language, grammar, writing, and finally the *mutammimat* or complementary ones among which are the Quran with its readings, exegesis, and the *usul al-fiqh*. Al-Ghazali also includes in this category the *kalam,* which he believes to be indispensable for the community, and interestingly enough, philosophy which includes geometry and arithmetic, logic, theology (*ilahiyyat*) and natural sciences (or *tabi'iyyat*). Ghazali insists that there is no opposition between the rational and the religious sciences, for they complement each other. Al-Ghazali "tried to prove [the religious beliefs revealed in the law] rationally with philosophical demonstration, considering this to be the true task of philosophy in opposition to the philosophers who were considered the adversaries of belief".[37] In his classification, rational sciences are compared to food and religious ones to medicine, both being indispensable for man's survival.

Ibn Khaldun (1332-1406), like al-Ghazali, divides human sciences into two categories. The first category is that of the *aqli* or rational sciences and the second that of the *naqli* or transmitted ones. According to Ibn Khaldun, the first one is that of natural or philosophic sciences with reason, reflection or investigation as a faculty of perception. This category is not the prerogative of any religion, community, or race but that of all rational human beings since the beginning of civilization. The so-called rational sciences are the following: logic, natural sciences, mathematics, metaphysics, and numerology. The sciences that were previously referred to as the science of the ancients (*'ulum al-awā'il*) or foreign sciences (*al-'ulum al-dakhila*) were 'integrated' as part of the natural or philosophical sciences into the rational sciences (*al-'ulum al-aqliyya*); this 'integration' marks another stage of development in the history of Islamic thought. Yet, if all nations partake of the rational sciences, the Islamic community is endowed with a knowledge that goes beyond the rational and that is based on transmitted knowledge (*al-'ulum al-naqliyya*), which Ibn Khaldun also calls positive sciences (*al-'ulum al-wad'iyya*). Its source is divine revelation and the Prophet-Legislator, its foundations are the Quran and the Sunna, and reason has no role to play in its foundations. According to his classification, religious sciences and literary arts (*dini* and *adabi*) are part of the second category, those which are *naqli* or transmitted, a fact that reflects a new understanding of sciences. This classification is not merely an expression of a grouping of different branches of sciences, but goes beyond that and becomes a system into which all human knowledge, whatever its source, is included along with the other Islamic sciences as part of an integral epistemological order. The Ottomans fully adopted the twofold classification of the sciences made by Ibn Khaldun. The system spread and the terms '*aqli*' and '*naqli*' or '*aqliyat*' and '*naqliyat*' are frequently encountered in Ottoman literature.

The evolution that is reflected in the terminology that applies to the teaching of the Ottoman *medreses* is also reflected in the epistemological system of Tashkopruzade (1495-1561). Tashkopruzade lived in a time that witnessed the peak of the development of the Ottoman *medreses*. He belonged to a family of ulema, among whom many were *medrese* teachers. His famous encyclopedia '*Mawdu'at al-'Ulum*', where he deals with the classification of sciences, is considered to be one of the finest sixteenth century literary works. For Tashkopruzade, the goal of learning in Islam is '*ma'rifatullah*', the knowledge of God. The reason behind the acquisition of a science is teleological or *ghayawi*. One should study all sciences, and it is only when our capacity limits us, advises Tashkopruzade, that we must concentrate on the study of the principles of religion, that is on theology. The principles of religion strengthen the foundations of faith (*iman*), while *fiqh* helps differentiate between the lawful (*halal*) and the unlawful (*haram*), and Sufism (*tasawwuf*), i.e. "the fruit of faith and the finality of Islam" leads to perfection (*ihsan*). *Tafsir* and *hadith* are contained in the above sciences. It is in this way that the acquisition of knowledge will grant man eternal happiness.

Following Farabi's tradition of thought, Tashkopruzade divides the existents into four categories: in the script (*fi'l-khutut*), the speech (*fi'l-alfaz*), the intellect (*fi'l-adhhan*), and the real (*fi'l-a'yān*). In his division, the realization of every existent leads to the next one, and finally to the ultimate Being. Script leads to speech, then to that which is in the intellect, and ultimately to certainty. Real existence is in the True and Real Being (*al-wujud al-haqiqi al-asil*).

Tashkopruzade recorded about 300 different branches of arts and sciences, mentioning the names of the famous authors who dealt with them and the works devoted to these. He divides sciences into seven main sections based on the above four categories of existence. Tashkopruzade's classification system based on this ontological grouping shows a mutation from epistemology to ontology. He calls each section a 'large tree' (or *dawha*). The first three *dawhas* correspond to the first three ontological categories 1. the script 2. the speech 3. the intellect. All the remaining *dawhas* of sciences belong to the fourth ontological category; thus in this category of existence, that of true existence, sciences are grouped in four sections:[38]

1-Theoretical sciences (theology, natural and mathematical sciences)

2-Practical sciences (ethics, politics, administration and domestic economy)

3-Sciences prescribed by the religious law

4-Esoteric sciences or knowledge of the inward (*'ilm al-batin*) (devotion, customs, actions that lead to perdition and actions that lead to salvation).

According to this classification, learning starts with script and speech and leads to its true finality which is man's own inner perfecting, the cleansing of his heart, his inclination towards good deeds and his spiritual advancement. It is interesting to note that Tashkopruzade's classification of the sciences regards religious and rational sciences as pertaining to the same level of existence and treats them in pedagogical perspective.

To appreciate the transformation of the concept of knowledge in Tashkopruzade, one can refer to Sabra's theory.[39] For Sabra, the transmission of Greek science to the world of Islam consists of two stages: appropriation and naturalization. According to him, the transmission of ancient science to Islam was an act of appropriation rather than mere 'reception', and ancient science entered the world of Islam not as an invading force but as an invited guest. In the naturalization stage, the type of thought and discourse found in the writings of philosophers like Farabi and Avicenna began to be practised in the context of *kalam*, and the philosopher-physician (represented by Razi) was replaced by the jurist-physician (represented by Ibn al-Nafis), the mathematician (*Ta'limi*) by the specialist in the Islamic law of inheritance or *Fara'idi*, and the astronomer-astrologer by the time keeper or *Muwaqqit*.

We can develop Sabra's theory even further and consider a third phase, namely a phase of 'integration' of natural and mathematical sciences. Tashkopruzade's understanding of knowledge can be seen as the culmination of this third stage, where all human knowledge–rational (*aqli*), transmitted (*naqli*), and Sufi (*yaqini*) -melt in one ontological unity of human experience and knowledge that leads to eternal happiness.

In the light of this, we can see Tashkopruzade's classification of sciences as a developed form which assimilates the Ghazalian and the Khaldunian systems, which in themselves were the starting points for a unified view of sciences and knowledge. For Ghazali the uniting factor between the various sciences was their complementarity, for Ibn Khaldun it was an epistemological one, and for Tashkopruzade it became ontological, a merging of human experiences. The rational, legal and mystical sciences are therefore part of one and the same category, *wujud al-a'yān* or true existence. Tashkopruzade has gone one step further, bringing the Ghazalian and

Khaldunian theory to completion, that is, to a stage of total integration. Thus, with this evolution of the concept of *ilm*, the Ottoman *medreses*, starting from the era of Mehmet II, included the rational sciences in their formal education, which was to take its definitive shape during the time of Suleyman the Magnificent. We can therefore regard the evolution of this concept as a catalyst for the change in the understanding of formal collegiate education.

The growing assimilation of the rational sciences to the official teaching system is also observed in Tashkopruzade's second important work, *Shaqaik-i Numaniya*, on the biographies of Ottoman scholars during the reign of the first ten Ottoman monarchs (from Osman I to Suleyman I). According to the statistics based on this important work, the distribution of the works authored by scholars in various disciplines during the reign of the first ten *padishahs* were as follows: rational sciences: 25.7%; history, literature and ethics: 25.7%; exegesis: 22.8%; jurisprudence: 14.2%; Sufism: 8.5%; scholastic theology and the tenets of faith *aqa'id*: 2.8%.[40] This study shows that the milieu of the *medreses* was clearly being transformed so that rational sciences occupied the first place, while jurisprudence came after exegesis.

5. CONCLUSION

In his definition of the rational sciences Ibn Khaldun tells us that the intellectual sciences are natural to man, in as much as he is a thinking being. They are not restricted to any religious group. They have existed and been known to the human species since civilisation has its beginning in the world. These sciences were more extensively cultivated by the two great pre-Islamic nations, the Persians and the Greeks. Muslim scientists assiduously studied the Greeks sciences. They became skilled in the various branches. The progress they made in the study of those sciences could not have been better. They surpassed their predecessors in the intellectual sciences.[41] Rational sciences or intellectual sciences were an integral part of the culture in the Islamic world and gradually became a component of the teaching activities of its official institutions. The *medrese* was an institution 'indigenous' to Islam, which in time integrated the sciences that were not native in the culture. The *medrese* has been seen by Makdisi as a third stage in the formation of the teaching institution in Islam, the first two being the masjid and the masjid-inn. *Medreses,* which were originally meant as institutions for the teaching of jurisprudence (*fiqh*), did not initially 'include' the teaching of sciences other than jurisprudence as such and the sciences related to jurisprudence. One may indeed argue that there were circumstances in the cultural history of Islam when a bias against philosophy is detected because of its use against the main Sunni trend, as we mentioned earlier. This 'sensitivity' should be placed in its own context, keeping in mind that *falsafa* was sometimes used by marginal groups as a doctrinal or political weapon against the established main schools. Furthermore, as Sabra says, *falsafa* could be perceived as a 'competitor' to *kalam*. Greek philosophy or *falsafa* was nevertheless appropriated and integrated into the culture of Islam. The sciences of the ancients including *falsafa* were transmitted through personal channels rather than through official institutions. We have argued that the rational sciences became an integral part of the cultural life not only on an individual level, but also on an institutional one. The rational sciences were made an integral part of the teaching program in the *medreses*, that is, the official institutions for teaching in Islam, as shown in the pre-Ottoman Anatolian *medreses* and in the Ottoman *medreses* from the time of Mehmet II onwards. The examples delineate a definite ascending curve in the gradual inclusion of the rational sciences into official teaching from the time of Nizam al-Mulk through the Seljuks to the Ottomans.

IRCICA, Istanbul

6. NOTES

[1] For the general framework of this subject see Ihsanoglu (2002a).
[2] On cultural life in Anatolia before the Ottomans: see Cahen (1988), Turan (1969), Turan (1980), Cetin (1990), Şeker (1997), Yildiz (1994) and Mercil (1994).
[3] Cahen (1988, pp. 136, 150, 223-4), Cetin (1990, pp. 202-3).
[4] Cahen (1988, pp. 150, 160, 165).
[5] Lapidus (1994, p. 305).
[6] Cahen (1988, pp. 347-8).
[7] Sozen (1972, pp. 86-89).
[8] Kuran (1969, p. 21).
[9] Ibid p. 21.
[10] Ibid p. 42.
[11] Hillenbrand (pp. 1144-5).
[12] Makdisi (1981, p. 282).
[13] Gimaret (pp. 783-4).
[14] Lewis (1985, pp. 70, 133-135).
[15] The pre-Islamic sciences that were transferred to the Muslim world and appropriated through translation were first called the sciences of the ancients or the *ulum al-awāil*. As they started being absorbed and integrated into Muslim culture, the appellation changed and they were referred to as the rational sciences (*al-ulum al-aqliyya*) or designated through one of their branches, namely the philosophical sciences (*al-ulum al-falsafiyya or falsafa*).
[16] Ihsanoglu (2002b), pp. 368–390).
[17] For the writer's manuscript see Sesen, Izgi and Akpinar (1986, pp. 486-487).
[18] Ibid p.486.
[19] Bayram and Karabacak (1981, pp. 52-69).
[20] Sayılı (1960, pp. 253-4).
[21] Ibid pp. 253-54.
[22] Ibid p. 257.
[23] *OALT*, 1 (pp. 22-24).
[24] Sayılı (1948, pp. 655-677).
[25] Seljuk medical tradition was continued and developed in a similar fashion by the Ottomans. Theoretical medicine was transmitted through the traditional teacher-student relationship while applied medicine was practised in the hospitals built by the Seljuks in various cities of Anatolia. The Seljuk hospitals which were conceived according to the established pre-Anatolian tradition and were generally known as *bimaristans* carried different names: *Dar al-Shifa, Dar al-Sihha, Shifahane, Bimarhane* or *Maristan*. The first hospital, *Gevher Nesibe Sultan Dar al-Shifa*, was built by the Seljuks in Kayseri in 1206. The other hospitals were located in the cities of Sivas, Divrigi, Cankiri, Konya, Tokat, Erzurum, Erzincan, Mardin and Amasya. The *Dar al-Shifa* served for health services and the dissemination of medical education in Anatolia. An assertion made by Suheyl Unver, which for lack of clear evidence is still a subject of dispute, is that independent medical schools (*medreses*) affiliated to the *Dar al-Shifa* were founded at the time. However historical records corroborate the existence of independent colleges of medicine or medical educational

facilities attached to *medreses* as in the case of the Mustansiriyya *Medrese* in Baghdad founded between 1227-1233. In the Turkish context, the evidence for the existence of an independent medical school (*medrese*) is found in Istanbul in the Süleymaniye medical *medrese*, a complex built by Süleyman the Magnificent in 1556.

[26] "*Wa shurita an la yushtaghal bil-funun al-falsafiyya aslan.*" by Ayverdi (1972, p. 270).

[27] Bilge (1984, p. 229). We have some important clues concerning the later inclusion of the rational sciences in the teaching activities of the Edirne *Dar al-Hadith* in the works on mathematics and astronomy copied in that institution. Also, the lists of the professors appointed in the fifteenth and sixteenth century institutions include many famous mathematicians and astronomers, such as:

i) Sinan Pasha (d. 1486), who was a famous statesman and scholar during the Mehmet II era. His works on astronomy and mathematics are extant. *OALT 1*, pp.45-8; *OMLT 1*, pp. 17-28.
ii) Molla Lutfi (d. 1495), who was Sinan Pasha's student and wrote works on mathematics. *OMLT 1*, pp. 34-40.
iii) Mirim Kose (d. 1550), who was a famous Ottoman mathematician and astronomer, a descendant of Bursali Kadizade Rumi and Ali Kushju. He wrote works on astronomy. *OALT 1*, pp. 129-30.
iv) Perviz Abdullah (d. 1579), who wrote works on medicine and astronomy. *OALT 1*, pp. 189-90.

[28] Sabra (1994, p. 11).

[29] Ibid p. 23.

[30] Ibid p. 23.

[31] *"Medaris-i Semaniyeden her biri icin ki usul-i erkani kavaid-i hikemiye uzere muessese ve furu-i muhassenat bunyani mevazin-i hendesiyyeden muktebestir. Balayi kitab-i celide serd ve tafsil olundugu uzere seccade nishin-i sadr-i ifade olmaga istihkaki zahir, mebadi ve mukaddimat ve akliyyat ve nakliyatta naziri nadir, esbab-i liyakat-i makam-i tedrisi. Cami' ulum-i nafia tahsiline sarfi omr-i aziz eylemish bir Muderris-i bari' tayin oluna."Fatih Mehmet II Vakfiyeleri,* (pp. 262-3).

[32] See note 31.

[33] See note 31.

[34] *"Talaba-i 'ilm... ayam-i tahsilde dershaneye hazir olup tedris-i kutub-i mutadawila-i makbule ve muzakara-i funun-i maqula ve manqula'ye istigalde ihtimami... "* Kurkcuoglu (1962, fcs.83-4/32).

[35] Katip Çelebi (d. 1658) claims that, in his time, the interest in rational sciences decreased and that in time they were excluded from the *medreses*' teaching. However his viewpoint should be re-examined in the light of historical facts and in the broader context of the development of the cultural and intellectual life in the Ottoman capital, especially in the seventeenth century. The ample evidence we have in the rich Ottoman scientific literature surveys published by IRCICA indicates a progressive curve in the inclusion of the rational sciences and does not confirm Katip Celebi's statement. For a critical evaluation of Katip Çelebi's words, see Ihsanoglu (1996, p. 39-84) and Ihsanoglu (2004).

[36] For the copies of the first book and the copy records see *OALT* 1, pp.9-21; for the second book *OMLT* 1, pp. 7-18.

[37] Mahdï (1964, p. 105).

[38] Tashköprüzade (1313 and 1968).

[39] Sabra (1996, pp. 3-27).

[40] Lekesiz (1989, p. 169).

[41] Ibn Khaldun (1980, pp. 111-116).

7. REFERENCES

Ayverdi, Ekrem Hakkı. *Osmanli Mimarisinde Çelebi ve II Sultan Murad Devri.* Istanbul, 1972.

Bayram, Sadi, and Ahmed Karabacak. "Sahib Ata Fahruddin Ali'nin Konya Imaret ve Sivas Gokmedrese Vakfiyeleri." *Vakıflar Dergisi* 13 (1981): 31-69.

Bilge, Mustafa. *Ilk Osmanli Medreseleri.* Istanbul, 1984.

Cahen, Claude. *La Turquie pré-Ottomane.* Istanbul/Paris, 1988.

Cetin, Osman. *Anadolu'da Islamiyetin Yayilisi.* Istanbul, 1990.

Fatih Mehmet II Vakfiyeleri. Ankara, 1938.

Gimaret, D. "Mu'tazila." *EI*^{2}. 783-93.
Hillebrand, R. "Madrasa." *EI*^{2}. 1123-1154.
Ibn Khaldun, *The Muqaddimah an Introduction to History*. Translated by F. Rosenthal. Princeton, 1980.
Ihsanoglu, Ekmeleddin. "Fatih Kulliyesi Medreseleri Ne Degildi! Tarih Yaziciligi Bakimindan Tenkit ve Degerlendirme Denemesi" in *Buyuk Cihad'dan Frenk Fodulluguna*. Istanbul, 1996: 39–84.
Ihsanoglu, Ekmeleddin et al. *Osmanli Astronomi Literaturu Tarihi / History of Astronomy Literature During the Ottoman Period* (refered as *OALT*). 2 Vols Istanbul, 1997.
Ihsanoglu, Ekmeleddin et al. *Osmanli Matematik Literaturu Tarihi / History of Mathematical Literature During the Ottoman Period* (refered as *OMLT*). 2 Vols. Istanbul, 1999.
Ihsanoglu, Ekmeleddin. "Il Ruolo Delle Istituzioni" in *Storia Della Scienza*. Vol. III. Roma, 2002 a: 110–139.
Ihsanoglu, Ekmeleddin. "Ottoman Educational and Scholarly-Scientific Institutions" in *History of the Ottoman State, Society and Civilization*. Vol 2, Istanbul, 2002 b: 357–515.
Ihsanoglu, Ekmeleddin. "The Initial Stage of Historiography of Ottoman Medreses (1916–1965): The Era of Discovery and Construction" in *Science, Technology and Learning in the Ottoman Empire*. Article VI. Hampshire: Ashgate, 2004: 41–85. (Variorum Collected Studies Series)
Kuran, Abdullah. *Anadolu Medreseleri*. Vol 1. Ankara, 1969.
Kurkcuoglu, Kemâl Edip (ed.) *Suleymaniye Vakfiyesi*. Ankara, 1962.
Lapidus, Ira. *A History of Islamic Societies*. Cambridge, 1994.
Lekesiz, Hulusi. *Osmanlı Ilmi Zihniyetinde Degisme*. Unpublished master thesis, Ankara 1989.
Lewis, B. *The Assassins: a Radical Sect in Islam*. London, 1985.
Mahdi, Muhsin. *Ibn Khaldun's Philosophy of History*. London, 1964.
Makdisi, George. *The Rise of Colleges, Institutions of Learning in Islam and the West*. Edinburgh, 1981.
Mercil, Erdogan. "The Anatolian Principalities" in *A Short History of Turkish-Islamic States (Excluding the Ottoman State)*. Ankara, 1994: 185-210.
Sabra, A. "Science and Philosophy in Medieval Islamic Theology, the Evidence of the Fourteenth Century." *Zeitschrift fur Geschichte Der Arabisch-Islamischen Wissenschaften* 9 (1994): 1-42.
Sabra, A. "Appropriation and Subsequent Naturalisation of Greek Science" in *Tradition, Transmission, Transformation, Proceedings of Two Conferences on Pre-Modern Science Held at the University of Oklahoma*. Edited by. F. Jamil Ragep and Sally P. Ragep, Leiden/New York/Koln, 1996: 3-27.
Sayılı, Aydin. *The Observatory in Islam and its Place in the General History of the Observatory*. Ankara, 1960.
Sayılı, Aydın. "Vacidiye Medresesi, Kütahya'da bir Ortaçağ Türk Rasathanesi." *Belleten* 12/47 (1948): 655-677.
Seker, Mehmet. *Fetihlerle Anadolu'nun Türklesmesi ve Islamlasmas*. Ankara, 1997.
Sesen, Ramazan, Cevat Izgi, and Cemil Akpinar. *Catalogue of Manuscripts in the Koprulu Library*. Vol 1. Istanbul, 1986.
Sözen, Metin. *Anadolu Medreseleri. Selşuklu ve Beylikler Devri*. 2. Istanbul, 1972.
Tashköprüzade. *Miftah al-Siyada wa Misbah al-Siyada fī Mawzu' al-Ulum*. 3 Vols. Cairo 1968.
Tashköprüzade. *Mavzuat al-Ulum*. 2 Vols. Translated by Kemaleddin Mehmed Tashkopruluzade. Istanbul 1313.
Turan, Osman. *Turk Cihan Hakimiyeti Mefkuresi Tarihi: Türk Dünya Nizaminin Milli İslami ve İnsani Esaslar* Vol 1-2. Istanbul, 1969.
Turan, Osman. *Selcuklular Tarihi ve Turk Islam Medeniyeti*. Istanbul, 1980.
Yıldız, Hakkı D. "Seljuks of Anatolia." *A Short History of Turkish-Islamic States (Excluding the Ottoman State)*. Ankara, 1994: 120-135.

OSMAN RECEP BAHADIR
AND H. H. GÜNHAN DANIŞMAN

LATE OTTOMAN AND EARLY REPUBLICAN SCIENCE PERIODICALS

Center and Periphery Relationship in Dissemination of Knowledge

1. THE NATURAL SCIENCES UNDER THE OTTOMAN DYNASTY UNTIL THE 19TH CENTURY

The available evidence indicates that from the founding of the Ottoman state in the later part of the 13th century to the middle of the 15th century, the Ottoman scholars who were trained in such medieval subjects as *sheriah* (Islamic law), rhetoric and logic were not very interested in the study of the natural sciences. With the accession of Mehmet II in *1451,* however, a new era of learning began in which philosophical and scientific thought played a predominant role. Particularly at the new higher educational *medreses* (colleges) at Ayasofya and at those the Conqueror established as part of his *Külliye* (the Fatih Complex) constructed on the fourth hill of the city, eminent philosophers and scientists of his day were ardently supported by the young Sultan. A scholar in his own right and fluent in several western and eastern languages, Mehmet II was especially interested in both Aristotelian and Stoic philosophy according to his biographer Kritovoulos. The royal library at Topkapi Palace contained nearly 600 manuscripts in non-Islamic languages, 75 of which dealing with mathematics and sciences are believed to have been compiled by the Conqueror himself. Among the most important of these treatises are Euclid's *Geometry,* Ptolemy's *Geography* and *Almagest,* Apolonyos's *Konika* and Serenos's two treatises on mathematics, as well as a number of treatises on astronomy. In addition to the treatises mentioned above, Hesiod's *Theogony,* Homer's *Iliad,* Diogenes's work on the Lives of Philosophers, as well as a 13th century copy of Plutarch's *Biography of Renowned Men* translated from Greek to Turkish on the Sultan's command and a copy of Francesco Berlinghieri's Italian translation of Ptolemy's *Geography* published in 1480 and dedicated to Mehmet II are part of the palace library collection.[1]

During this golden age of Ottoman science in the second half of the 15th century, Mehmet II was able to attract to his court and then enthusiastically patronize the eminent mathematicians and astronomers of his day. Among them the astronomer Ali Kuşçu, who was born in Samarkand during the first quarter of the 15th century as Aladdin Ali bin Mohamed Kushdji, and studied mathematics and science under the famous astronomer Uluğ Bey and Kadi-zâde at the Samarkand observatory

G. Irzık & G. Güzeldere (eds.), Turkish Studies in the History and Philosophy of Science, 285-308.

established by Uluğ Bey (1393-1449), was appointed by the Conqueror as the *müderris* (professor) of the Ayasofya (Hagia Sofia) *medrese,* which position he retained until his death in 1474. Important treatises on mathematics and geometry were written by Molla Lütfi from Tokat and his teacher Sinan Paşa (1440-86), and a number of treatises on mathematics and the astrolabe were written by Mahmud bin Mehmed, also known as Mirim Çelebi, who was also a student of Sinan Paşa. In addition to mathematics and astronomy, two important books on medicine date from this era, the first one of these called *Kitab-ı Tıb* (Book of Medicine) by Mohamed lbn Hamza Akşemseddin, and the other one written in 1465 and called *Cerrahname-i İlhanî* (Book of Ilkhanid Surgery) by Şerefeddin Ali bin el Hajj Ilyas, the chief surgeon of the Amasya *Darüşşifa* (the Amasya Hospital).

Although Süleyman the Magnificent assigned one of the four *medreses* he built at his *külliye* (the Süleymaniye complex) to the study of mathematics and another to medicine, Ottoman science seems to have stagnated and declined during the 16th century. On the other hand, Ottoman fleets were able to go beyond the Mediterranean to sail in the Indian and Atlantic oceans during the 16th century, and as a result of these new excursions, two Ottoman captains, Seyit Ali Reis and Piri Reis, produced important works on naval geography. One of these works was a copy of the Christopher Columbus map of 1489 presented to Sultan Selim I in Egypt by Piri Reis (1470-1554), and another was a book by him called *Kitab-ı Bahriye* (Book of the Navy), written in 1521 and presented to Süleyman the Magnificent four years later. In the introduction of this book, the author has written that the earth is like a globe and that he has seen a model of such a globe made by a Portuguese priest, and that America was discovered by Christopher Columbus.

During the last quarter of the 16th century, an astronomer named Taki a-Din bin Mehmed bin Ahmed (1520-1585) presented a report to his teacher Sadeddin Efendi observing that it had become necessary to modify Uluğ Bey's system of astronomy, which did not always produce accurate readings. Sadeddin Efendi, who was well respected by Sultan Murat III, took this matter to the court and obtained the permission of the Sultan to build an observatory with all the appropriate instruments, on the hills above the Tophane area of the Galata region in Istanbul.

Takiyüddin's contact with European science was probably his Jewish assistant from Salonica. Unfortunately, this venture had a very short life, because the *ulema* (the orthodox Islamic teachers) reacted strongly to the establishment of such an observatory on the grounds that the privacy of the angels in the sky was being violated, and the place was demolished upon the insistence of the Sheikh ül-Islam.

The recession that plagued Ottoman science in the 16th century continued in the following century as well. The *medreses* of this era produced only a handful of encyclopedic scholars, while courses relating to scientific topics were replaced by non-scientific subjects. It is true that mathematics, astronomy and philosophy remained in the curriculum, but they were no longer considered to have much weight. During this era, publication of medical books compiled from European sources indicates that the Ottoman men of medicine were aware of advances in this field in the West.

The first half of the 17th century is marked by the efforts of one Ottoman intellectual who complained bitterly in his writings of the religious fanaticism of his time and stressed the need for scientific thinking, as well as the importance of restoring to their proper place the science and philosophy courses which had been downgraded in *medrese* education. His name was Katip Çelebi (1608-1656) and he obviously studied western science when important discoveries were taking place both in the macrocosmos and the microcosmos as a result of important developments in lens technology leading to the use of the first telescopes and microscopes. In spite of these efforts at pulling down the barriers which separated Ottoman science from the science of the west, the prejudice against science and philosophy in Ottoman society was still strong in the first quarter of the 18th century, as was made clear from *a fatwa* (a religious jurisdiction) issued by the office of the Sheikh ül-Islam (the chief religious officer or the Mufti of Istanbul) Ebu Ishak Ismail Efendi in 1716 banning the donation of books on science and philosophy to the public libraries from the private collection of the Grand Vizier Damad Ali Paşa, which had been taken over by the state. However, this conservative attitude was to change for a brief period with the rise of liberal policies in the Tulip Period under Sultan Ahmet III during which the number of original and translated books showed a big increase, especially after Ibrahim Müteferrika opened the first Turkish printing press in Istanbul.[2] An important book published by Ibrahim Müteferrika was Katip Çelebi's *Cihannüma* (World Mirror), a general work on geography with many illustrations and maps.

The Tulip Period's importance also lies in the fact that the first initiative for military reform in the Ottoman Empire took place when a French officer called Rochefort submitted to the Grand Vizier Damad Ibrahim Paşa a report on establishing a "corps of foreign military engineers." Although Rochefort's report was not implemented due to the violent end of the Tulip Period by a popular uprising against the Grand Vizier in 1730, the proposal by another French officer during the reign of Mahmut I was fruitful, and an artillery (*humbaracılar*) company was founded under the command of the proposer, Comte de Bonneval. Also of importance was the founding in 1734 of a school of engineering (*hendesehane*) in the Toptaşı district of Üsküdar, which lasted for a short period because of threats of uprising by the regular army troops.

In the second half of the 18th century the reign of Sultan Mustafa III saw a new interest develop in mathematics and astronomy. Finding his astrologers to be totally useless, the Sultan asked help from the French Government and in return received from the French Academy of Science a number of works on astronomy. In addition to these, Mustafa III found in his own palace library several more books on astronomy which were brought to Istanbul by Yirmisekiz Çelebi Mehmet Efendi, an attaché to the first Ottoman embassy in Paris during the time of Ahmet III, approximately fifty years earlier. Again, during the reign of Mustafa III, an Imperial College of Naval Engineering (*Mühendishane-i Bahri-i Hümayun*) was founded by Baron de Tott near the Kasımpaşa district along the Golden Horn.

The last reigning Sultan at the end of the 18th century was Selim III, who was also an enlightened sovereign and was interested in modern military engineering. In 1790 he decided to establish a new school of mathematics and artillery in Hasköy called the Imperial College of Military Engineering (*Mühendishane-i Berri-i Hümayun*), and its curriculum included arithmetic, geometry, geography, trigonometry, algebra, topography, military history, mechanics, natural science, and civil engineering. The rebellion in 1807 by the Janissaries which led to the assassination of Selim III and the abolition of his New Army (*Nizam-i Cedid*) did not suppress the spirit of reform, because the military colleges were not shut down and others were opened, e.g. the Military College of Medicine (*Tıphane-i Amire*) in 1838, following the accession of Mahmud II that marked the end of the Janissary Corps in 1826. This process of modernization of Ottoman higher education would reach its high point with the founding of an Ottoman university of sciences (*Darülfünun*) during the second half of the 19th century.

2. OTTOMAN SCIENTISTS AND OTTOMAN SCIENCE PERIODICALS IN THE SECOND HALF OF THE 19TH CENTURY UP TO THE GREAT WAR

Parallel with the initiation of modern scientific education in the Ottoman Empire, modern scientific research and a conscious pursuit of world scientific developments started through the efforts of the scientists teaching at the military colleges of engineering and medicine. One of the early examples of this was an important pamphlet on logarithmic tables written by Gelenbevi Ismail Efendi (1730-1791), a teacher at the Imperial College of Military Engineering (*Mühendishane-i Berri-i Hümayun*). One of the headmasters of the short lived engineering college (*mühendishane*), Ishak Efendi (d. 1836), published a four-volume book entitled *Mecmua-i Ulum-i Riyaziye* (Collected Works of Mathematical Sciences), which contained both translated material and original research. On the other hand, Hüseyin Rıfkı Tamani (d. 1871), the headmaster of Imperial College of Military Engineering (*Mühendishane-i Berri-i Hümayun*), was one of the first Ottoman scientists to transfer detailed scientific knowledge from Europe in a systematic manner, particularly about mathematics, astronomy and physics. Şanizade Ataullah Efendi (1771-1826) was the first Ottoman physician to write a modern textbook on anatomy. Tevfik Paşa of Vidin (1832-1901), contributing to the new discipline of linear algebra, wrote a book in English entitled *Linear Algebra* in 1882. And Salih Zeki Bey (1864-1921) was the prominent member of the last generation of Ottoman scientists who introduced modern mathematics and physics to the curriculum of the high schools and the universities in a comprehensive manner.

In line with the increasing number of Ottoman scientists being influenced by the latest research in western science, and as a result of the creation of a significant atmosphere of scientific education, the Ottoman community of scientists began to publish the first indigenous science periodicals. *Vakayi-i Tıbbiye* (Medical Events) was the first periodical of medicine in the Ottoman Empire, started in 1849. Twenty-eight issues of the journal were published by the Imperial College of Medicine (*Mekteb-i Tıbbiye-i Şahane*) for the duration of two years and ten months. The First Ottoman periodical of science was *Mecmua-i Fünun* (Journal of Science), published in 1862 by the Ottoman Scientific Society (*Cemiyet-i Ilmiye-i Osmaniye*), which was headed by Münif Paşa. Following the publication of 47 issues, the journal ceased to exist, but was re-published in 1883, to be closed down once again after a single issue. The importance of this journal was that it attempted to cater to a wide range of Ottoman society as the first popular science periodical by printing original and translated articles on natural sciences like physics, chemistry, geology, geography, astronomy and medicine, as well as social sciences such as economics, history, history of art, forestry, transportation and public works, philosophy, ethnology, literature, education, languages, urban planning, insurance, rare books, politics and military science.

Publication of numerous other periodicals of science continued until the end of the 19th century. Among the more important periodicals during this era, *Rehber-i Fünun* (Guide to Sciences) published 11 issues in 1882, *Medrese-i Fünun* (College of Sciences) published 8 issues in 1884, *Hazine-i Fünun* (Treasure of Sciences) published 1 issue in 1885, *Kevkeb-ül Ulum* (Star of Sciences) published 16 issues in 1886, *Numune-i Terakki* (Example of Progress) published 9 issues in 1887 and 1888, and *İrtika* (Rising High) published 23 issues in 1897.

During the politically troubled decade prior to the Second Constitutional Period, there were no periodicals of science in print. However, the situation changed following the Young Turks revolution, and from 1908 to 1918 under the regime of the Union and Progress Party a large number of weekly and monthly science periodicals were published. Among major journals of this period, *Genç Mühendis* (Young Engineer) published 62 issues during 1909 and 1910, *Darüşşafaka* (Imperial Orphanage) published 12 issues in 1911, *Fen Gazetesi* (Science Newspaper) published 13 issues in 1913 and 1914, *Riyaziyat* (Mathematics) published 9 issues in 1917, and *Bilgi Yurdu Işıgı* (The Light of the Land of Knowledge) published 17 issues in 1917 and 1918. The important point about all these journals was that none of them had a long publication life and none was able to continue from the Ottoman Empire period into the Republican era.

3. THE CHANGING ENVIRONMENT OF SCIENCE IN THE FIRST YEARS OF THE REPUBLIC

It is obvious that Mustafa Kemal Paşa and his companions had very clear ideas about the importance of science and the need for scientific education long before their declaration of the Republic in 1923. The following quote from his speech delivered to a group of teachers in Bursa at an evening meeting held at *Şark Tiyatrosu* (Eastern Theatre), in order to celebrate the Grand Victory of a few weeks earlier at the Battle of Dumlupınar on 30 August 1922, gives explicit insights into his thinking on this subject:

> Ladies and gentlemen; do you know what is the secret of victory, beating the enemy that had trampled on the most sacred, most pleasant, most beautiful places of our country with their dirty feet? It is, accepting principles of science and technology as our guide in directing and administering the armies. We will follow the same objective in establishing schools and faculties which are essential to the training of our nation. Yes, science and technology will be our guide in the political and social life of our nation, in the intellectual education of our nation. With the help of schools, with the help of scientific and technological education that the schools will *give*, the Turkish nation, Turkish art, economics, Turkish poetry and literature would develop with all their novelties.[4]

On 22 September 1924, at a speech delivered at Samsun Independence Commerce School (*Istiklal Ticaret Mektebi*), again to a gathering of the teachers, Mustafa Kemal said the following historic words:

> Gentlemen, for everything in the world, for civilization, for life, for success, the most real guide is science, technology. Looking for guidance other than science and technology is delusion, ignorance, going astray. But it is mandatory to understand every scientific development in its stages and follow its progress in time.[5]

In September 1925, in his opening speech of the First Congress of Medicine, the definition of the Republic that the Prime Minister Ismet Inönü gave was similarly interesting:

> The Republic that is based on the mathematical principles of the science of technology, and the propositions of the scientists...[6]

Mustafa Kemal was not a philosopher, nor had he a proper education in science. Yet he understood the concept of science, both in its modern sense and in its integrity. In his efforts towards modernizing his nation, including economic development, to which he attributed major importance, he assigned the biggest role to science and technology. For that reason, he underlined the importance of science and technology on every occasion, and he strove tirelessly to formalize scientific education in the country. Even in the midst of the Independence War, before the Battle of Sakarya was fought, he held the Second Congress of Education in 16-20 July 1921, and he himself joined the congress by coming in straight from the battlefront. Until his death, the leader of the young Republic insisted on enforcing scientific education and popularizing scientific thinking amongst the masses. During the first decade of the Republic, his 'call to science' attitude is especially significant. It can be said that in no other period of her history did Turkey aim at being a society of science, as she did during the period from 1923 to 1928.

A sweeping set of social reforms was implemented during the first years of the Republic. In 1924, *medreses* (religious colleges), *tekkes* (dervish lodges) and *zaviyes* (dervish cells) were closed down. In 1927 religious education was abolished at all the schools. With an amendment ratified in the Parliament on 10 April 1928, the phrase "The religion of the Turkish State is Islam", was removed from the Constitution. There was a steady increase in the number of students registered at the primary, secondary and high schools, as well as the new universities. The number of students attending the Engineering School (*Mühendis Mektebi*) was 83 in 1923, while it had increased to *255* by 1928. On 26 December 1925, European time and calendar systems were instituted. In 1928 the Latin alphabet replaced the Arabic script with the aim of increasing the literacy rate in the country within a short period of time. From 1927 onwards, successful students were given scholarships by the state to continue their education in Europe. During the academic year of 1928-29, the total number of students abroad was 170, among which 36 were female students.[8] Most of the doctoral students in this period were majoring in the positive sciences. Turkish students undertaking their doctoral studies abroad in social sciences increased only after the Second World War.[9] Compared to the previous five-year period, the number of science books published increased three-fold during the first five years of the Republic.[10]

The leaders of the new nation were able to achieve major advances in the fields of transportation and communications within the first years of the Republic. There was a national campaign of railway construction between 1923 and 1929 when the length of the railways increased from 5 km to 7 km per 1000 km^2 and from 3 km to 4 km per

10,000 people.[11] Turkey became one of the first countries to begin radio broadcasting. Two radio transmitters in Ankara and Istanbul began long wave broadcasting in 1927, while the earliest national radio service, that of the British Broadcasting Corporation (BBC) began public broadcasting only one year earlier in 1926.[12] Finally, the first automatic telephone exchange came into service in 1926.[13]

It was within this new atmosphere of science that specific technology and industry periodicals such as the monthly *Demiryolları Mecmuası* (Journal of Railways), the weekly *Telsiz* (Wireless), and *Türk Mühendisleri Ocağının Fen Mecmuası* (Technology Journal of the Chamber of Turkish Engineers) were published between 1925 and 1928. Yet perhaps the most important characteristic of this early period of the Republic was the publication of the first popular science periodicals, *Fen Alemi* (World of Science), which published 24 issues during 1925 and 1926, and *Tabiat Alemi* (World of Nature), which published 14 issues from 1925 to 1927.

The importance of these two journals can be better understood when it is realized that there were no other popular science periodicals in Turkey for the next 40 years, from the closing down of *Tabiat Alemi* (World of Nature) in 1927 until the publication *of Bilim ve Teknik* (Science and Technics) journal in 1967 by TÜBITAK (Turkish Academy of Science and Technology). The sustained existence of popular periodicals of science, without any state subsidies and relying entirely on subscription income and newspaper stand sales, indicates the public's interest in science at a very crucial time in the life of the nation. An article that appeared on 15 March, 1926, in *Fen Alemi* (World of Science) by the journal's owner and editor-in-chief Mehmet Refik Bey, announcing that more than 425 people from different age and career groups had registered for the evening classes of practical electricity and mechanics started at *Darülfünun Fen Fakültesi* (Ottoman University Faculty of Science) illustrates this surprisingly vigorous atmosphere of science in Istanbul at the end of the first quarter of the 20th century.[14]

4. THE CONTENTS OF THE FIRST PERIODICALS OF SCIENCE IN THE REPUBLICAN PERIOD

During the first five years of the Turkish Republic, when the Arabic script was still in use, there were six periodicals of science in publication. The earliest scientific periodical was *Mualimler Mecmuası* (Journal of Teachers), a professional periodical, first published in 1922, during the Armistice. *Darülfünun Fen Fakültesi Mecmuası* (Journal of Ottoman University Faculty of Science), which started its publication life in 1924, was the most academic periodical of this period. From 1925 to 1927, the two popular periodicals of science, *Fen Alemi* and *Tabiat Alemi* were in print. During 1927 and 1928, another academic science periodical, *Mühendis Mektebi Mecmuası* (Journal of the School of Engineering), and finally, *Kimya ve Sanayi Mecmuası* (Journal of Chemistry and Industry), a professional science journal were published.

Mualimler Mecmuası (Journal of Teachers) was the only scientific periodical published in the country during the period of ten months following the declaration of the Republic on 29 October, 1923, until the appearance of *Darülfünun Fen Fakültesi Mecmuası* in September, 1924. The journal targeted the community of high school teachers who were receiving special attention from the Republican government. The first issue of the journal was published on 22 September, 1922, with the following subtitle appearing on its cover, "Journal of education and profession, published monthly for the present time." The last published issue was No. 54 in October, 1924; hence 41 issues of *Mualimler Mecmuası* were published during the Republican period.[15] Although the main content of the journal consisted of articles on the problems of high school teachers, nevertheless, there were 30 specific articles on pure science in these 41 issues of the journal published after October, 1923. Nine of these articles were on history of science, 5 were on geology, 4 were on physics, and 2 each were on chemistry, mathematics, and astronomy, while there were single articles on geography, zoology, medicine, philosophy of science, and policy of science, with one book review. While the articles on history, philosophy and policy of science covered 76 pages, a total of 147 pages were reserved for the 17 articles on natural sciences. Six of the articles were translations from the works of foreign scientists (3 on physics and one each on astronomy, chemistry, and history of science), equal to 20% of both the total number of articles and total pages. Some selected topics of articles published in the journal were as follows: "Development of the Theory of Atoms from Antiquity to the 1900s" (translation of an article by Prof. Artad of Paris Catholic Institute by Ahmed Tevfik Bey); "Developments in Mathematics during the Last Century" (an article by Hüsnü Hamid Bey); "Life and Works of Salih Zeki Bey" (a biographical article by Ahmed Fahri); "Physiological Characteristics of Ants" (an article by Ali Haydar); "Discovery of the South Pole" (an article by Abdülkadir Sadri); and "Looking at an Apple Fall" (an article by Henri Poincaré on the general theory of gravity translated by Harun Elreşit).

The first truly scientific periodical of the Republican period was *Darülfünun Fen Fakültesi Mecmuası* (Journal of O.U. Faculty of Science),[16] the first issue of which was published in September 1924. It was a quarterly journal and the first issue carried the clause "Year 2," because the journal considered itself as the successor of *Darülfünun Fünun Fakültesi Mecmuası,* which was published between March 1916 and August 1917. There is no information available on the editorial board of the journal, which carried the subtitle of "Mathematics, Physics, Chemistry, and Applied Sciences". Until the issue dated June-July-August 1928 (Year 5, Number 4), the journal was published in Arabic script. From the issue dated September-November-December 1929 (Year 6, Number 1) onwards, the journal appeared in Latin script and was renamed as *Istanbul Darülfünun Fen Fakültesi Mecmuası*. The last issue of the journal was published in December 1933, just after the "University Reform" Law formally ended *Darülfünun* and established Istanbul University. Between September 1924 and August 1928, 47 articles were published in the first 14 issues of *Darülfünun Fen Fakültesi Mecmuası*. Of these articles 17 were on physics, 8 each were on mathematics and chemistry, 4 each were on geology and zoology, 2 each were on history of science and philosophy of science, and one each were on astronomy and botany. As for the number of pages, physics covered 500, geology 164, chemistry 123, mathematics 118, philosophy of science 43, zoology 39, history of science 27, astronomy 23, and botany 11 pages. Articles on physics comprised 51% of all articles on natural sciences. A sample of article titles in the journal are as follows: "On the Characteristics of

Gluceron Acid" (an article by Ömer Şevket); "The Scientific Values of Einstein's Theories" (two articles in series by Hüsnü Hamid Bey); "Solutions of Two Problems by Planuade" (an article by Mehmed Nadir); "Ninhydrine Reaction and Pregnancy Tests" (an article by Ömer Şevket and Dr. Burhan Şevket); "Hypotheses in Physics" (an article by Henri Poincare translated by Hüsnü Hamid); "On Meteorites" (an article by Ahmed Müştak Kargılı); "Leibniz and Newton" (an article on the lives of these two scientists and their work on calculus and differential equations by Mehmed Nadir); "La Grotte de Yarım Bourgas" (an article in French by Raymond Hovass); "On the Theory of Evolution" (an article in Turkish by Raymond Hovass); and "Obtention des Lois de la Réflexion et de la Réfraxion par une Méthode Particulière" (an article in French, and Turkish, by Müderris Tevfik proposing a new method of derivation of the laws of reflection and refraction of light).

Mühendis Mektebi Mecmuası (Journal of the School of Engineering)[17] was a monthly science periodical published by *Mühendis Mektebi Heyet-i Talimiyesi* (Committee of Instructors of the School of Engineering). Its first issue was published in June 1927, and the journal continued in print till the last issue numbered 74-76 and dated December-January 1934. It was published in Arabic script until the 18th issue dated November 1928, and then the remaining issues were published in Latin script. The first issue of the journal carried an inaugural article entitled *"Ifade-i Meram"* (Expression of Intention), and three aims of the journal were listed here: a) to reflect the developments of theoretical analyses and practical applications of technological problems published in foreign periodicals, b) to inform the readers on public works and construction projects in different parts of the country, and c) to help the experts in technology who want to carry out research in their fields of interest, as libraries in the country were insufficient. During the period between June 1927 to November 1928, when the journal was published in Arabic script, 101 articles appeared in *Mühendis Mektebi Mecmuası,* covering a total of 528 pages. Of these totals 30 articles of 197 pages were on construction of roads, bridges and manholes, making 30% of the total number of articles and 37% of the total number of pages. The next popular subject area was general construction theory with 22 articles covering 126 pages, making 22% of the total number of articles and 24% of total pages. Book reviews with 11 articles constituted the next largest group, and these were entitled *"Asar-i Münteşire"* (published works), in which 52 different books were reviewed. Forty-one of these books were on civil engineering, while 3 were on architecture, 2 each were on mechanical engineering, electricity and electrical motors, and mathematics, with one each on hydrodynamics and mining engineering. Thirty-eight books were in French, 8 in German, 3 in Italian, 2 in English and one in Spanish, with almost all of them being recent publications. The remaining articles consisted of 5 articles on housing projects, public squares and urban planning, while 8 articles with a total of 75 pages were direct translations from European languages. Some article titles from this journal are as follows: "On the Number and Dimension of the Caissons and the Settlements in the Brick-and-Stone Piers of the Two Ends of Karaköy Bridge in Istanbul" (an article by Mühendis Fikri); "Cartography through the Use of Photographs" (an article by Subhi Kemal); "Calculations for Reinforced Concrete Water Towers" (an article by Muallim Ihsan); "German Concrete Code dated September

1925" (translation by Muallim Ihsan); "Pouring Concrete in Cold Weather" (an article by Mühendis Fikri); "The Athletic Stadium of the City of Lyon" (an article by Mühendis Ihsan); "On Acoustics in Architecture" (an article by Salih Murad); "On Pneumatic Drills for Rock Excavation" (an article by Dr. Ing. Hohage); "Measurement of Pressures Less than an Ohm" (an article by Mehmed Emin); and, "On Stevenson Dikes" (an article by Mühendis Nebil).

Kimya ve Sanayi Mecmuası (Journal of Chemistry and Industry) was the first journal of chemistry in the country with its first issue appearing in September 1927. It was the official publication of the Association of Turkish Chemists (*Türk Kimyagerler Cemiyeti*), and was published quarterly. Only two more issues of the journal appeared after September 1927, in January and April 1928, before it ceased publication. In an article entitled *"Ifade-i Meram"* (Expression of Intention) in the first issue of the journal, it was indicated that the journal needed at least 200 subscribers in order to remain in print, and the difficulty in acquiring these subscribers could have been the reason for closing down the journal after the third issue. A total of 25 articles were published in the three issues of the journal, each issue being 32 pages. The major area of the articles was the investigation of the practical problems of the chemical industry, covering 37 pages out of a total of 96 pages. The only theoretical article was on the concentration of hydrogen ions, covering 31 pages in two separate issues. There were three articles introducing famous chemists (one of them Marcellin Berthelot) totaling 21 pages out of 96. Finally, in each issue there was a section entitled "new books," in which important books on chemistry published in foreign countries were reviewed. Some sample titles from the journal are: "A Stabilization Experiment for Nitrocellulose" (an article by chemist A. Kemal); "On Measurement of Small Amounts of Water in Oils" (an article by M.W. Boller-H.V. translated by Hikmet Vefik); "On Catalysis" (an article by Prof. M. Faillebin); "On the Life of Svante Arrhenius (1859-1927)" (an article by Fazlı Faik); and "Rapid Separation of Lead and Silver" (a translation of an article in a foreign journal).

Fen Âlemi (World of Science) was one of the two popular science periodicals, and it appeared for 24 issues during the period between January 1925 and December 1926.[18] The monthly journal consisted of 24 pages, and was illustrated. In the first issue, under the title, *"Ifade-i Meram"* (Expression of Intention), it stated:

> This periodical is published to introduce the scientific and technological developments of the century, and to help our readers in overcoming the difficulties they encounter in science. The articles it would publish, keeping up with the new technologies, will be written in a clear language and will be illustrated as much as possible, to allow the general public to benefit from them. We hope that we will be useful to the friends of science and to the artisans.

Thus the periodical's aim was to introduce the scientific and technological developments in the world and in Turkey to the general public, and to create public interest in the basic concepts and problems of science and technology. The articles were written briefly and in a language easily understandable by non-experts. On average 8 to 10 articles were published in each issue, and on the first page of every issue, under the title of *"Küçük Fen Haberleri"* (Short Science News Items), 4 or 5 notices on recent developments in science, technology and scientific education from around the world were included. Most of the articles in the journal were on applied sciences. Only 8 articles in 24 issues were on the theory of natural sciences, and 5 of these were commentaries on the theories of Einstein, and the remaining articles were on the new schools of physics, theories of light, and explanations of chemical reactions. The most popular subject area was electricity, and more specifically the radio-telegraph, telegraph and telephone. There were a total of 16 articles on these subjects with a total of 107 pages, amounting to one fifth of the total number of articles. The next most popular subject area was chemistry, with 12 articles covering 99 pages. The other subjects, in order of decreasing importance, were mechanics and civil engineering, transportation technology, history of science, medicine, and book reviews. A selection of titles from this journal are as follows: "The Scientific Bases of Einstein's Theories" (an article by Doctor Kerim); "A New Measure Against Fires in Airplanes" (Short News Items); "White Coal" (an article by Mehmed Refik); "Application of Electricity in Homes" (an article by Salih Murad); "On Space and Time Notions before Einstein" (an article by Muallim Kerim); "Galileo and His Times" (an article by Salih Murad); "The Basis of Wireless and Telephone" (an article by Mühendis Abdüllatif); "On the Life of Students at American Universities" (Short News Items); and "Analysis of Sugar Content in Urine" (Short News Items).

Tabiat Âlemi (World of Nature) was the second popular science periodical, of which 14 issues were published between December 1925 and February 1927.[19] The journal was subtitled "Illustrated Monthly Turkish Periodical on the Progress of Science and Arts." Each issue was 36 pages and contained illustrations. The largest number of articles, a total of 29 articles covering 90 pages, was on electricity, the radio-telegraph, telegraph and telephone. The next popular title was "Short Science News Items," covering a total of 78 pages, and more than 100 such entries were published. The third popular subject area consisted of history and philosophy of science and medicine, with 24 articles covering 70 pages. Two articles on natural sciences totaling 63 pages, 14 articles on health and medicine totaling 36 pages, and 12 articles on photography totaling 24 pages were the other major areas covered by the journal. Some typical article titles in this journal are as follows: "Darwin's Theory of Evolution" (no author); "An Automatic Telephone: Holmtan-Ericsson System" (an article by Mehmed Emin); "Wireless at the North Pole" (Short News Items); "Light Waves, or Particles?" (translation of an article on the experiments made at the University of Chicago on whether light has the characteristics of waves or of particles); "Locomotive Engine-Driver's Guide" (a book review); "Airplanes Without Pilots" (an article by Salih Murad); and, "On Einstein's Theories" (an article by Salih Murad).

5. THE SCIENTISTS WHO CONTRIBUTED TO THE FIRST SCIENCE PERIODICALS OF THE REPUBLIC

The managing-director of *Mualimler Mecmuası* (Journal of Teachers) was Hüseyin Besim Bey, who was a member *of Istanbul Mualimler Birliği* (Association of Istanbul Teachers), which was the publisher of the journal. The two main contributors to the journal were Ahmet Tevfik Bey, a teacher of natural sciences at *Istanbul Erkek Lisesi* (The Istanbul High School for Boys), and Hüsnü Hamid Bey, the director of *Darülfünun Fen Fakültesi* (O.U. Faculty of Science).

Hüsnü Hamid Bey was a professor of mathematics, who contributed frequently to *Darülfünun Fen Fakültesı Mecmuası,* 6 articles with a total of 206 pages, in addition to three articles he published in *Mualimler Mecmuası.* Born in the town of Elmalı in Antalya Province in 1890, Hüsnü Hamid Bey registered at the primary school in his home town, and as a result of the exceptional performance he showed here, he was sent to attend the high school in Konya with a full scholarship from the government.[20] In 1906 he sat for the entrance examinations for *Mühendis Mektebi* (School of Engineering) in Istanbul, and was accepted. During his third year at this school, he was successful at an examination initiated by the Constitutional Ottoman Government, and was sent to the University of Lausanne with a full scholarship to study mathematics. After receiving his bachelor's degree in mathematics in 1912, he returned to Istanbul and was sent initially to the Salonica High School as a mathematics teacher. Following the Greek invasion of Salonica during the Balkan War, he had to return to Istanbul once again, and was reappointed as a teacher of mathematics in Beirut. Here he met his future wife Enise Hanım, and they were married in late 1913. In 1914 he was appointed to Van High School, but the beginning of the First World War and the Russian invasion of the city of Van prevented him from assuming his duties there. Returning to Istanbul in 1915, he was appointed as assistant professor at *Darülfünun Fen Fakültesi* (O.U. Faculty of Science), and started to work for Salih Zeki Bey. Upon Salih Zeki Bey's death in 1921, he became a full professor and was elected as the director of the Faculty. As director, he was instrumental in the creation of the Institute of Electro-Mechanics during 1926 and 1927, where for the first time education in electrical and mechanical engineering was offered. This institute was transferred to *Mühendis Mektebi* (School of Engineering) during the "university reform" of 1933. Hüsnü Hamid Bey was reelected as the director of the Faculty of Science in 1928, and contributed to the "university reform" studies together with Prof. Malche, the head of the reform committee. He was a close friend of professors Fuat Köprülü and Neşet Ömer Irdelp, and was expected to be made the dean of the Faculty of Science after the reform. However, he was unexpectedly removed from the university in 1933. After he left the university, Hüsnü Hamid Bey worked as a teacher of mathematics at the Haydarpaşa High School for a period of three years, and also lectured on perspective at the Department of Architecture of the Academy of Fine Arts. Following the death of the Academy's teacher of mathematics, he was appointed to this post, and until his retirement in 1955, he lectured at the Academy on mathematics, advanced mathematics, and geometric design. He died in Istanbul in 1975. Hüsnü Hamid Bey was the author of several books on mathematics, and he undertook research on mathematics and history of mathematics. He presented a paper at the annual congress of L'Association Française pour L'Advancement des Sciences in 1928 entitled "Sur la Caractéristique et la Sous-Caractéristique du Paraploide des Normales aux Surfaces Réglés," and again in the same year, at a science conference held in Italy, he delivered a paper entitled "On the History of mathematics in Turkey." Hüsnü Hamid Bey was also the first person to introduce insurance mathematics in the country, and he lectured on the subject at the Commercial Vocational College in the 1940s.

Mehmet Tevfik Bey was the next most frequent contributor to *Darülfünun Fen Fakültesi Mecmuası,* but there is no biographical information available on him, except that he was a professor of physics at *Darülfünun Fen Fakültesi* (O.U. Faculty of Science). He contributed 8 articles to the journal with a total of 184 pages.

Three other scientists who were important contributors to this journal were Ahmet Müştak (Kargılı) Bey, a geologist, with 33 articles and a total of 136 pages, then Ömer Şevket (Öncel) Bey, a professor of chemistry, with 5 articles and 98 pages, and finally Mehmet Nadir Bey, a professor of mathematics, with 4 articles and 52 pages.

Ahmet Müştak (Kargılı) Bey was born in 1865. He studied at *Askeri Tıbbiye* (Army Medical College), and graduated as the school's valedictorian. Afterwards, he began working as an assistant to Mazhar Paşa, who was lecturing on anatomy at the same college. He was made a prisoner of war during the Balkan War. Following his release after the war, he began to lecture on *tabakat-ül arz* (geology) at *Darülfünun* (Ottoman University) until the beginning of the Second Constitutional Period. He published a book entitled *Darülfünun Ilm-i Arz Dersleri* (Geology Lectures at Ottoman University) in 1925. He died in Istanbul in 1938.[21]

Ömer Şevket (Öncel) Bey was born in Salonica in 1880. He graduated from *Mekteb-i Tibbiye-i Mülkiye Eczacilik Sınıfı* (Imperial College of Medicine, Department of Pharmaceutics) as the school's valedictorian in 1901, and worked as a pharmacist for a period of time. In 1909 he was sent to Germany to be educated as a professor of analytical chemistry, which was a new subject to be taught at *Eczacı Mekteb-i Âlisi* (The Higher College of Pharmacists). In Berlin he studied with Prof. Dr. Carl Neuberg, and during 1912 and 1913 he published 11 articles, three of them co-authored with Neuberg, in *Biochemistiche Zeitschrift*. Returning to Istanbul in 1913, he began to lecture on analytical chemistry at *Eczacı Mektebi Âlisi* (The Higher College of Pharmacists), and on organic chemistry at *Darülfünun Fen Fakültesi* (O.U. Faculty of Science). With the "university reform" of 1933, he was removed from the university. Ömer Şevket (Öncel) Bey had written several books on analytical and organic chemistry, the most important one being *Kimya-yi Uzvi Tatbikati* (Application of Organic Chemistry), published in 1917. He became the first scientist to use Latin script for the formulae of organic components in this book. He died in Istanbul in 1950.[22]

Mehmet Nadir Bey was born at the island of Chios in 1856. He attended *Bursa Askeri Idadisi* (Bursa Military Junior High School) and *Kuleli Askeri Lisesi* (Kuleli Military High School), and then went on to *Harbiye* (The War Academy). He was an honor student at all levels. He graduated from *Deniz Harb Okulu* (Naval War Academy) as *erkân-i harbiye-i bahriye mülâzimi* (naval staff first lieutenant) and was appointed to a clerkship at *Divanhane Bahriye Meclisi Başkanlığı* (Presidency of the Naval Assembly of the Consulate). Then he became an assistant lecturer of mathematics at *Deniz Harb Okulu* (Naval War Academy) at Heybeliada, as well as a teacher of mathematics at *Darüşşafaka Lisesi* (The High School of the Imperial Orphanage), where Salih Zeki was one of his students. Mehmet Nadir quit his job at *Darüşşafaka* (Imperial Orphanage) without permission and went to London with the purpose of attending advanced classes to improve his mathematics. Upon his return to Istanbul, he was imprisoned for one year and then expelled from the military. From then on, he continued to teach mathematics at private schools in order to earn his living. Between 1884 and 1896, he operated his own private school, *Numune-i Terakki* (Model of Progress). For a period of five years until 1902, he worked as the director of *Aşiret Mektebi* (Tribal School) and then was appointed as the Director of Education in Aleppo until 1908. Having fallen out with the regime of the Union and Progress Party, he was exiled to Tripoli in 1908, and could not return to Istanbul until the Italian invasion of 1911. With the help of Mustafa Salim Bey, a professor of mathematics at *Darülfünun* (Ottoman University), he began teaching advanced algebra at the newly established *Inas Darülfünunu* (Ottoman University for Girls) in 1915. In 1919 the chair of numerical theory was established at *Darülfünun* by its director Salih Zeki Bey, and Mehmet Nadir Bey was appointed to this chair. He held this post until his death in December 1927.[23] Mehmet Nadir Bey was the author of a textbook on numerical theory, which incorporated his original research in this field. Between 1900 and 1914, in a Parisian journal called *L'Intermédiane des Mathématiciens*, his problems and solutions on numerical theory were published as a series of 12 articles. He also developed an important new rule of divisibility, which received a special compliment from the famous German mathematician Felix Klein when the two scientists were in communication with each other during the First World War. Finally, Mehmet Nadir Bey was the publisher and editor-in-chief of a weekly science periodical entitled *Numune-i Terakki* (Model of Progress), which ran for 9 issues during 1887 and 1888.

During the period between June 1927 till May 1929, for Volumes 1 and 2 of *Mühendis Mektebi Mecmuası* (Journal of the School of Engineering), the members of *Tahrir Heyeti* (Editorial Board) were as follows:

1. Engineer Burhanettin Bey (editor-in-chief), a teacher at *Mühendis Mektebi* and a professor at *Darülfünun;*
2. Salih Murad, a teacher at *Mühendis Mektebi* and at Robert College;
3. Engineer Suphi Kemal, a teacher at *Mühendis Mektebi*;
4. Engineer Fikri Santur, the Director of *Mühendis Mektebi*;
5. Engineer Ahmet Ihsan, a teacher at *Mühendis Mektebi*;
6. Engineer Yusuf Razi, a teacher at *Mühendis Mektebi;* and
7. Engineer Dr. Kerim, a teacher at *Mühendis Mektebi.*

The editor-in-chief of the journal, Burhanettin (Berkan) Bey, was the professor of mathematical analysis at *Darülfünun* (Ottoman University) and also taught the same subject at *Mühendis Mektebi* (School of Engineering).[24] Born in Istanbul in 1886, he graduated from *Hendese-i Mülkiye* (Civil Engineering School) in 1908, and then went to Paris to attend Ecole Nationale des Ponts et Chaussées, receiving his bachelor's degree in engineering in 1912. After his return, he worked as a hydraulics engineer at the Ministry of Public Works. In 1913 he joined *Mühendis Mektebi* as an assistant teacher of *islâh-i enhar* (improvement of rivers). In November 1915, he became the teacher of mathematical analysis at *Darülfünun Riyaziyat Şubesi* (O.U. Department of Mathematics), and at the same time continued to lecture at *Mühendis Mektebi.* By 1925 he was lecturing on a wide range of subjects including *idrolik* (hydraulics), *islâh-i enhar* (improvement of rivers), *kârgir köprüler* (brick-stone bridges), *tevzi-i meyah* (distribution of rivers), *tefcir* (drainage), and *kuvvay-i meyahiye* (forces of water). When the School of Engineering was transformed into Istanbul Technical University in 1946, Burhanettin (Berkan) Bey became Emeritus Professor under the Chair of Hydraulics and Forces of Water at the Faculty of Civil Engineering, and in this position he was instrumental in establishing the hydraulics and forces of water laboratory at the university. He published two important textbooks, one called *Idrolik* (Hydraulics) and the other *Büyük Bentler* (Large Dams). He was one of the professors dismissed from the Faculty of Science as a result of the "university reform" of 1933. He died in Istanbul in 1953.[25]

Salih Murad (Uzdilek) was born in Istanbul in 1891. After graduating from *Bahriye Mektebi* (Naval School) in 1908 as first lieutenant, he worked for two years in the Naval Command and in 1911 was sent to study electrical engineering at the University of London. Salih Murad was given an award by the Ministry of Navy for the remarkable success he showed at the University of London and for the excellent paper he delivered at the Congress of the Third Century of Logarithms in Edinburgh. Upon his return to Istanbul at the beginning of the First World War, he worked as a teacher of mathematics at *Bahriye Mektebi* and at Robert College. After joining *Mühendis Mektebi* (School of Engineering) as a teacher, he became professor in 1928, and was made Emeritus Professor under the Chair of Physics at the Faculty of Mechanical Engineering of Istanbul Technical University in 1944. He was appointed as the Dean of the Faculty of Mining in 1956 and remained in office until 1958. His major publications were *Tarih-i Riyaziyat* (History of Mathematics), published in 1909, *Malumat-i Fenniye* (Scientific Knowledge), published in 1915, *Değişen Dünyanin Sırrı* (Secret of the Changing World), published in 1947, and *Nükleer Ilim ve Teknoloji Terimleri Sözlüğü* (Dictionary of Nuclear Science and Technology), published in 1963. Salih Murad Uzdilek died in Istanbul in December 1967.[26]

Engineer Suphi Kamil (Tanığ) was born in Cerrahpaşa, Istanbul, in 1888. He graduated from *Davutpaşa Arakiyeci Ahmet Ağa Mektebi* (Davutpaşa Arakiyeci Ahmet Ağa Primary School) in 1894 as the school's valedictorian, from *Davutpaşa Merkez Rüştiyesi* (Davutpaşa Central Junior High School) in 1902, and finally from *Mercan Mülki Idadisi* (Mercan Civilian High School) in 1907. He then attended *Hendese-i Mülkiye Mektebi* (Civil Engineering School) and graduated as an engineer in 1914. During 1915 to 1918 he worked for Hijaz Railways, and was a site engineer for Samsun railway construction in 1918 and 1919. Between 1919 and 1924, he was a member of *Istanbul Tapu Müdürlüğü Fen Heyeti* (Istanbul Province Title-Deeds Directorate's Technical Commission). In 1923 he joined *Mühendis Mektebi* (School of Engineering) as a teacher of geometry. In the following years he lectured on general principles of construction, topography, and road construction. He worked as the director of the School of Engineering twice, between November 1929 and September 1932, and between June 1935 to September 1939.[27]

Engineer Fikri (Santur) was born in Salonica in 1876. He graduated from *Hendese-i Mülkiye Mektebi* (Civil Engineering School) in 1899 and started to lecture there immediately. He lectured on geometry, hydraulics and steam engines, and worked at the Ministry of Public Works at the same time. He was the director of the School of Engineering twice, between May 1927 and October 1929, and between September 1932 and May 1935. He played an important role in the establishment of Istanbul

Technical University, and he is the author of numerous articles and books on geometry, mechanics, iron and wooden bridges, hydraulics and reinforced concrete. Fikri (Santur) Bey died in Istanbul in June 1961.[28]

Engineer Ahmet Ihsan (Inan) graduated from *Mühendis Mektebi* (School of Engineering) in 1920, and he started working at the same school in 1924. Initially, he lectured on *islâh-i enhar* (improvement of rivers), but later was appointed to lecture on *betonarme* (reinforced concrete). When the University Law was passed, he became Emeritus Professor under the Chair of Reinforced Concrete at Istanbul Technical University. Between 1946 and 1948, he served as the dean of the Faculty of Civil Engineering. His major books are entitled *Sulama Kurutma* (Irrigation and Drainage), *Akarsu Hidroliği* (River Hydraulics), and *Betonarme* (Reinforced Concrete).[29]

Engineer Yusuf Razi Bey was born in Istanbul in 1870. He attended *Hendese-i Mülkiye Mektebi* (Civil Engineering School) after graduating from *Istanbul Sultanisi* (Istanbul Imperial Lycee). He then went to Paris and graduated from Ecole Nationale des Ponts and Chaussées. After working as an engineer in various places around the country, he started to lecture at *Hendese-i Mülkiye Mektebi* (Civil Engineering School) in 1892 on steam machines. At the same time he was the assistant director of *Demiryolları Idaresi* (Director of Railways). In addition to steam machines, he lectured on *resm-i hatti* (technical drawing), on ports and French. He worked as the Prefect of the City of Istanbul for a brief period at the end of the 1920s. He became professor in 1928. The first covered tramway stops of the city were built by Yusuf Razi Bey.[30]

Engineer Dr. Kerim (Erim) was born in Istanbul in 1894. He was the grandchild of the famous mathematician Abdurrahman Paşa of Kazan. After completing *Mühendis Mektebi* (School of Engineering), he took his doctorate degree in mathematics from the University of Erlangen in Germany. He was the first mathematician with a Ph.D. Degree in Turkey. After his return, he became the teacher of *hesab-i nazarî* (theoretical calculus) and *tahlil-i hendese* (analytical geometry) at *Mühendis Mektebı* (School of Engineering). He became professor in1929. He wrote numerous articles on mathematics and philosophy of mathematics in *Mühendis Mektebi Mecmuasi* (Journal of School of Engineering), especially in the issues published in Latin script. After the "university reform" of 1933, he became the dean of the Faculty of Science. In 1934 he headed the commission for mathematical terms and worked on the formation of a new terminology. He established the periodical *Fakülte* (Faculty), publishing original articles in foreign languages. He presented his works in various congresses held in Europe and Pakistan. His published books are entitled *Mihanik* (Mechanics), *Nazari Hesap* (Theoretical Calculus), *Analitik Geometri* (Analytical Geometry), *Analiz* (Analysis), and *Riyazi Mekanik* (Mathematical Mechanics). Dr. Kerim Erim also worked on philosophy of mathematics and physics, and some of his articles on this subject were entitled, *"Matematik ve Gerçek"* (Mathematics and Reality) and *"Matematik ve Mantık"* (Mathematics and Logic). In 1930 he interviewed Einstein, covering various subjects, and published his impressions in *Mühendis Mektebi Mecmuası*. In the conferences he gave in 1930s, he spoke on the philosophical conclusions of Einstein's theory of relativity. Determinism and probability were among the subjects he investigated. Dr. Kerim Erim died in Istanbul in December 1952.[31]

Chemist A. Kemal Bey was the editor-in-chief of the first issue of *Kimya ve Sanayi Mecmuası* (Journal of Chemistry and Industry), while chemist M. Ilhami Cıvaoğlu became the editor-in-chief for the second and third issues of the journal. The members of the editorial board were Prof. Fazlı Faik, Prof. Ligor, chemist Kemal Hikmet Bey and Ihsan Bey.[32] There is no biographical information available on Kemal Hikmet and Ihsan Beys.

Fazli Faik (Yeğül) was born in Salonica in 1882. He attended *Mülkiye Baytar Mekteb-i Âlisi* (College of Veterinary Medicine) in 1899, and graduating in 1903, he started to teach at the same school in 1906. Fazli Faik Bey was sent to Germany to study chemistry in 1909, and worked with famous chemists of the period in Berlin until 1911. Joining *Darülfünun Fen Fakültesi* (O.U. Faculty of Science) in 1916 as the assistant to Fritz Arndt, he began to lecture on chemical analysis. During the Armistice, he taught inorganic chemistry at *Mülkiye Baytar Mekteb-i Âlisi* (Higher College of Veterinary Medicine). When *Yüksek Ziraat Enstitüsü* (Higher Institute of Agriculture) was established, he joined it as assistant professor in 1933. He became professor in 1936, and emeritus professor in 1943, retiring in 1947. He died in Istanbul in 1965. His textbook entitled *Tahlil-i Mikdarî Tatbikati* (Application of Quantitative Analysis) was published in 1916.[33]

Prof. Ligor (Kimyacı) was born in Istanbul in 1875. After he graduated as a pharmacist from *Mekteb-i Tıbbiye-i Mülkiye* (Imperial School of Medicine) in 1897, he started to work at *Şam Tıbbiye Mektebi* (Damascus Medical School) as a teacher of chemistry and *fenn-i ispençiyarî* (science of pharmaceutics). He joined *Darülfünun Fen Fakültesi* (O.U. Faculty of Science) in 1919 as a professor of inorganic and analytical chemistry, and continued to work there until his removal with the "university reform" of 1933. After that date, he worked as a teacher of chemistry at *Izmir Kiz Lisesi* (Izmir High School for Girls) until his retirement in 1942. He changed his family surname "Taranakidis," firstly to "Urut" and then to "Kimyacı" (Chemist) following the Surname Law. Ligor Bey published 12 textbooks on inorganic and analytical chemistry and pharmaceutics. His research was published in French journals. He published 11 articles in French and Turkish. Prof. Ligor (Kimyacı) died in November 1956.[34]

Ilhami Cıvaoğlu was born in Istanbul in 1898. He started studying chemistry at *Darülfünun Fen Fakültesi* (O.U. Faculty of Science) in 1918, and physics in 1919. In 1920 he graduated from the chemistry department, and in 1921 from the physics department, and began to work as the director of a cement factory in Bakırköy. In 1923 he was sent to Paris to major in physical chemistry and radioactivity. Upon the suggestion of the French physicist Jean Perrin, he started to work at the laboratories of Marie Curie. There, he carried out a series of research projects on counting alpha particles of radium. Returning to Istanbul in 1926, Ilhami Cıvaoğlu started working as an assistant at the newly established chair of physical chemistry of Faillebin at *Darülfünun* (Ottoman University). He established the laboratory for radioactivity measurement at the Faculty of Science, and carried out research on the radioactivity levels of underground waters and thermal springs at Yalova, Bursa, Armutlu, Kütahya, and Alemdağ. He became assistant professor in 1928, and then professor and head of the chair after Faillebin in 1930. Soon after he transferred to Istanbul

Technical University and became emeritus professor in 1956. He was appointed as the dean of the Faculty of Mechanics in 1949, and then as the rector of the University in 1957. He supervised the establishment of the Faculty of Chemical Engineering and served as its dean for two terms until 1967. He retired from Istanbul Technical University in 1973, but continued to work as the holder of the chair of chemistry at Galatasaray Engineering School until 1979. In addition to his academic career, Ilhami Cıvaoğlu was known for his contributions to numerous scientific and learned institutions. He was instrumental in the foundation of Societies of Physics and Chemistry in 1919, served as the vice-president of *Türk Dil Kurumu* (Turkish Language Society) between 1941 and 1962, and contributed to the opening of three new universities: Atatürk University in Erzurum, Karadeniz Technical University in Trabzon, and the Aegean University in Izmir. He received *Légion d'Honneur* in 1951, with his research entitled "Relations of Distributions of Ions in Solvent Molecules," and was given honorary membership in the Society of Industrial Chemistry in France in 1969. Some of his important publications were *Kimyanın Temel Prensipleri* (Basic Principles of Chemistry), published in 1967, *Temel Operasyonlar* (Basic Operations), in 1966, *Genel ve Teknik Kimya Dersleri* (Lectures in General and Technical Chemistry), in 1965, and *Tabii Yakacaklar* (Natural Combustibles) and *Yanma Kimyası* (Combustion Chemistry) in 1949.[35]

The founder and editor-in-chief of *Fen Âlemi* (World of Science) was Mehmet Refik (Fenmen) Bey. Born in Preveze in 1882, he was the grandson of the famous Hero of Freedom, Mithat Paşa. Mehmet Refik completed his higher education in Europe, graduating from the Department of Physics at the University of Lausaune. He also received the Degree of High Distinction in electrical engineering from Liège University in Belgium. Returning to Istanbul, he started to work as a teacher of electricity at *Hendese-i Mülkiye Mektebi* (School of Civil Engineering). Later he worked as an electrical engineer at the Ministry of Public Works. When it was decided to transfer the School of Civil Engineering from the military, in order to appoint a civilian director the Ministry of Public Works requested from the *Osmanli Mühendis ve Mimar Cemiyeti* (Society of Ottoman Engineers and Architects) the names of two candidates to be chosen by secret ballot. Mehmet Refik Bey received the majority of votes and was appointed as the director of the new *Mühendis Mektebi* (School of Engineering). During his tenure of office, important steps were taken towards the modernization of the School of Engineering, lectures on new topics were introduced, and professors from Europe were invited to teach at the school. In 1910 Mehmet Refik Bey was sent to Europe to inspect foreign schools, and he visited engineering schools in Germany, Belgium, France and Switzerland. On his return, he undertook the establishment of new laboratories of physics, chemistry, and construction materials. A new drafting studio was added, new topographical equipment was bought, and the students were given the chance to do field training courses. Furthermore, summer training of students in institutions of industry and construction was made obligatory, and technical visits were planned for senior students to visit factories and railroad bridges in France and other European countries. Plans were made to initiate a museum of engineering, while hundreds of books on engineering and mathematics were purchased for the school library, mainly from France. French language courses were started at this period, and soon the students were able to follow lectures by foreign professors without the need for translators. At the same time, Mehmet Refik Bey was very much in favor of the students organizing themselves to defend their rights, and he supported the establishment of *Mühendis Mektebi Talebe Cemiyeti* (School of Engineering Students Association) in 1910. However, not everyone appreciated the renovations carried out by Mehmet Refik Bey. A student protest at the school was used as a pretext, and he was dismissed from the directorship by the Ministry of Public Works in September 1913, without any due investigation. Consequently, Mehmet Refik Bey resigned from the School of Engineering, and joined the department of physics and electrics at *Darülfünun Fünun Medresesi* (O.U. Faculty of Science) in 1919. In 1926 he was appointed to the directorship of *Zonguldak Yüksek Mühendis Mektebi* (Zonguldak Higher College of Engineering) and in 1928 to the general directorship of *Zonguldak Kömür İşletmeleri* (Zonguldak Coal Works). He was elected a deputy to the Turkish Parliament between 1943 and 1946 from Kocaeli. He died in 1957. His most important book, entitled *Einstein Nazariyesi:Mekân, Zaman ve Kütle Mefhumlarının Tebdili* (Theory of Einstein: The Changes in the Concepts of Space, Time and Mass) was published in 1924.[36]

Mehmet Refik Bey, with 11 articles, was the most frequent contributor to *Fen Âlemi* (World of Science). Other scientists who made substantial contributions to the journal were Dr. Kerim (Erim), Salih Murad (Uzdilek), Fikri (Santur), telegraph engineer Abdullatif Bey, engineer Yusuf Ziya Bey, Ali Hikmet Bey, and Ömer Şevket Bey. In addition to the above, the following authors also wrote on various subjects: Dr. Kemal Cenab, professor Tevfik, engineer Halim Celil, A. Hamdi (a teacher at the Naval School), engineer Şuarte (of Siemens Company), Rezvan Ziya, telephone engineer Dominik, Ahmed Ahmedi, engineer Suphi, Halid Ziya, and engineer Donsiyer.

The founder and editor-in-chief of *Tabiat Âlemi* (World of Nature) was Salih Murad (Uzdilek) Bey, who was also a member of the editorial board of *Mühendis Mektebi Mecmuasi* (Journal of School of Engineering).[37] Two thirds of the total number of articles in this journal belonged to Salih Murad Bey (34 articles covering 82 pages), which were essentially on subjects relating to electricity, the radio, telegraph, or telephone. The other authors of the journal were Mehmet Emin Bey (5 articles), medical doctor Hakki Bey of Rusçuk (8 articles), agriculturalist Cevat Şükrü Bey (2 articles), geometry teacher Ali Erten Bey (1 article), and teacher of reinforced concrete at the School of Engineering Ahmet Ihsan Bey (1 article).

6. ASSESSING THE IMPACT OF THE FIRST SCIENCE PERIODICALS OF THE REPUBLICAN PERIOD

Mualimler Mecmuası (Journal of Teachers), strictly speaking, was not fully a science periodical, but a journal for professional teachers. Nevertheless, its contribution during a period when no other science periodical was in print was crucial. Important articles on science were published in this journal for the first ten months of the Republic. and the fact that these articles were aimed at educating and informing the high school teachers at the beginning of an educational reform movement in the country increases the value of the journal's contribution to the society in general.

In evaluating the role played by *Darülfünun Fen Fakültesi Mecmuası* (Journal of O.U. Faculty of Science), it is important to point out that it was essentially a journal of scientific education, rather than a journal of pure science. The articles written with the purpose of giving information on various aspects of science amounted to almost two-thirds of the total number of published articles. These either transmitted information regarding scientific developments in certain fields of European science within an historical framework, as well as along a chronological narrative, or they related news of recent scientific discoveries and developments of new theories in certain subject areas. As for the articles summarizing results of basic research, they were highly original and constituted a major portion of the articles of a purely scientific nature, in spite of the fact that some of them referred to research carried out by the authors more than ten years earlier. For example, most of the articles by Ömer Şevket Bey were ones that he had already published in *Biochemistiche Zeitschrift* while he was in Germany in 1912 and 1913. Another important characteristic of the articles published in this journal was that they referred to specific, rather than general, problems of science. However, the articles by Mehmet Nadir Bey and Hüsnü Hamid Bey, and the article by Ali Yar criticizing Wronski, as well as the article by Mehmet Tevfik Bey on the refraction and reflection of light, were exceptions to this trend. In particular, Mehmet Tevfik Bey's article entitled *"Inikas ve Inkisar Kanunlarının Hususi Bir Prensip ile Istihracı* (Derivation of the Rules of Reflection and Refraction with a Special Principle), where he compared Newton's particle theory of light with Huygens's wave theory of light, was highly original. In conclusion, it can be said that this journal of the Faculty of Science encouraged new scientific research and resulted in the betterment of scientific education in the country.

All seven members of the editorial board of *Mühendis Mektebi Mecmuası* (Journal of School of Engineering) were teachers at this school, and the journal had the professed mission to become the voice of the civil engineering profession which was assuming a leading role in a young country determined to develop swiftly. The editorial board wrote in the first issue of the journal that publications are the essential outcomes of scientific and technological developments, and that such publications had to communicate these developments to the community on a regular basis. The editorial board further observed that as a result of developments in the country due to the government's efforts to improve the construction sector and public works, this journal should present theoretical analyses, as well as information about the technological applications. The journal also had to help those involved in applications to become aware of the developments in their fields, and learn about the changing construction methods of the developed countries. The Ministry of Public Works also publicly supported the publication of the journal, and this information was included in the inaugural article by the editorial board. It is important to add here that the journal carried out a reader survey in order to revise and enrich its contents during the second half of the first year of its publication, through a questionnaire directed at a chosen and limited group of its readers, consisting of the former and current engineers-in-chief and general directors in various branches of the Ministry of Public Works. The results of this sounding were published in the 12th issue of the journal dated May 1928, and they indicated that the respondents asked for more articles giving practical information on construction engines and machinery, and on new developments in electricity. Furthermore, analyses of the new construction projects and techniques in Europe, as well as the solutions to the problems encountered with indications of their applicability to Turkey were requested. The contributors to the questionnaire also requested information on major construction projects in the country, and test results of various construction materials. The editorial board responded positively to these demands from the readers in the later issues of the journal.

Kimya ve Sanayi Mecmuası (Journal of Chemistry and Industry) was published in 1927 and 1928 in a country where the chemical industry was still not very developed. Judging from the contents of the articles that appeared in the journal, it is evident that the Association of Turkish Chemists, as the owner of the journal, attempted to provide practical and useful information rather than translation of articles from foreign journals. It is difficult to assess the full impact of the journal due to its short publication life of 3 issues.

Fen Âlemi (World of Science), as the first popular periodical of the Republican period, was aiming to introduce the scientific and technological developments in the world and in Turkey to the general public, as well as to popularize interest in the basic concepts and problems of science and technology. 1925 and 1926 were years during when the Republican administration attached great importance to science and scientific education. During these years, science was very prestigious in the rest of the world, as well, and Einstein's general theory of relativity was still very influential. At the same time, this was the period when electricity, mechanization, telephone and telegraph, and other devices of modern technology were becoming the essential parts of social life. In such an environment, *Fen Âlemi* obviously filled an important gap by helping to diffuse information on new developments to the general public, facilitating their appreciation, understanding and even liking of science.

Tabiat Âlemi (World of Nature) was different from *Fen Âlemi* in certain aspects. This second popular science journal tended to present simplified subjects related to problems of everyday life, supported with illustrations, rather than address issues of pure science directly. It followed a different policy of publication by targeting the masses who had never been in touch with science. Thus, *Tabiat Âlemi* can be considered more like a "magazine" of science, rather than a journal of science. The restricted number of contributors was consistent with this policy, and the journal depended on numerous articles and extra efforts of its owner Salih Murad Bey. *Fen Âlemi*, on the other hand, was rich both in the number and the academic quality of its contributors. However, both journals were very successful commercially; for example there were only a handful of unsold copies at newsstands from the first four issues of *Tabiat Âlemi*, as was announced by the owner of the journal in an article in the 12th issue evaluating the first year of publication.

Both of these popular journals catered to the positive atmosphere of science that was created in the country during the first years of the Republic, they were modern looking and they followed the world agenda in science. It is not clear, however, to what exactly was the circulation for each of them. It may be reasonable to assume that both journals printed about 500 copies. Considering that the daily circulation of *Cumhuriyet* (The Republic) newspaper in 1924 was around 7000 copies, and that *Kimya ve Sanayi Mecmuası* (Journal of Chemistry and Industry) attempted to reach 200 subscribers in order to be able to stay in print, it can safely be assumed that these two popular science journals had reached a circulation somewhere between 300 and 500, within the total population of under 13 million in the country at the end of 1923.

7. OBSERVATIONS ON A UNIQUE CASE OF KNOWLEDGE TRANSFER FROM CENTER TO PERIPHERY

The Ottoman Empire during the 19th century, especially in the period leading to the First World War, implemented a determined policy of modernization through westernization. Between 1923 and 1928 the Republican leaders of Turkey set a policy of modernization as the goal for the young nation. In both cases science was one of the prime movers behind modernization. However, the degree of success achieved during the first five years of the Republic was much greater than during the last few decades of the Ottoman Empire. Some of the reasons for this success seem to lie in the degree of importance attributed to scientific education and application of the principles of science to technological transformation by the Republican statesmen. The speed with which this transformation was put into practice seems to have run parallel with the ability and willingness to transfer scientific knowledge from Western Europe to Republican Turkey. The main points in this unique case of knowledge transfer can be summarized as follows:

1. Creation of an active atmosphere of science in the country, through the government's determined policy to reform the national educational system by establishing institutions of scientific learning and as a result of tacit encouragement offered by the leaders of the Republican regime to the scientists and to scientific publications, undoubtedly was a prerequisite for this rapid transfer of scientific knowledge and for its effective reception;

2. In spite of the persistent continuation of conservative education at *medreses* (religious colleges), the liberal military, and civilian schools of engineering and medicine established during the 19th century by the Ottoman government helped to lay a strong foundation for the Republican administration's new system of scientific education;

3. The Republic inherited from the Ottoman period a fairly respectable tradition of science, in which some positive contributions were made on an international scale, for example research and publications by Ottoman scientists like surgeon Cemil Topuzlu, radiologist Esat Feyzi, bacteriologists Osman Nuri and Mustafa Adil,[38] while Ottoman educators like Salih Zeki Bey (1864-1921) whose contribution to the education of modern mathematics and science and to the formation of scientific thought, not only by lectures delivered at universities and high schools, but also through publications of books like *Kamus-u Riyaziyat* (Dictionary of Mathematics) and translations of works like Henri Poincaré's *Ilim ve Faraziye* (Science and Hy-

pothesis) or *Ilimin Kiymeti* (Value of Science) which were used as text books for the first year students at the School of Engineering, played a crucial role in the creation of leading scientists of the Republican Turkey, such as Mustafa Inan (1911-1967), who was one of the most brilliant scientists working in applied mechanics at Istanbul Technical University;[39]

4. Furthermore, the owners, as well as the members of the editorial boards, including the contributors to the first periodicals of science in the early Republican period, had quite naturally completed their educational formation, and their views on modern science, during the last decades of the Ottoman Empire, when most of these scientists had received their bachelor or doctorate degrees in France, England, Germany or Belgium, thus facilitating a direct transfer of the latest level of European research, including their own research results, to the colleges of higher education where the first generation of Republican scientists and engineers were being trained; and finally,

5. As a result of an environment in which every obstacle to the independent development of science was effectively eliminated, the science periodicals published in the early Republican period between 1923 and 1928 contributed to the intellectual and material modernization of the country, in addition to their role as transmitters of the ideal of creation of a society of advanced science, which was instrumental in the government's decision to implement the "university reform" of 1933, in spite of certain mistakes in application, which led to the final transformation of Turkish universities as modern institutions of scientific research and learning.

8. NOTES

[1] Information made public in a report written in 1929 by Prof. Adolf Deismann of Berlin University after he was invited by the Republican government to investigate the manuscript collection at the Topkapi Palace Library. For the biography of Mehmet II, see:
Kritovoulos of Imbros, *History of Mehmed the Conqueror by Kritovoulos (1451-1467)*, trans. Charles Riggs (Princeton, N.J., 1954); and
For more information on Ottoman scientists and scholars mentioned below, refer to relevant articles in *The Encyclopedia of Islam*, E.J. Brill and Luzac Co., Leiden and London, as well as *Islam Ansiklopedisi*, Istanbul; and finally, for the Tophane Observatory built by Sultan Murad II, see Rasadhane, *Islam Ansiklopedisi*, Vol. 9, 1964, p. 631; and Sevim Tekeli's articles on Takiyüddin, Belleten, Vol XXV, 1961, pp. 213-38; as well as Islam Tetkikleri Enstitüsü Dergisi, *Istanbul, 1960, pp. 1-30.*

[2] It is useful to note here that the Sephardic Jews introduced the first printing press in the city in 1494, while in 1567 an Armenian, and in 1627 a Greek printing press began publications in those languages, respectively. The first Turkish printing press was put into operation in 1727; see: Ertuğ, H. R., *The Turkish Press*, The Turkish Press Institute, Istanbul, 1964.

[3] Dinç, G., "Arap Harfleri ile Basılmıs Tıbbı Süreli Yayinlar Üzerine bir Inceleme", *Tıp Tarihi Araştırmaları*, No. 5 (Istanbul: 1993), 118-120.

[4] Edited/ *Atatürk'ün Söylev ve Demeçleri*, Vol. II, Ankara: T.T.K. Basımevi, 1952; 43-44.

[5] *Ibid.*, 197

[6] Bilsel, C., *Istanbul Üniversitesi Tarihi*, Istanbul: I.Ü. Yayınları, No.182, Kenan Matbaasi, 1943; 27.

[7] *Mühendis Mektebi Mecmuası*, No. 29 (Istanbul, 1933), Graph No. 1231.

[8] Dölen, E., "Cumhuriyetin Ilk Onbeş Yılında Istanbul Üniversitesi'nde Kız Öğrenciler", N. Yıldırım, (ed.), *Sağlık Alanında Türk Kadını*, Istanbul: I.Ü. Yayini, 1998; 8-47.

[9] Seyhan, M., "Doktora Yapan Ilk Türkler", *Meydan* (October, 1978), 54-56.

[10] Bahadir, O.R., "Cumhuriyetin Ilk Yillarinda Bilim Kitapları", *Virgül*, No. 26 (January, 2000), 72-3.
[11] Tezel, Y., S., *Cumhuriyet Döneminin Iktisat Tarihi*, Istanbul: Tarih Vakfi Yurt Yayinlari, 3rd Ed., 1994; 129.
[12] Dervişoğlu, A., "Türkiye Cumhuriyeti'nin 75. Yılında Elektrik-Elektronik Mühendisliğindeki Gelişmelerin Bilime ve Ülkenin Gelişimine Katkıları", *Türkiye Cumhuriyeti'nin 75. Yılında Bilim, "Bilanço 1923-1998' Ulusal Toplantisi*, I. Kitap, I. Cilt, Ankara: Türkiye Bilimler Akademisi, 1999; 395.
[13] *Ibid.*, 394.
[14] Refik, M., "Fen Fakültesinin Ameli Elektrik ve Makine Dersleri", *Fen Alemi*, No. 15, (March, 1926), 258.
[15] The complete collection of *Mualimler Mecmuası* (Journal of Teachers) is available at the Atatürk Library of the Greater Istanbul Municipality.
[16] The complete collection of *Darülfünun Fen Fakültesi Mecmuası* (Journal of O.U. Faculty of Science) is available at the Atatürk Library of the Greater Istanbul Municipality.
[17] The complete collection of *Mühendis Mektebi Mecmuası* (Journal of the School of Engineering) is available at the Mustafa Inan Library of Istanbul Technical University.
[18] The complete collection of *Fen Alemi* (World of Science) is available at the State Library of Beyazit in Istanbul.
[19] The complete collection of *Tabiat Alemi* (World of Nature) is available at the Hakki Tarik Us Library in Istanbul.
[20] Biographical information on Hüsnü Hamid Bey was supplied by his surviving son Demir Sayman, a civil engineer, in an interview conducted on 11 April, 1999.
[21] "Ahmet Müştak Kargılı", *Büyük Larousse Sözlük ve Ansiklopedisi*, Vol. 1 (Istanbul, 1994), 214.
[22] Dölen, E., *Osmanlılarda Kimyasal Semboller ve Formüller*, Istanbul: TMMOB Kimya Mühendisleri Odası Yayını, 1996; 102.
[23] Inönü, E., *Mehmet Nadir, Bir Eğitim ve Bilim Öncüsü*, Ankara: TÜBITAK Bilim Adamı Yetistirme Grubu Yayınları, 1997; 12-24.
[24] Ayni, M. A., *Darülfünun Tarihi*, Istanbul: Pinar Yayınları, 1995; 78.
[25] Kazancıgıl, A., "Fahir Yeniçağ ve Türkiye'de Atom Fiziği", *Bilim Tarihi*, No. 4, (February 1992), 4.
[26] "Uzdilek, Salih Murad", *Ana Britannica Genel Kültür Ansiklopedisi*, Vol. 21 (Istanbul, 1990), 457.
[27] Çağatay, U. & Karatekin, E., *Yüksek Mühendis Mektebi*, Istanbul: I.T.Ü. Kütüphanesi No. 389, Berksoy Matbaasi, 1958; 383-385.
[28] *Ibid.*, 353-355.
[29] *Ibid.*, 582-583.
[30] *Ibid.*, 342-344.
[31] Bahadir, O.R., "Matematikçi Kerim Bey ve Einstein", *Toplumsal Tarih*, No. 72 (December, 1999), 22-25.
[32] Dölen, E., *Açıklamalı Türkiye Kimya Dergileri Bibliografyasi (1911-1990)*, Istanbul: TMMOB Kimya Mühendisleri Odası Istanbul Şubesi Yayini, 1990; 20.
[33] Erk, N., "Ord. Prof. Dr. Fazlı Faik Yeğül (1882-1965)", *Ankara Üniversitesi Veteriner Fakültesi Dergisi*, No. 12 (3) (Ankara, 1965), 250-256.
[34] Baytop, T., *Türk Eczacılık Tarihi*, Istanbul: I.Ü. Eczacilik Fakültesi Yayınları, 1985; 421-422.
[35] "Civaoğlu, Ilhami", *Ana Britannica Genel Kültür Ansiklopedisi*, Vol. 5 (Istanbul, 19787), 616.
[36] Çağatay & Karatekin, *op. cit.*, 210-218.
[37] See above under *Mühendis Mektebi mecmuası* (Journal of the School of Engineering) for the life story of Salih Murad (Uzdilek).
[38] Bahadir, O.R., *Osmanlılarda Bilim*, Istanbul: Sarmal Yayınevi, 1996; 53.
[39] Atay, O., *Bir Bilım Adamının Romanı*, Istanbul: Iletişim Yayınları, 1990; 64.

Kolibri 360 GmbH
Ferdinand-Jühlke-Straße 7
99095 Erfurt, Deutschland
produktsicherheit@kolibri360.de